Lecture Notes in Physics

The Lecture Notes in Physics

The series Lecture Notes in Physics (LNP), founded in 1969, reports new developments in physics research and teaching – quickly and informally, but with a high quality and the explicit aim to summarize and communicate current knowledge in an accessible way. Books published in this series are conceived as bridging material between advanced graduate textbooks and the forefront of research and to serve three purposes:

- to be a compact and modern up-to-date source of reference on a well-defined topic

- to serve as an accessible introduction to the field to postgraduate students and nonspecialist researchers from related areas

- to be a source of advanced teaching material for specialized seminars, courses and schools

Both monographs and multi-author volumes will be considered for publication. Edited volumes should, however, consist of a very limited number of contributions only. Proceedings will not be considered for LNP.

Volumes published in LNP are disseminated both in print and in electronic formats, the electronic archive being available at springerlink.com. The series content is indexed, abstracted and referenced by many abstracting and information services, bibliographic networks, subscription agencies, library networks, and consortia.

Proposals should be sent to a member of the Editorial Board, or directly to the managing editor at Springer:

Christian Caron
Springer Heidelberg
Physics Editorial Department I
Tiergartenstrasse 17
69121 Heidelberg / Germany
christian.caron@springer.com

W. Hillebrandt
F. Kupka (Eds.)

Interdisciplinary Aspects
of Turbulence

 Springer

Wolfgang Hillebrandt
MPI für Astrophysik
85741 Garching
Germany
wfh@mpa-garching.mpg.de

Friedrich Kupka
MPI für Astrophysik
85741 Garching
Germany
fk@MPA-Garching.MPG.DE

Hillebrandt, W., Kupka, F. (Eds.), *Interdisciplinary Aspects of Turbulence*, Lect. Notes Phys. 756 (Springer, Berlin Heidelberg 2009), DOI 10.1007/978-3-540-78961-1

ISBN: 978-3-540-78960-4   e-ISBN: 978-3-540-78961-1

DOI 10.1007/978-3-540-78961-1

Lecture Notes in Physics ISSN: 0075-8450

Library of Congress Control Number: 2008933388

Cover design: eStudio Calamar S.L., F. Steinen-Broo, Pau/Girona, Spain

Printed on acid-free paper

9 8 7 6 5 4 3 2 1

springer.com

Preface

What do combustion engines, fusion reactors, weather forecast, ocean flows, our sun, and stellar explosions in outer space have in common? Of course, the physics and the length and time scales are vastly different in all cases, but it is also well known that in all of them, on some relevant length scales, the material flows that govern the dynamical and/or secular evolution of the systems are chaotic and often unpredictable: they are said to be *turbulent*.

In fact, the term "turbulence" is used for an enormous variety of phenomena in very different fields, including geophysics, astrophysics, and engineering. Unfortunately, these communities do not talk to each other too often. Therefore, back in 2005, we organized a workshop on "Interdisciplinary Aspects of Turbulence" at the Ringberg Castle in the Bavarian Alps, to discuss topics such as the basic concepts of turbulence, the different approaches of modelling and simulations used in the various areas, and also possible tests. This workshop was a great success and the proceedings can be found on the Internet (www.mpa-garching.mpg.de/mpa/publications/ proceedings/proceedings-en.html) as well as pdf-files of several of the talks presented (www.mpa-garching.mpg.de/hydro/Turbulence/).

However, we felt that it would be a good idea to ask a few of the participants to contribute to an edited book on this subject. Most of them agreed, and the outcome is this *Lecture Notes in Physics* (LNP) volume. It covers several subjects on which considerable progress was made during the last decades, from questions concerning the very nature of turbulence to some practical applications. After an introductory chapter an approach to relate turbulence to statistical mechanics and nonlinear dynamics is presented, followed by a discussion of turbulent convection in our Sun and in other stars. Next, in the chapter "Turbulence in Astrophysical and Geophysical Flows", a basic introduction to turbulence in the context of geophysics and astrophysics is provided, showing that the same methods, i.e. Reynolds stress models, can be applied to a variety of problems successfully. The chapter "Turbulence in the Lower Troposphere: Second-Order Closure and Mass-Flux Modelling Frameworks" is devoted to the modeling of atmospheric turbulence in the context of numerical weather predictions and in the chapter "Magnetohydrodynamic Turbulence" the focus is on magnetohydrodynamic turbulence, the application in mind being the design of magnetically confined plasmas in fusion reactors. In contrast, the chapter "Turbulent Combustion in Thermonuclear Supernovae" deals with turbulent combustion in premixed flames with an application taken again from astrophysics, namely thermonuclear supernova explosions. Finally, in the chapter "One-Dimensional Turbulence Stochastic Simulation of Multi-Scale Dynamics", a new method is presented to deal with the multi-scale character of turbulence by means of stochastical processes which, besides its value on its own right, has the potential to serve as a "subgrid-scale" model in large-scale numerical simulations.

We are grateful to our colleagues who contributed with enthusiasm to an unusual volume and we hope that some of the readers will share our elation.

Garching, Germany   *Wolfgang Hillebrandt*
Garching, Germany   *Friedrich Kupka*

Contents

Turbulence in the Lower Troposphere: Second-Order Closure and Mass–Flux Modelling Frameworks

Magnetohydrodynamic Turbulence

Turbulent Combustion in Thermonuclear Supernovae

One-Dimensional Turbulence Stochastic Simulation of Multi-Scale Dynamics

An Introduction to Turbulence

W. Hillebrandt and F. Kupka

Max-Planck-Institut für Astrophysik, Garching, Germany
whillebrandt@mpa-garching.mpg.de
fkupka@mpa-garching.mpg.de

In fluid dynamics, turbulence is commonly defined as a flow regime characterized by chaotic, stochastic property changes, such as rapid variation of
pressure and velocity in space and time. Turbulence sets in if the dimensionless (scale-dependent) Reynolds number, measuring the relative importance
of inertial forces to viscous forces, exceeds a certain value, roughly 2300, as
found in experiments (see [5] for the case of pipe flows – lower values are
sometimes found for other types of flow such as plane Couette flow [26]).

Although, in a sense, the dynamics of fluid flows is well understood since
about 200 years, described by an innocently looking nonlinear partial differential equation, the Navier–Stokes equation, even giants of modern physics,
such as Werner Heisenberg, are quoted saying that the two major unsolved
problems of physics are the unification of quantum physics and gravity and
turbulence, and that they are more optimistic that the former will be solved
some day.

So, what is the problem with turbulence? The problem is that with increasing Reynolds-number the flow patterns become so complicated on so many
different length scales that one has to resort to statistical methods to describe
the empirical findings. Kolmogorov's 'theory' of fully developed homogeneous
turbulence (discussed in some detail in this chapter) is the best-known example. However, Kolmogorov's hypotheses only hold exactly in the limit of
infinite Reynolds number, and 'a solution to the problem of turbulence' would
be, roughy speaking, to derive them, for all kinds of flows, directly from the
Navier–Stokes equation. In fact, at present we are very far from being able to
do this, but several modern attempts to come closer to a solution, at least for
subclasses of complex fluid flows, are subject of this book.

The main emphasis of this chapter is to introduce the basic concepts of turbulence. Before expounding on the meaning of turbulence we shall derive the
hydrodynamic equations from statistical mechanics, since statistical methods
are used in several chapters of this book and most derivations in textbooks
limit themselves to the traditional, continuum mechanics approach. Next, various aspects (including shortcomings) of Kolmogorov's theory are discussed,

Hillebrandt, W., Kupka, F.: *An Introduction to Turbulence*. Lect. Notes Phys. **756**, 1–20 (2009)
DOI 10.1007/978-3-540-78961-1_1   © Springer-Verlag Berlin Heidelberg 2009

followed by an outline of more modern developments. We conclude with an outlook on the other contributions to this book.

1 The Fundamentals of Turbulence

1.1 The Boltzmann and Maxwell–Boltzmann Equations

The equations governing the dynamics of fluids and dense gaseous systems can be derived from classical statistical mechanics. To this end one considers a classical mechanical system of N structureless particles without internal degrees of freedom. For given initial conditions their motions are governed by the Hamilton equations

$$\frac{\partial \boldsymbol{q}_i}{\partial t} = \frac{\partial H}{\partial \boldsymbol{p}_i}, \qquad \frac{\partial \boldsymbol{p}_i}{\partial t} = -\frac{\partial H}{\partial \boldsymbol{q}_i} \qquad i = 1, \ldots, N, \tag{1}$$

where $\boldsymbol{q}_i = (q_1, q_2, q_3)$ are the generalized coordinates and $\boldsymbol{p}_i = (p_1, p_2, p_3)$ the generalized momenta of the particles, and $H = H(\boldsymbol{q}_1, \ldots, \boldsymbol{q}_N, \boldsymbol{p}_1, \ldots, \boldsymbol{p}_N, t)$ is the Hamiltonian of the system.

Of course, for large numbers of particles these equations of motions cannot be solved directly, because the initial conditions are not known well enough and/or the computational demands are prohibitive. The way out is to recourse to a statistical description.

At any instant, the system of N particles under consideration can be represented by a point in the $6N$-dimensional phase space. In terms of the multi-particle distribution function $F(\boldsymbol{q}_1, \ldots, \boldsymbol{q}_N, \boldsymbol{p}_1, \ldots, \boldsymbol{p}_N, t)$, the probability of finding the system in the volume element $d\Omega \equiv d\boldsymbol{q}_1 \ldots d\boldsymbol{q}_N d\boldsymbol{p}_1 \ldots d\boldsymbol{p}_N$ of phase space is

$$F(\boldsymbol{q}_1, \ldots, \boldsymbol{q}_N, \boldsymbol{p}_1, \ldots, \boldsymbol{p}_N, t) d\Omega \tag{2}$$

and the phase–space integral is normalized to 1.

For a system of noninteracting particles F can be written as a product of single-particle distribution functions $f_i(\boldsymbol{q}_i, \boldsymbol{p}_i, t)$, and for identical particles it is simply

$$F(\boldsymbol{q}_1, \ldots, \boldsymbol{q}_N, \boldsymbol{p}_1, \ldots, \boldsymbol{p}_N, t) = \prod_{i=1}^{N} f_i(\boldsymbol{q}_i, \boldsymbol{p}_i, t) = [f(\boldsymbol{q}, \boldsymbol{p}, t)]^N. \tag{3}$$

In order to derive evolution equations for the distribution functions, one uses the fact that for a canonical ensemble the phase–space volume element $d\Omega$ is constant in time which leads directly to the Liouville equation

$$\frac{dF}{dt} \equiv \frac{\partial F}{\partial t} + \sum_{j=1}^{3N} \left(\frac{\partial F}{\partial q_j} \frac{\partial q_j}{\partial t} + \frac{\partial F}{\partial p_j} \frac{\partial p_j}{\partial t} \right) = 0 \tag{4}$$

which, in the case of noninteracting identical particles, reduces to the Vlasov equation:

$$\frac{\partial f}{\partial t} + \sum_{k=1}^{3} \left(\frac{\partial f}{\partial q_k} \frac{\partial q_k}{\partial t} + \frac{\partial f}{\partial p_k} \frac{\partial p_k}{\partial t} \right) = 0. \tag{5}$$

With increasing density, collisions between particles can no longer be ignored. Under the assumption of only two-body collisions (which is a fair approximation for short-range interactions and "moderate" densities) and molecular chaos, one can integrate the Liouville equation successively over the coordinates of the N-particle distribution function building up a hierarchy of integro-differential equations, commonly called the BBGKY (Born, Bogoljubov, Green, Kirwood, Yvon) hierarchy [3, 22]. For example, integrating over all-but-one particle coordinates and momenta one gets

$$\frac{\mathrm{d}F_{(1)}(\boldsymbol{q}_1, \boldsymbol{u}_1, t)}{\mathrm{d}t} \equiv \frac{\partial F_{(1)}}{\partial t} + \boldsymbol{u}_1 \frac{\partial F_{(1)}}{\partial \boldsymbol{q}_1} + \boldsymbol{b} \frac{\partial F_{(1)}}{\partial \boldsymbol{u}_1} \tag{6}$$

$$= \frac{1}{m_1} \int \frac{\partial \Psi(\boldsymbol{q}_1, \boldsymbol{q}_2)}{\partial \boldsymbol{q}_1} \frac{\partial}{\partial \boldsymbol{u}_1} F_{(2)}(\boldsymbol{q}_1, \boldsymbol{q}_2, \boldsymbol{u}_1, \boldsymbol{u}_2, t) \, \mathrm{d}\boldsymbol{q}_2 \, \mathrm{d}\boldsymbol{u}_2 .$$

where $F_{(1)}(\boldsymbol{q}_1, \boldsymbol{p}_1, t)$ is the probability to find a particle at $(\boldsymbol{q}_1, \boldsymbol{p}_1)$, and we have used for the acceleration

$$\frac{1}{m} \frac{\mathrm{d}\boldsymbol{p}}{\mathrm{d}t} \equiv \frac{\mathrm{d}\boldsymbol{u}}{\mathrm{d}t} = -\sum_j \frac{1}{m_j} \frac{\partial \Psi}{\partial \boldsymbol{q}_j} + \boldsymbol{b}. \tag{7}$$

Here $\Psi(\boldsymbol{q}_i, \boldsymbol{q}_j)$ is the interaction potential between particles i and j, $\boldsymbol{u}$ the velocity, and $\boldsymbol{b}$ the accelerations caused by external forces.

In Eq. (7) the two-particle function $F_{(2)}(\boldsymbol{q}_1, \boldsymbol{q}_2, \boldsymbol{u}_1, \boldsymbol{u}_2, t)$ appears as a new unknown for which by integrating the Liouville equation over $(N-2)$ particle coordinates and momenta one obtains the next-order equation with a new unknown function $F_{(3)}$, and so on. Because this system of equations cannot be solved in closed form, the BBGKY-hierarchy has to be truncated or "closed", e.g., by making certain assumptions about $F_{(i)}$ at a certain order.

Under the additional assumption that $F_{(1)}$ is constant during collisions and for identical particles the Boltzmann equation is obtained which reads

$$\frac{\partial f}{\partial t} + \boldsymbol{u} \, \mathrm{grad}_q f - m \, \mathrm{grad}_q \Phi \, \mathrm{grad}_p f = \left[\frac{\partial f}{\partial t} \right]_{\mathrm{c}} \tag{8}$$

or

$$\frac{\partial f}{\partial t} + \boldsymbol{u} \, \mathrm{grad}_q f - \mathrm{grad}_q \Phi \, \mathrm{grad}_u f = \left[\frac{\partial f}{\partial t} \right]_{\mathrm{c}} . \tag{9}$$

Here, Φ is an external potential (such as gravity) and grad_q stands for $\left(\frac{\partial}{\partial q_1}, \frac{\partial}{\partial q_2}, \frac{\partial}{\partial q_3} \right)$, etc. The collision term $[\frac{\partial f}{\partial t}]_{\mathrm{c}}$ describes the changes of the single-particle distribution function caused by collisions in a statistical way and represents irreversible processes such as viscosity and diffusion. It is this term which makes the otherwise linear Boltzmann equation difficult to solve and it is the main problem when deriving macroscopic equations from the kinematic ones.

The next step towards hydrodynamics is to derive macroscopic equations by appropriately averaging the Boltzmann equation. The standard way towards this goal is to define macroscopic quantities as velocity moments of the distribution function, the r-th moment being

$$\left\langle \Theta^{(r)} \right\rangle \equiv \frac{1}{n} \int \Theta^{(r)} f \, \mathrm{d}\boldsymbol{p}, \tag{10}$$

with

$$\Theta^{(\tau)} \equiv m \boldsymbol{u}^{\tau} \qquad \tau = 0, 1, 2, \ldots \tag{11}$$

and the mean particle-number density n is defined as

$$n \equiv \int f \, \mathrm{d}\boldsymbol{p} \tag{12}$$

for which the identity

$$\langle n\Theta \rangle = \frac{\int n\Theta f \, \mathrm{d}\boldsymbol{p}}{n} = \int \Theta f \, \mathrm{d}\boldsymbol{p} = n \langle \Theta \rangle \tag{13}$$

holds.

The first three moments, $\tau = 0$, 1, and 2 in Eq. (11), are of particular importance because of their direct physical meaning. They read

$$\rho = \int m \, f(\boldsymbol{q}, \boldsymbol{p}, t) \, \mathrm{d}\boldsymbol{p} \tag{14}$$

$$\rho \boldsymbol{v} = \int m \boldsymbol{u} \, f(\boldsymbol{q}, \boldsymbol{p}, t) \, \mathrm{d}\boldsymbol{p} \tag{15}$$

$$\rho \varepsilon = \int \frac{m}{2} |\boldsymbol{w}|^2 f(\boldsymbol{q}, \boldsymbol{p}, t) \, \mathrm{d}\boldsymbol{p}. \tag{16}$$

Here ρ and $\rho \boldsymbol{v}$ are the density and the momentum density, respectively, and $\rho = mn$. In order to get Eq. (16) we have introduced $\boldsymbol{w}$ as the difference between the mean velocity $\boldsymbol{v}$ and the velocity $\boldsymbol{u}$ of a particle, i.e., $\boldsymbol{w}$ is the (statistical) velocity fluctuation due to thermal motions and, thus, $\rho \varepsilon$ is the internal (or thermal) energy density.

Next the Boltzmann equation (Eq. 8) is multiplied by $\Theta^{(r)}$ and integrated over momentum space which gives

$$\int \Theta^{(r)} \left[\frac{\partial f}{\partial t} + \boldsymbol{u} \, \mathrm{grad}_q f - \mathrm{grad}_q \Phi \, \mathrm{grad}_u f \right] \mathrm{d}\boldsymbol{p} = \int \left[\frac{\partial f}{\partial t} \right]_c \Theta^{(r)} \, \mathrm{d}\boldsymbol{p}. \tag{17}$$

Since $\Theta^{(r)}$ does not depend on $\boldsymbol{q}$ and t explicitly, Eq. (17) can be rewritten in a more convenient way as

$$\frac{\partial}{\partial t} \int \Theta^{(r)} f \, \mathrm{d}\boldsymbol{p} + \int \boldsymbol{u} \operatorname{grad}_q (\Theta^{(r)} f) \, \mathrm{d}\boldsymbol{p} - \int \Theta^{(r)} \operatorname{grad}_q \Phi \operatorname{grad}_u f \, \mathrm{d}\boldsymbol{p}$$

$$= \int \left[\frac{\partial f}{\partial t} \right]_c \Theta^{(r)} \, \mathrm{d}\boldsymbol{p} \tag{18}$$

which after averaging and some algebraic manipulations, making use of Eqs. (10) and (13), transforms into the Maxwell–Boltzmann (transport) equation for the moment $\Theta^{(r)}$:

$$\frac{\partial}{\partial t} \left\langle n\Theta^{(r)} \right\rangle + \operatorname{div}_q \left\langle n\Theta^{(r)} \boldsymbol{u} \right\rangle + \operatorname{grad}_q \Phi \left\langle n \operatorname{grad}_u \Theta^{(r)} \right\rangle$$

$$= \int \left[\frac{\partial f}{\partial t} \right]_c \Theta^{(r)} \mathrm{d}\boldsymbol{p}. \tag{19}$$

Some general properties of Eq. (19) are worth to be mentioned. Firstly, the right-hand side is identical to zero, if $\Theta^{(r)}$ (or any linear combination of the $\Theta^{(r)}$'s) is conserved in collisions. Secondly, in the nonrelativistic limit and for elastic collisions mediated by short-range forces there exist exactly five conserved quantities: the mass m, the three components of momentum $m\boldsymbol{u}$, and the kinetic energy $\frac{1}{2}m|\boldsymbol{u}|^2$.

As before, the Maxwell–Boltzmann equations are a hierarchical system of equations because the collision term is not simply a function of the r-th moment, and again, to solve them closure conditions have to be invented. For a discussion of the physics of the equations derived in this and the following subsection we refer to [10, 23]. A more general treatment can be found in [18].

1.2 The Equations of Hydrodynamics

In deriving the Maxwell–Boltzmann equations, we have assumed that it makes sense to perform averages in momentum space, and the necessary conditions were stated explicitly. This allowed us to introduce macroscopic quantities such as the density or the energy density and to separate the forces between particles into short-range ones, written as a collision term, and long-range forces, put in explicitly as an external potential Φ. In physical applications, Φ could be the gravitational or the Coulomb potential and would have to be determined from some field equation.

The derivation of the hydrodynamic equations is now straight-forward. They are obtained from the Maxwell–Boltzmann equations by taking the first three moments and by introducing an equation of state for closure. Moreover, we will use the fact that mass and momentum are conserved quantities and, hence, the corresponding collision terms vanish.

Thus, the zeroth moment is simply $\Theta^{(0)} = m$ and the corresponding moment equation, known as the equation of mass conservation or the continuity equation, reads

$$\frac{\partial}{\partial t} \langle nm \rangle + \operatorname{div} \langle nm\boldsymbol{u} \rangle = 0, \tag{20}$$

or, with $nm = \rho = \int mf \, d\boldsymbol{p}$ in a more convenient form:

$$\frac{\partial \rho}{\partial t} + \text{div}\,(\rho \boldsymbol{v}) = 0. \tag{21}$$

The first moment is $\Theta^{(1)} = m\boldsymbol{u}$ and with $\boldsymbol{u} = \boldsymbol{v} - \boldsymbol{w}$ and $\langle \boldsymbol{w} \rangle = 0$ the corresponding moment equation follows as

$$\frac{\partial}{\partial t}(\rho \boldsymbol{v}) + \text{div}\,[\rho \langle \boldsymbol{u} \otimes \boldsymbol{u} \rangle] + \rho \,\text{grad}\, \Phi = 0, \tag{22}$$

where $\boldsymbol{u} \otimes \boldsymbol{u}$ is the dyadic product of $\boldsymbol{u}$ with itself, for which because of $\langle \boldsymbol{w} \rangle = 0$ the relation

$$\langle \boldsymbol{u} \otimes \boldsymbol{u} \rangle = \langle \boldsymbol{v} \otimes \boldsymbol{v} \rangle + \langle \boldsymbol{w} \otimes \boldsymbol{w} \rangle \tag{23}$$

holds. Therefore one can rewrite Eq. (22) as

$$\frac{\partial}{\partial t}(\rho \boldsymbol{v}) + \text{div}\,[\rho(\boldsymbol{v} \otimes \boldsymbol{v})] + \text{div}\,\boldsymbol{\Pi} = -\rho \,\text{grad}\, \Phi \tag{24}$$

which is the standard form of the Navier–Stokes equation, the equation of motion of fluid dynamics. The pressure tensor $\boldsymbol{\Pi}$ appearing in Eq. (24) is defined as $\boldsymbol{\Pi} \equiv \rho\,(\boldsymbol{w} \otimes \boldsymbol{w})$ and is commonly rewritten as

$$\boldsymbol{\Pi} = p\boldsymbol{I} - \boldsymbol{\pi}, \tag{25}$$

with $\boldsymbol{I}$ being the unit tensor, $p \equiv \frac{1}{3} \rho \,\langle |\boldsymbol{w}|^2 \rangle$ the isotropic pressure, and $\boldsymbol{\pi} \equiv \rho \langle \frac{1}{3} |\boldsymbol{w}|^2 \boldsymbol{I} - \boldsymbol{w} \otimes \boldsymbol{w} \rangle$ the (tensor) viscosity.

For an ideal gas or fluid the viscosity vanishes and the Navier–Stokes equation becomes

$$\frac{\partial}{\partial t}(\rho \boldsymbol{v}) + \text{div}\,[\rho(\boldsymbol{v} \otimes \boldsymbol{v})] + \text{grad}\, p = -\rho \,\text{grad}\, \Phi, \tag{26}$$

known as the Euler equation, which with the help of Eq. (21) is often written as

$$\frac{\partial \boldsymbol{v}}{\partial t} + (\boldsymbol{v}\,\text{grad})\boldsymbol{v} + \frac{1}{\rho}\text{grad}\, p = -\text{grad}\, \Phi. \tag{27}$$

The Euler equation is a hyperbolic non-linear partial differential equation. The non-linearity enters via the $(\boldsymbol{v}\,\text{grad})\boldsymbol{v}$ term, and this term is responsible for turbulence to occur, as we shall see later.

Finally, the moment equation for the second moment $\Theta^{(2)} = \frac{1}{2}m|\boldsymbol{u}|^2$ gives us a conservation law for the energy density. Note that $\Theta^{(2)}$ is the total energy of a particle and therefore $\langle n\Theta^{(2)} \rangle = \left[\frac{\rho}{2} \langle |\boldsymbol{w} + \boldsymbol{v}|^2 \rangle\right]$ is the total energy density, i.e., the sum of kinetic and thermal energy.

Its change with time is given by

$$\frac{\partial}{\partial t}\left\langle n\Theta^{(2)} \right\rangle = \frac{\partial}{\partial t}\left[\frac{\rho}{2} \langle |\boldsymbol{w} + \boldsymbol{v}|^2 \rangle\right] = \frac{\partial}{\partial t}\left[\frac{\rho}{2}|\boldsymbol{v}|^2 + \frac{\rho}{2} < |\boldsymbol{w}|^2 > \right] \tag{28}$$

and the second term of the Maxwell–Boltzmann equation (Eq. (19)) reads

$$\operatorname{div}\left\langle n\Theta^{(2)}\boldsymbol{u}\right\rangle = \sum_i \frac{\partial}{\partial x_i}\left[\frac{\rho}{2}\sum_j \langle u_j^2 u_i\rangle\right] \tag{29}$$

which after some lengthy manipulations can be written as

$$\operatorname{div}\left\langle n\Theta^{(2)}\boldsymbol{u}\right\rangle = \tag{30}$$

$$\sum_i \frac{\partial}{\partial x_i}\left\{\frac{\rho}{2}\left[|\boldsymbol{v}|^2 v_i + v_i\left\langle|\boldsymbol{w}|^2\right\rangle + 2\sum_j v_j\left\langle w_i w_j\right\rangle + \left\langle w_i|\boldsymbol{w}|^2\right\rangle\right]\right\}.$$

With the definitions $\varepsilon \equiv \frac{1}{2}\left\langle|\boldsymbol{w}|^2\right\rangle$ (internal specific energy) and $\boldsymbol{h} \equiv \rho\left\langle \boldsymbol{w}\frac{1}{2}|\boldsymbol{w}|^2\right\rangle$ (energy flux by heat conduction) one obtains the energy equation

$$\frac{\partial}{\partial t}(\rho E) + \operatorname{div}\left[(\rho E + p)\boldsymbol{v}\right] + \operatorname{div}\boldsymbol{h} - \operatorname{div}(\boldsymbol{\pi}\boldsymbol{v}) = -\rho\boldsymbol{v}\operatorname{grad}\Phi \tag{31}$$

where we have used $E \equiv \frac{1}{2}|\boldsymbol{v}|^2 + \varepsilon$ as an expression for the total specific energy density (in units of energy per mass). For adiabatic motions of an ideal fluid without external forces (or gravity) Eq. (31) reduces to

$$\frac{\partial}{\partial t}(\rho E) + \operatorname{div}\left[(\rho E + p)\,\boldsymbol{v}\right] = 0 \tag{32}$$

or, in terms of specific entropy (entropy per mass) S,

$$\frac{\partial \rho S}{\partial t} + \operatorname{div}(\rho S\boldsymbol{v}) = 0\,. \tag{33}$$

Together with an equation of state, $p = p(\rho, E)$, and expressions for the heat flux and the viscosity tensor the system of equations (21), (24), and (31) is closed. If the diffusion approximation holds (see also [15]), a possible choice for $\boldsymbol{h}$ is the phenomenological ansatz

$$\boldsymbol{h} = -\kappa\operatorname{grad}T\,, \tag{34}$$

with the heat conduction coefficient κ. The viscosity tensor reads [15]

$$\pi_{ik} = \eta\left(\frac{\partial v_i}{\partial x_k} + \frac{\partial v_k}{\partial x_i} - \frac{2}{3}\delta_{ik}\operatorname{div}\boldsymbol{v}\right) + \zeta\delta_{ik}\operatorname{div}\boldsymbol{v}\,, \tag{35}$$

where η and ζ are the molecular and the bulk viscosity coefficients, respectively. This linear dependence of π_{ik} on $\boldsymbol{v}$ holds for a *Newtonian fluid* [1, 16]. For emulsions or fluids of long-chained molecules the generation of internal forces through shear and compression can be non-linear. However, for the basic properties of turbulent flows and their statistical treatment, which are of interest in this volume, such peculiarities are less important.

1.3 Kolmogorov's Theory of Fully Developed Homogeneous Turbulence

There are many cases in daily life where one can observe chaotic flows of fluids or gases, with rapid stochastic variation of velocity and pressure, properties we have used to characterize turbulence in the introduction. Boiling water in a pot is an example, the jet flow from a nozzle into a quiescent environment is another one, or terrestrial atmospheric circulations, or sun spots.

In order to show why some fluid flows are laminar and others are turbulent, we rewrite the Navier–Stokes equation (Eq. 24) slightly:

$$\frac{\partial}{\partial t}\,(\rho \boldsymbol{v}) \;=\; -\,\nabla\,(\boldsymbol{T} + \boldsymbol{\pi}) + \mathbf{F}, \tag{36}$$

where now $\boldsymbol{T} = T_{ij} = p\delta_{ij} + \rho v_i v_j$ is the stress tensor with isotropic pressure p, $\mathbf{F} = -\rho\,\mathrm{grad}\,\Phi$ is a vector describing external forces, and $\boldsymbol{\pi}$, as before, is the viscosity tensor which is zero for ideal fluids. For an "incompressible" fluid, defined through $\mathrm{div}\,\boldsymbol{v} = 0$, in physical applications this means that pressure and density are approximately constant over a characteristic length scale of the fluid (which is a fair approximation for most low Mach-number flows, for which $|\boldsymbol{v}|/c_\mathrm{s} \ll 1$ and c_s is the speed of sound in the fluid, which is obtained from the equation of state $p = p(\rho, E)$). In that case and without external forces Eq. (36) reduces to

$$\rho\frac{\partial \boldsymbol{v}}{\partial t} + \rho(\boldsymbol{v}\,\mathrm{grad})\boldsymbol{v} + \mathrm{div}\,\boldsymbol{\pi} = 0 \tag{37}$$

and because of Eq. (35), with $\mathrm{div}\,\boldsymbol{v} = 0$, the viscosity is given by

$$\pi_{ik} \;=\; \nu\rho\left(\frac{\partial v_i}{\partial x_k} + \frac{\partial v_k}{\partial x_i}\right) \tag{38}$$

where we have introduced the kinematic viscosity as $\nu := \eta/\rho$.

As can be seen from Eq. (37), two terms regulate the velocity change of the fluid. The first term is the inertia of the fluid given by the extremely non-linear expression $\rho(\boldsymbol{v}\mathrm{grad})\boldsymbol{v}$ and the second is the dissipation term $\mathrm{div}\,\boldsymbol{\pi}$ which is linear in $\boldsymbol{v}$. If the kinematic viscosity is high and/or the characteristic length scale is small, the inertial term has little influence on the fluid which, consequently, will not show the non-linear patterns of chaotic motions. In contrast, if the viscosity is low and/or the characteristic length scales are large, the non-linear term dominates over dissipation and the motion can become turbulent. One should note, however, that for any given value of the viscosity there exist length scales for which a fluid flow is likely to become turbulent. Therefore, the phenomenon of turbulence is related to properties of the flow rather than those of the fluid.

An example of a flow which changes its characteristics depending on length scales involved is convection. We shall discuss it a bit more in detail to provide an illustration of the importance of scales. In laboratory fluids as well as in stars at the onset of convection large plumes of hot material form and float upwards while heavier cool surface material sinks downwards. This instability is called the Rayleigh–Taylor (RT) instability and the growth rate of small perturbations is given by the dispersion relation

$$\omega_{RT}^2 \;=\; g\,k\,\frac{\rho_2 - \rho_1}{\rho_2 + \rho_1},\tag{39}$$

where g is the (local) gravitational acceleration and k the wave number of the perturbation (see [4] for a general discussion).

However, due to the strong nonlinearity of the Navier–Stokes equation (the $v\nabla v$ term in particular), velocity fluctuations created on large scales couple to all scales down to the viscous dissipation scale η_k, the Kolmogorov scale [see Eq. (41)]. In between the RT scale L and η_k, there exists a region where the hydrodynamic behaviour of the fluid is dominated by the inertial term in the Navier–Stokes equation. It is therefore called the "inertial range". In this range, fluid properties should be statistically independent of both the geometry on the RT scale and the microscopic viscosity. Therefore, one expects a universal scaling of the velocity fluctuations in that range.

In such a cascade of turbulent velocity fluctuations, kinetic energy is transported from the large (RT) scales down to the Kolmogorov scale where it is dissipated into heat. If the assumption is made that this transport is fully described by its constant mean value $\bar{q}$ (which is a reasonable assumption if convective overturn times are short compared with the thermal evolution time and if turbulence is homogeneous and isotropic, i.e, its statistical properties are invariant under translation and rotation in space), a simple scaling law follows, commonly called Kolmogorov scaling [12] (see also the independent work of [19, 21, 29] and the refinements in [8, 9]):

$$v(l) \;=\; v(L)\,\left(\frac{l}{L}\right)^{1/3} \; ; \; \eta_k \ll l \ll L.\tag{40}$$

Here, η_k is related to the Reynolds number defined as $\mathrm{Re} := v(L)L/\nu$ by

$$\eta_k \;=\; L\,\mathrm{Re}^{-3/4}\tag{41}$$

and the energy flux through scale space is given by

$$\bar{q} \;=\; \frac{v(l)^3}{l} \;=\; \frac{v(L)^3}{L} \;=\; \frac{\nu\,v(\eta_k)^2}{(\eta_k)^2}.\tag{42}$$

(Here and in what follows $v(l)$ are the mean (turbulent) velocity fluctuations on length scale l.)

In stars such flows are highly turbulent always, mainly because of the huge values of L and despite the rather "normal" viscosities. Reynolds numbers are typically around 10^{12} or even 10^{14}. This in turn means that turbulence spans about 10 orders of magnitude in scale space. This is also the reason why, in contrast to most laboratory flows, in astrophysics in general the Euler rather than the Navier–Stokes equation is used to describe the flow.

One can state the assumptions which go into Kolmogorov's scaling law a bit more precisely. To that end we start by defining the probability function for the velocity difference of two fluid elements separated by a distance l:

$$\delta v(\boldsymbol{l}, \boldsymbol{x}) = |\boldsymbol{v}(\boldsymbol{x} + \boldsymbol{l}) - \boldsymbol{v}(\boldsymbol{x})|, \tag{43}$$

with the definition that $v(l) = \langle \delta v(|\boldsymbol{l}|) \rangle$ is the space average of δv for fixed length scale l.

Given this definition Kolmogorov's first assumption is that for $l \ll L$ $\delta v(\boldsymbol{l}, \boldsymbol{x})$ is a function only of $l = |\boldsymbol{l}|$, $\bar{q}$, and ν. This essentially requires the statistically averaged properties of the flow field to be independent of location, $\boldsymbol{x}$, and direction (dependence only on l), thus indirectly introducing the requirements of homogeneity and isotropy for $\delta v(\boldsymbol{l}, \boldsymbol{x})$. His second assumption is that for $\eta_k \ll l$, $\delta v(\boldsymbol{l}, \boldsymbol{x})$ is independent of ν. With both assumptions being fulfilled $\delta v(\boldsymbol{l}, \boldsymbol{x})$ can depend only on two variables: l and $\bar{q}$. This uniquely defines the functional form of the m-th moments of $\delta v(\boldsymbol{l}, \boldsymbol{x})$, the so-called "structure functions", which read

$$\langle \delta v(\boldsymbol{l}, \boldsymbol{x})^m \rangle \propto (\bar{q} l)^{m/3}. \tag{44}$$

Setting $m = 1$ and using the definition of $v(l)$ one obtains Eqs. (40) and (42), or for the turbulent energy spectrum in wave number (k-space, with k related to l through a Fourier transform with respect to l):

$$E(k) = C \bar{q}^{2/3} k^{-5/3}, \tag{45}$$

with the Kolmogorov constant C, and by rewriting Eq. (42) as

$$\bar{q} = \nu \frac{\left[v(L)(\eta_k/L)^{1/3} \right]^2}{\eta_k^2} \tag{46}$$

we recover for the viscous dissipation length η_k the relation

$$\frac{L}{\eta_k} = \mathrm{Re}^{3/4}. \tag{47}$$

1.4 Some Problems of Kolmogorov's Theory

While Kolmogorov's law, expressed by the first structure function, has been confirmed by experiments as well as by numerical simulations for many different high Reynolds-number flows, for the scaling exponents ζ_m for higher order moments,

$$\langle \delta v(\boldsymbol{l}, \boldsymbol{x})^m \rangle \propto l^{\zeta_m}, \tag{48}$$

deviations from the linear relationship $\zeta_m = m/3$ of Eq. (40) are well known. Experimentally, one finds $\zeta_m \leq m/3$, the deviations increasing with increasing m. These findings indicate that the assumptions leading to Kolmogorov scaling are not exactly fulfilled, which is not that surprising. In fact, if one abandons the assumption that the energy flux from large to small scales is sufficiently described by its mean value $\bar{q}$, one can explain the experiments in an easy way. Values $\zeta_m \leq m/3$ can be obtained if some rare but strong fluctuations are present in the energy dissipation rate. This phenomenon is commonly called "intermittency" [6, 7, 24], signalling a break-down of Kolmogorov's mean field theory for higher order structure functions where fluctuations of q become increasingly more important.

Several modifications of Kolmogorov's original theory have been suggested to account for intermittency. The first attempt is due to Kolmogorov and Oboukhov who independently invented the so-called "log-normal" model [13, 20]. Their hypothesis, often called "Kolmogorov's third hypothesis", is that the logarithm of the energy dissipation rate q is a normal random variable of variance equal to $A(\boldsymbol{x}, t) + \mu \log(L/l)$, where the term $A(\boldsymbol{x}, t)$ depends on the characteristics of the large-scale motion and μ is a parameter, possibly a universal constant. In addition, often the assumption is made that for $\log q$ the following relation holds:

$$\log q = -(\mu/2)\log(L/r) - A(\boldsymbol{x}, t)/2. \tag{49}$$

In more general terms, the ideas of homogeneous dissipation can be modified by calculating corrections $\delta\zeta_m$ to the Kolmogorov $\zeta_m = m/3$, for instance, from approximate solutions of the Navier–Stokes equation or in so-called multi-fractal models (see also [7]). Of course, these corrections have to be scale dependent because, according to experimental findings, they are large in the viscous subrange (small l), but small to zero in the inertial subrange (intermediate l).

In general, for many practical applications the assumptions of fully developed homogenous (and isotropic) turbulence are not justified. Therefore, further modifications of Kolmogorov's theory have been investigated and some of them are subject of this book.

2 Statistics and Simulations of Turbulent Flows

2.1 Length and Time Scales of Hydrodynamics

The averaging process discussed in Sect. 1 implicitly assumes that the time scales t and the length scales l which are of interest for the dynamics of the flow are well separated from the time and length scales relevant for the equation of state $p = p(\rho, E)$ used to close the system (21), (24), and (31). This limits the region of applicability of these *macroscopic* equations, as it requires that

$$\lambda \ll l, \quad \tau \ll t, \tag{50}$$

where λ and τ are the mean free path and the thermodynamical relaxation time scale of the particles of which the fluid is made of. Averaging in physical space as discussed in Sect. 1 (see also [10, 23]) can take place on length scales λ_h and time scales τ_h for which

$$\lambda \ll \lambda_h \ll l, \quad \tau \ll \tau_h \ll t. \tag{51}$$

If the particles under consideration are atoms or molecules at standard terrestrial temperature and pressure, the inequalities (51) are easily fulfilled. For all hydrodynamical problems considered here, we assume that (51) holds and hence the physical processes we consider operate on length scales $l \gg \lambda_h$ and on time scales $t \gg \tau_h$. In particular, we also have that $\eta_k \gg \lambda_h$.

2.2 Statistical Interpretation of Solutions and Alternatives

From a mathematical point of view, one might simply attempt to solve Eqs. (21), (24), and (31) as an initial-boundary value problem. To do so $\rho(\boldsymbol{x}, t)$, $\boldsymbol{v}(\boldsymbol{x}, t)$, and $E(\boldsymbol{x}, t)$ are specified at some time t_0 for all locations $\boldsymbol{x}$ in a spatial domain D, the volume (or area) considered in a given physical problem. In addition to these *initial conditions*, the quantities $\rho(\boldsymbol{x}, t)$, $\boldsymbol{v}(\boldsymbol{x}, t)$, and $E(\boldsymbol{x}, t)$ are specified for all $t > t_0$ on locations $\boldsymbol{x}_b$ at the boundary ∂D of D. These *boundary conditions* are required to obtain a unique solution on D and ensure physical causality. Analytical solutions for this class of problems are known only for few, special cases and proofs of existence or even uniqueness of solutions for $\rho(\boldsymbol{x}, t)$, $\boldsymbol{v}(\boldsymbol{x}, t)$, and $E(\boldsymbol{x}, t)$ have remained a difficult mathematical challenge until today. The pragmatic approach to this challenge is the computation of approximate solutions. Exact solutions are assumed to exist because of the sound physical and mathematical basis of the derivation of the hydrodynamical equations and the successful comparisons of numerous approximate solutions with experimental data.

Because of the nonlinearity of $\rho(\boldsymbol{v} \, \mathrm{grad})\boldsymbol{v}$ the differences $\delta \rho = \rho_1 - \rho_2$, $\delta \boldsymbol{v} = \boldsymbol{v}_1 - \boldsymbol{v}_2$, and $\delta E = E_1 - E_2$ between initially similar solutions $(\rho_1, \boldsymbol{v}_1, E_1)$ and $(\rho_2, \boldsymbol{v}_2, E_2)$ can grow exponentially fast – and even faster – on time scales $t - t_0$, i.e. within a time interval of physical interest. This has two different implications: the growth of arbitrarily small or finite-sized perturbations (which is related to the onset of turbulence) and the limited long-term predictive power of the hydrodynamical equations per se. An elementary introduction into the stability analysis of turbulent flows is given in [15]. But even when we are able to resolve all length scales from the size of the physical system (some maximum distance within D) down to the Kolmogorov scale η_k in numerical solutions of Eqs. (21), (24), and (31), the predictive power of such solutions is limited to a finite interval $t_1 - t_0$, because initial conditions can only be known with finite accuracy. This restriction is inevitable whenever initial data are taken from experiments or approximate solutions have to be compared to experimental measurements. It is doomed to occur for any turbulent flow. Thus, even direct numerical simulations of turbulent flows, approximate numerical solutions to Eqs. (21), (24), and (31) which explicitly account for all length scales in D down to η_k, are often the subject of a statistical interpretation. This approach is by no means straightforward, since the general hydrodynamical equations are dissipative, despite they are derived from the conservative equations of motions for particles, (1), for which the concept of phase space is indeed well understood. For a historical overview on the subject of statistical predictability in turbulent flows we refer to [25]. A short introduction can also be found in [16].

As is illustrated, for instance, by the various contributions and references in [17], the usefulness of statistical concepts for the study of turbulent flows has been questioned in the scientific literature more than two decades ago. Much of this criticism was in fact motivated by the limitations of simple Reynolds stress models which only consider second order moments such as turbulent kinetic energy and neglect the relevance of higher order correlations on the averaged properties of the flow. We note that in deriving Reynolds stress models products of the unaveraged hydrodynamical equations, i.e. (21), (24), and (31) or some simplification thereof, are used to construct dynamical equations for products of the basic dynamical variables ρ, v, and E (or some alternative, dependent variables, see chapter "Turbulence in Astrophysical and Geophysical Flows"). Ensemble averaging those equations yields the dynamical equations of the *moments* of the hydrodynamical equations. This procedure in the end yields an entire hierarchy of equations which is unclosed similar to the Boltzmann and Maxwell–Boltzmann equations discussed in Sect. 1.1. But contrary to the latter a closure of the Reynolds-averaged moment equations is a much more difficult task, since it requires knowledge on the macroscopic behaviour of the flow. The most simple models are obtained, if that process is already stopped at second order moments. This includes many of the phenomenological models for turbulent flows such as mixing length theory. Information on the large-scale structures, or actually on any structures in the flow, is lost in such models, which is one reason for their limitations.

Until today such simple models are frequently used to parameterize properties of turbulent flows, although they have also been known for a long time to fail in various applications (see [17]). However, the criticism went much further: it was even argued that the very existence of large-scale coherent structures prohibits the success of any statistical method. Coherent structures are commonly referred to as spatial regions that at a given time show some organization with respect to any quantity related to the flow (cf. [16]). This concept is not new at all in the study of turbulent flows: already five decades ago an extensive description was given in [27] and also in some earlier literature. One of the weak points behind those arguments used to discourage further use of statistical methods for studying or modelling turbulence is that very often the coherent structures themselves behave in a manner complex enough to justify statistical methods. As was already pointed out in [17] other branches of physics have successfully used essentially statistical methods to study phenomena which feature the equivalent of coherent structures, for example, the theory of critical phenomena in ferromagnets [14, 30]. It was concluded that one important message behind those results is the indispensability of identifying the right quantities and objects on which to do statistics.

During this debate many new concepts for the study of the physical nature of turbulence have been proposed. In some cases hope was expressed that they could even replace statistical methods. Research efforts were devoted to cellular automata and lattice gas dynamics, to chaos theory and a dynamical systems approach to turbulence, and to more systematic studies of coherent structures in turbulent flows as well as of vortex dynamics (see [17] for some earlier references). Later on multifractal models based on wavelet transforms have been introduced to analyse the scale-dependent properties of turbulent flows (see, for instance, [11] and also [7]). The latter are in fact more akin to a statistical approach. A lot of physical insight was gained from these methods on how turbulence develops on the smallest scales. Research on coherent structures and vortex dynamics has also been useful to further our understanding of the dynamical properties of turbulent flows on large scales. However, the study of complex physical systems that are the subject of astro- and geophysical sciences, where the effects of turbulence can be observed on length scales differing among each other by many orders of magnitude and where complex boundary conditions interact with the flow, has seen much less benefit from this kind of research. In fact, the greatest advancements in these fields have definitely resulted from the dramatic refinement of numerical simulations over the last two decades, partially because of the now available 'raw computational power', and also due to refinements in algorithmic solution procedures and the microphysics used as input data for such simulations. The sheer amount of output data produced by hydrodynamical simulations is one reason why statistical methods have not disappeared. Instead, they have continued to be a standard tool for interpreting their results. Another reason is related to the fact that for large physical systems not all the length and time scales of interest can be included in the numerical scheme used to perform the simulation. Thus, hybrid approaches are necessary. They deal with separate ranges of length and time scales by different means, e.g., large eddy simulations combined with mass flux–based sub-grid-scale parametrizations. In principle, simulations could be performed separately for different scale ranges, but in general this approach is prohibitive both due to computational costs and conceptual difficulties such as coupling to larger scales by means of complex boundary conditions or direct exchange between scales not contained within the simulation. We note here that the one-dimensional turbulence (ODT) methods discussed in the chapter "One-Dimensional Turbulence Stochastic Simulation of Multi-scale Dynamics" address this problem from a new point of view. In the meantime, Reynolds stress and mass flux models have seen far more success when combined with numerical simulations than many of the approaches originally proposed to replace such methods altogether.

At the bottom line, no approach currently used for the study of turbulence can deal with complex flows of large physical systems on its own. This might explain why the past two decades have not seen new, alternative approaches completely taking over the field, while refined Reynolds stress and mass flux models have continued to be used widely in the parametrization of flow physics operating on scales not explicitly accounted for in numerical simulations.

2.3 Ensembles, Averages, and Multiscale Problems

With this background in mind it is clear that the physical interpretation of a quantity computed within a turbulence model or a numerical simulation can be subtle and requires to specify the context of the computation. The most important types of averages considered in the following chapters are ensemble averages, volume averages, and (mass) flux averages. For the latter the sign of a physical quantity at a given location is used to group different regimes of the flow field together as part of the averaging process. *Ensemble averages* assume the existence of long-term, quasi-stationary states or at least the possibility to collect related, physically plausible initial conditions together. In numerical simulations such ensembles are constructed from time sequences under the assumption of a quasi-ergodic hypothesis. In that case different locations in phase space are visited proportional to their realization probabilities within the simulation time. We note here that ensemble averages in Fourier space are actually performed for n-point correlation functions (cf. [16]) and additional assumptions (such as homogeneity and isotropy) are introduced to interpret them in terms of spectra in k-space. *Volume averages* are a much more straightforward concept directly derived from the conservation properties of the hydrodynamical equations (21), (24), and (31) and the limited number of computational degrees of freedom (resolution, number of grid points in a numerical method, etc.) and the ensuing limitations due to the size D of the domain to be modelled. Problems such as numerical weather prediction or certain types of combusting flows can challenge the definition of such averages, since the scales at which physical processes take place can change. This may even lead to the point at which parametrizations or certain averaging concepts break down. Other multiscale flows are more benign and allow robust definitions of averages useful in turbulence modelling. We conclude here with a practical remark: both the notation used for and the exact meaning of certain type of averages are not 'normalized' by some generally accepted convention. Hence, a careful inspection of definitions is always useful, even within the individual chapters of this book.

3 Summary

In the following we give a short summary of the different topics selected by the authors of the remaining chapters of this book. The chapters have been written in such a way that they can be read both successively as implied by the sequence in this book and also selectively. Cross references between chapters have been provided by the authors where appropriate.

The chapter "Nonextensive Statistical Mechanics and Nonlinear Dynamics" discusses some recent developments in statistical mechanics and nonlinear dynamics which have become more and more interesting for research on turbulence. It first introduces how Boltzmann–Gibbs statistics, the foundation of classical statistical mechanics, and its famous Boltzmann–Gibbs entropy functional should be generalized, in particular for physical systems which are characterized by long-range interactions. The generalizations preserve many key properties of the classical Boltzmann–Gibbs statistics and its associated entropy functional, but abandon strict additivity. Classical statistical mechanics is recovered as a special case [2, 28]. The new distributions and their associated entropy functionals require an additional physical quantity, for instance the extensivity parameter q, to be determined from measurements. The resulting framework is very useful to recover the distributions of velocity fluctuations in a variety of turbulent flows, as shown from comparisons with experimental data. This new methodology may hence provide us with a new possibility to better understand the nature of probability density functions of the dependent variables in turbulent flows. A well-founded theory of the statistical distribution of those variables would provide a major step forward towards a theory of turbulence which could replace current models.

The chapter "Turbulent Convection and Numerical Simulations in Solar and Stellar Astrophysics" provides a discussion of turbulent convection in the Sun and other stars as well as an introduction to the numerical simulation of the latter. After discussing the applicability of the hydrodynamical approach and motivating the astrophysical interest in this kind of work, solar convection is chosen to explain various fallacies when dealing with large-scale turbulent flows. An example is the laminar appearance of very high Reynolds-number flows visualized by the structures observed at the solar surface. This example is also used to demonstrate how straightforward measurement techniques such as Doppler broadening of spectral lines can remain inconclusive in some situations, while less direct ones (such as helioseismology) can indeed be decisive. The focus is then given on numerical simulations and how they deal with unresolved, small scales. The possible pitfalls in time integration of numerical simulations and in the gathering of reliable ensemble averages is discussed. Finite computational resources introduce compromises and impose the necessity to choose the length and time scales accounted for in the simulation very carefully. It is shown that grid refinement techniques can be useful to resolve shear-driven turbulence while at the same time a large enough sample of coherent up- and down-flow structures can be included within the simulation volume. The influence of different types of boundary conditions is discussed for the case of simulations of convection near the solar surface. It is shown that the simulations can yield robust results on averaged properties of the flow in the interior of a convective zone, even if artificial (closed, stress-free) boundaries are used in the simulations, at the expense of not considering a large domain close to the boundaries for interpretation. Finally, problems related to turbulent mixing are discussed including some of the difficulties caused by the finite parameter space accessible to even the most advanced numerical simulations.

The chapter "Turbulence in Astrophysical and Geophysical Flows" first provides a basic introduction into turbulence in astro- and geophysical flows. It demonstrates how the same methods can be applied in the different fields and explains step by step, how physically more and more complete Reynolds stress models of turbulent flows can be built using the same methodology to account for different physical effects. They include buoyancy driven turbulence (turbulent convection) due to a gradient in temperature, in mean molecular weight, or any combination thereof. The influence of rotation is discussed, particularly on the transport of angular momentum and on mixing in stars. The competing role of these different physical mechanisms in predictions of (turbulent) mixing is discussed as well. This provides the background for the discussion of a complete, state-of-the-art Reynolds stress model for turbulent convection, which is shown to work well for convection in the dry (cloudless) boundary layer found in the atmosphere of the Earth. A connection to convection models based on plumes is made. It demonstrates how coherent structures can be accounted for in the Reynolds stress approach. Convection and mixing in the ocean is then used as another example how the same methodology can be used for the study of different physical systems. The chapter concludes with some of the limitations introduced by how small-scale dissipation is treated in all these models.

The chapter "Turbulence in the Lower Troposphere: Second-Order Closure and Mass-Flux Modelling Frameworks" deals with the modelling of turbulence in the lower atmosphere of the Earth in numerical weather prediction models. Such models have to deliver reliable predictions in a well-defined amount of time on available resources. They are a good example for what kind of compromises are chosen in simulations of turbulent flows when the goal is reliability of the entire model using available, finite resources. More complete physical models have to be discarded, if they do not blend well with other parts of the model or are simply unaffordable (it is of very limited practical use to receive a weather prediction for today not until tomorrow !). In such modelling, small-scale turbulence is commonly dealt with by Reynolds stress models, while mass flux models are used for the large-scale, coherent structures created by convection. Both methods are hence introduced in detail and then compared to each other. It is explained how the different physical mechanisms (creation and destruction of turbulence, etc.) are accounted for. This comparison is useful to translate between two types of averages: ensemble averaging and (mass) flux averaging. The differences between formally similar equations obtained by these two approaches are discussed, as well as a number of simplifications used in actual weather forecast programmes. Since weather is a very non-linear, multi-scale phenomenon, a practical issue is that an increase in computational power and thus in the number of length and time scales that can be accounted for in a numerical model does not always lead to improvements. This happens when a new "regime" is reached and a parametrization, valid at course resolution, breaks down. Thus, the different components of a forecast model should work together also when minimum and maximum scales within the numerical simulation change. Hybrid schemes, even based on different types of averaging, are thus highly interesting for numerical weather prediction, and the possibilities for such improved schemes are evaluated.

In the chapter "Magnetohydrodynamic Turbulence" the focus is changed to magnetohydrodynamic (MHD) turbulence. The latter is of interest not only to astrophysics, but also to the study of thermonuclear fusion for energy production on Earth, since magnetic confinement has developed into the most promising device to store hot plasmas for the reaction process. After introducing the dynamical equations, numerical simulations are used to study several MHD problems. Differences between the case of two and three spatial dimensions and between the magnetic and the non-magnetic case are explained. The predictions of kinetic and magnetic energy distribution according to several phenomenological models is probed with numerical simulations which resolve all scales down to the dissipation range. Anisotropy and the macroscopic structure of MHD turbulence are investigated. Classical statistical tools of turbulence modelling such as ensemble averaged two-point correlation functions and structure functions are used to explain the results and complement the visualizations.

The chapter "Turbulent Combustion in Thermonuclear Supernovae" discusses the subject of turbulent combustion in type Ia (thermonuclear) supernovae. Again this problem is characterized by an enormous range of scales and transitions between different flow regimes are expected to occur at least during long-term simulation runs. Since the flame surface itself cannot possibly be resolved by a simulation of an entire exploding star (not even, if the number of spatial dimensions is reduced to two or the computations are limited to sectors assuming additional symmetries in the flow), a modelling concept is required. This is taken from engineering flows (level-set technique). Then, an introduction into sub-grid-scale modelling is given, since the turbulent flow speed determines the rate of energy production in the complex and highly wrinkled flame front. This difference is essentially due to a larger area of a turbulent flame compared to a laminar one. The role of sub-grid-scale modelling is illustrated by a number of simulations. Finally, a comparison with astrophysical experimental data is made demonstrating that many of the observed properties such as change of total visual brightness of supernovae as a function of time and the production of ^{56}Ni isotopes can be reproduced by the hydrodynamical simulations.

The chapter "One-Dimensional Turbulence Stochastic Simulation of Multi-scale Dynamics" introduces a new approach for modelling the multi-scale behaviour of turbulent flows that we deal with in numerical weather prediction, turbulent combustion, and other phenomena described in the preceding chapters. The stochastic simulation of turbulent processes proposed in the 'ODT' approach resolves some of the problems encountered in current large-eddy simulations, where the unresolved, small scales are left to various kinds of parametrizations. This puts limitations on the accuracy with which we can predict the interchange of energy between resolved and unresolved scales and is of major concern, whenever the flow is directly coupled to dissipation (or other unresolved) scale processes. The stochastic simulations in the ODT framework share a number of concepts with lattice-gas hydrodynamics and lattice Boltzmann models proposed earlier. In the end, the new approach aims at performing a simulation of turbulent flow processes on length and time scales not accounted for in the simulation. However, no assumptions are made such as the volume-averaged flow quantities being equivalent to some kind of ensemble average represented by mean quantities, with a functional form claimed to be known in advance. Rather, such information is obtained only during the numerical simulation and in this way provides a feedback between scales contained within the hydrodynamical simulations and those only considered within the stochastic simulation. The advantage of this approach is the much higher resolution which can be achieved in one dimension. Some applications are presented which illustrate the potential of this method.

References

1. Batchelor, G.K.: An Introduction to Fluid Dynamics. Cambridge University Press, Cambridge (1967)
2. Beck, C., Cohen, E.G.D.: Physica A **322**, 267 (2003)
3. Bellomo, N., Gerasimenko, V., Petrina D.Ya.: BBGKY hierarchy and nonlinear kinetic theories. In: Series on Advances in Mathematics for Applied Sciences, World Scientific, Singapore (1994)
4. Chandrasekhar, S.: Hydrodynamic and Hydromagnetic Stability. Clarendon Press, Oxford (1961)
5. Draad, A.A., Kuiken, G.D.C., Nieuwstadt, F.T.M.: J. Fluid Mech. **377**, 267 (1998)
6. Friedrich, R., Peinke, J.: Physica D **102**, 147 (1997)
7. Frisch, U.: Turbulence. Cambridge University Press, Cambridge (1995)
8. Heisenberg, W.: Z. Phys. **124**, 628 (1948)
9. Heisenberg, W.: Proc. Roy. Soc. London, Series A **195**, 402 (1948)
10. Huang, K.: Statistical Mechanics. John Wiley & Sons, Inc., New York (1963)
11. Kestener, P., Arneodo, A.: Phys. Rev. Lett. **93**, 044501 (2004)
12. Kolmogorov, A.N.: Dokl. Akad. Nauk SSSR **30**, 301 (1941)
13. Kolmogorov, A.N.: J. Fluid Mech. **12**, 82 (1962)
14. Kosterlitz, J.M., Thouless, D.J.: J. Phys. **C6**, 1181 (1973)

15. Landau, L.D., Lifshitz, E.M.: Fluid Mechanics. Vol. 6 of Course of Theoretical Physics. Pergamon Press, Reading, Mass., (1963)
16. Lesieur, M.: Turbulence in Fluids, 3rd ed. Kluwer Academic Publishers, Dordrecht (1997)
17. Lumley, J.L. (ed.): *Whither Turbulence? Turbulence at the Crossroads*, Lect. Notes Phys. **357**. Springer Verlag, New York (1989)
18. Montgomery, D.C., Tidman, D.A.: Plasma Kinetic Theory, Advanced Physics Monograph Series, McGraw-Hill, New York (1964)
19. Oboukhov, A.M.: Dokl. Akad. Nauk SSSR **32**, 19 (1941)
20. Oboukhov, A.M.: J. Fluid Mech. **12**, 77 (1962)
21. Onsager, L.: Phys. Rev. **68**, 286 (1945) (abstract only)
22. Petrina, D.Ya., Gerasimenko, V.I., Malyshev, P.V.: Mathematical foundations of classical statistical mechanics: Continuous systems. In: Advanced Studies in Contemporary Mathematics, 2nd edn. Taylor & Francis, London (2002)
23. Shu, F.H.: The Physics of Astrophysics, Vol. II, Gas Dynamics. University Science Press, Mill Valley (1992)
24. Sreenivasan, K.R., Antonia, R.A.: Annu. Rev. Fluid Mech. **29**, 435 (1997)
25. Thompson, P.D.: A review of the predictability problem. In: Predictability of fluid motions, La Jolla Institute 1983, Holloway, G., West B.J. (eds.) AIP Conference Proceedings Vol. p. 1 106, American Institute of Physics, New York, (1984)
26. Tillmark, N., Alfredsson, P.H.: J. Fluid Mech. **235**, 89 (1992)
27. Townsend, A.A.: The Structure of Turbulent Shear Flow, 2nd edn. Cambridge University Press, Cambridge (1976)
28. Tsallis, C.: J. Stat. Phys. **52**, 479 (1988)
29. von Weizsaecker, C.F.: Z. Phys. **124**, 614 (1948)
30. Wilson, K.G.: Rev. Mod. Phys. **47**, 773 (1975)

Nonextensive Statistical Mechanics
and Nonlinear Dynamics

C. Tsallis

Santa Fe Institute, 1399 Hyde Park Road, Santa Fe, NM, USA
tsallis@santafe.edu
Centro Brasileiro de Pesquisas Fisicas, Xavier Sigaud 150, 22290-180 Rio de
Janeiro-RJ, Brazil
tsallis@cbpf.br

1 Introduction

This chapter addresses turbulence and related questions from a statistical–
mechanical viewpoint. More precisely, as one of the many existing realizations
of nontrivial–nonlinear dynamical phenomena, which stands at the grounding
of statistical mechanics itself. Consistently, we shall first review some general
aspects of the statistical approach of mechanical phenomena, and only later
make a connection with turbulence, in Sect. 4.4.

A mechanical foundation of statistical mechanics from *first principles*
should essentially include, in one way or another, the following main steps.

(i) Adopt a *microscopic dynamics*. This dynamics is typically deterministic, i.e. without any phenomenological noise or stochastic ingredient, so that
the foundation may be considered as *from first principles*. This dynamics could
be Newtonian, or quantum, or relativistic mechanics (or some other mechanics
to be found in future) of a many-body system composed by say N interacting elements or fields. It could also be conservative- or dissipative-coupled
maps, or even cellular automata. Consistently, time t could be continuous
or discrete. The same is valid for space. The quantity which is defined in
space-time could itself be continuous or discrete. For example, in quantum
mechanics, the quantity is a complex continuous variable (the wave function)
defined in a continuous space-time. On the other extreme, we have cellular
automata, for which all three relevant variables—time, space and the quantity
therein defined—are discrete. In the case of a Newtonian mechanical system
of particles, we may think of N Dirac delta functions localized in continuous
spatial positions which depend on a continuous time.

Langevin-like equations (and associated Fokker–Planck-like equations) are
typically considered not microscopic, but *mesoscopic* instead. The reason, of
course, is the fact that they include at their very formulation, i.e. in an essential manner, some sort of noise. Consequently, they should not be used as a

Tsallis, C.: *Nonextensive Statistical Mechanics and Nonlinear Dynamics.* Lect. Notes
Phys. **756**, 21–48 (2009)
DOI 10.1007/978-3-540-78961-1_2   © Springer-Verlag Berlin Heidelberg 2009

starting point if we desire the foundation of statistical mechanics to be from first principles.

(ii) Then assume some set of *initial conditions* and let the system evolve in time. These initial conditions are defined in the so-called *phase space* of the microscopic configurations of the system, for example Gibbs' Γ space for a Newtonian N-particle system (the Γ space for point masses has $2dN$ dimensions if the particles live in a d-dimensional space). These initial conditions typically (but not necessarily) involve one or more constants of motion. For example, if the system is a conservative Newtonian one of point masses, the initial total energy and the initial total linear momentum (d dimensional vector) are such constants of motion. If the masses have some spatial extent, then the total angular momentum is also a constant of motion. It is quite frequent to use coordinates such that both total linear momentum and total angular momentum vanish.

If the system consists of conservative-coupled maps, the initial hypervolume of an ensemble of initial conditions near a given one is preserved through time evolution. By the way, in physics, such coupled maps are frequently obtained through Poincaré sections of Newtonian dynamical systems.

(iii) After some *sufficiently long evolution time* (which typically depends on both N and the spatial range of the interactions), the system might approach some *stationary* or *quasi-stationary* macroscopic state (when the system has some kind of *ageing*, the expression *quasi-stationary* is preferable to *stationary*). In such a state, the various regions of phase space are being visited with some probabilities. This set of probabilities either does not depend anymore on time, or depends on it very slowly. More precisely, if it depends on time, it does so on a scale much longer than the microscopic time scale. The visited regions of phase space, that we are referring to, typically correspond to a partition of phase space with a degree of (coarse or fine) graining that we adopt for specific purposes. These probabilities can be either insensitive or, on the contrary, very sensitive to the ordering in which the $t \to \infty$ (*asymptotic*) and $N \to \infty$ (*thermodynamic*) limits are taken. This can depend on various things such as the range of the interactions, or whether the system is on the ordered or on the disordered side of a continuous phase transition. Generically speaking, the influence of the ordering of the $t \to \infty$ and $N \to \infty$ limits is typically related to some kind of breakdown of symmetry .

The simplest dynamical situation is expected to occur for an isolated many-body short-range-interacting classical Hamiltonian system (microcanonical ensemble); later on we shall qualify when an interaction is considered *short*-ranged in the present context. In such a case, the typical microscopic (nonlinear) dynamics is expected to be *strongly chaotic*, in the sense that the maximal Lyapunov exponent is positive. Such a system would necessarily be *mixing*, i.e. it would quickly visit virtually all the accessible phase space (more precisely, very close to almost all the accessible phase space) for almost *any* possible initial condition. Furthermore, it would necessarily be *ergodic* with respect to some measure in the full phase space, i.e. *time averages* and *ensemble averages* would coincide. In most of the cases this measure is expected to be uniform in phase space, i.e. the *hypothesis of equal probabilities* would be satisfied.

A slightly more complex situation is encountered for those systems which exhibit a continuous phase transition. Let us consider the simple case of a ferromagnet which is invariant under inversion of the (hard) axis of magnetization, e.g. the three-dimensional ($d = 3$) XY classical nearest-neighbour ferromagnetic model on a simple cubic lattice, i.e. coupled planar rotators localized on a lattice. If the system is in its disordered (paramagnetic) phase, the limits $t \to \infty$ and $N \to \infty$ commute, and the entire phase space is expected to be equally well visited. If the system is in its ordered (ferromagnetic) phase, the situation is expected to be more subtle. The $\lim_{N\to\infty} \lim_{t\to\infty}$ set of probabilities is, as before, equally distributed all over the entire phase space *for almost any initial condition*. But this is not expected to be so for the $\lim_{t\to\infty} \lim_{N\to\infty}$ set of probabilities. The system probably lives, in this case, only in half of the entire phase space. Indeed, if the initial condition is such that the initial magnetization is *positive*, even infinitesimally positive, then the system is expected to be ergodic *but only in the half phase space associated with positive magnetization*; the other way around occurs if the initial magnetization is *negative*. This illustrates, already in this simple example, the importance that the ordering of those two limits can have.

A considerably more complex situation is expected to occur if we consider a *long*-range-interacting model, e.g. the same $d = 3$ XY classical ferromagnetic model on a simple cubic lattice as before, but now with a coupling constant which decays with distance as $1/r^\alpha$, where r is the distance measured in crystal units, and $0 \le \alpha < d$ (the nearest-neighbour model that we just discussed corresponds to $\alpha \to \infty$, which is the extreme case of the *short*-ranged domain $\alpha > d$). The $0 \le \alpha/d < 1$ model also appears to have a continuous phase transition. In the disordered phase, the system is possibly ergodic over the entire phase space. But in the ordered phase the result possibly strongly depends on the ordering of the two limits. The $\lim_{N\to\infty} \lim_{t\to\infty}$ set of probabilities corresponds to the system living in the entire phase space. In contrast, the $\lim_{t\to\infty} \lim_{N\to\infty}$ set of probabilities for the same (conveniently scaled) total energy might be considerably more complex. It seems that, for this ordering, phase space exhibits at least two macroscopic basins of attraction. One of them leads essentially to half of the same phase space where the system lives in the $\lim_{N\to\infty} \lim_{t\to\infty}$ ordering, i.e. the half phase space which is associated with a sign for the magnetization which coincides with the sign of the initial magnetization. The other basin of attraction could correspond well to living in a very complicated, hierarchical-like, geometrical structure. This structure could be a zero Lebesgue measure one (in the full multidimensional phase space), somewhat similar to that of an airlines company, say Varig, whose main hub is located at Sao Paulo, or Continental Airlines, whose main hub is located at Houston. The specific *location* of the structure in phase space would depend on the particular initial condition within that special basin of attraction, but the *geometrical nature* of the structure would be virtually the same for any initial condition within that basin of attraction. The scenario that we have depicted here is conjectural, not yet proved. It is however based on various numerical evidences (see [1–3] and references therein). It is expected to be caused by a possibly *vanishing* maximal Lyapunov exponent. In other words, one would possibly have, instead of strong, *weak chaos*.

(iv) We have to focus now further on the specific role played by the *initial conditions*. If the system is strongly chaotic, hence mixing, hence ergodic, this point is irrelevant. We can make or not averages over initial conditions, we can take almost any initial condition, the outcome for sufficiently long times will be the same, in the sense that the set of probabilities in phase space will be the same. But if the system is only weakly chaotic, the result can drastically change from initial condition to initial condition. If both initial conditions belong to the same basin of attraction, the difference at the macroscopic level could be quite irrelevant. If they belong however to different basins of attraction, the results can be sensibly different. For some purposes we might wish to stick to a specific initial condition within a certain class of initial conditions. For other purposes, we might wish to average over all initial conditions belonging to a given basin of attraction, or even over all possible initial conditions of the entire phase space. The macroscopic result obtained after averaging might considerably differ from that corresponding to a single initial condition.

(v) Last but not least, the mathematical form of the *entropy functional* must be addressed. Strictly speaking, if we have deduced (from microscopic dynamics) the probabilities to be associated with every cell in phase space, we can, in principle, calculate useful averages of *any* physical quantity of interest which is defined in that phase space. In this sense, we do not need to introduce an entropic functional which is defined precisely in terms of those probabilities. Especially if we take into account that *any* set of physically relevant probabilities can be obtained through *extremization* (typically *maximization*) of an infinite number of entropic functionals (monotonically depending one onto the other), given any set of physically and mathematically meaningful constraints. However, if we wish to make contact with classical thermodynamics, we certainly need to know the mathematical form of such an entropic functional. This functional is expected to match, in the appropriate limits, the classical, macroscopic, entropy *à la Clausius*. In particular, one expects it to satisfy the Clausius property of extensivity, i.e. essentially to be proportional to the weight or mass of the system. In statistical–mechanical terms, we expect it to be proportional to N for large N.[1]

---

[1] Let us anticipate that it has been recently shown [4, 5] that, if we impose a Poissonian distribution for visitation times in phase space, in addition to the first and second principles of thermodynamics, we obtain the Boltzmann–Gibbs functional form for the entropy. If a conveniently deformed Poissonian distribution is imposed instead, we obtain the S_q functional form. These results in themselves cannot be considered as a justification from first principles of the

The foundations of any statistical mechanics are, as already said, expected to cover basically all of the above points. There is a wide-spread vague belief among physicists that these steps have already been satisfactorily accomplished since long for the standard, Boltzmann–Gibbs (BG) statistical mechanics. *This is not so!* Not so surprising after all, given the enormity of the corresponding task! For example, at this date, there is no available deduction, from and only from microscopic dynamics, of the validity, for thermal equilibrium with a thermostat, of the celebrated BG exponential weight

$$p_i = \frac{e^{-\beta E_i}}{\sum_{j=1}^{W} e^{-\beta E_j}}, \tag{1}$$

where β is proportional to the inverse temperature T of the thermostat, and $\{E_i\}$ are the eigenvalues of the Hamiltonian of the system (with a given set of boundary conditions). Neither exists the deduction from microscopic dynamics of the BG entropy

$$S_{BG} = -k \sum_{i=1}^{W} p_i \ln p_i \quad \left(\sum_{i=1}^{W} p_i = 1 \right), \tag{2}$$

where W is the total number of possible microscopic states and k is a conventional positive constant (typically taken to be either the Boltzmann constant k_B, or simply unity).

For standard systems, there is not a single reasonable doubt about the correctness of the expressions (1) and (2), and of their relationships. But, from the logical-deductive viewpoint, there is still pretty much work to be done! This is, in fact, a kind of easy to notice. Indeed, all the textbooks, without exception, introduce the BG factor and/or the entropy S_{BG} in some kind of phenomenological manner, or as self-evident, or within some axiomatic formulation. None of them introduces them as (and only as) a rational consequence of Newtonian, or quantum mechanics, using theory of probabilities. This is in fact sometimes referred to as the *Boltzmann program*. Boltzmann himself died without succeeding its implementation. Although important progress has been accomplished in these last 130 years, the Boltzmann program still remains in our days as a basic intellectual challenge. Were it not the genius of scientists like Boltzmann and Gibbs, were we to exclusively depend on mathematically well-constructed arguments, one of the monuments of contemporary physics—BG statistical mechanics—would not exist!

Many anomalous natural, artificial and social systems exist for which BG statistical concepts appear to be inapplicable. Typically because they live in peculiar stationary or quasi stationary states that are quite different from

---

Boltzmann–Gibbs, or of the nonextensive, statistical mechanics. Indeed, the visitation distributions are phenomenologically introduced, and the first and second principles are just imposed. This connection is nevertheless extremely clarifying, and can help producing a full justification.

thermal equilibrium, where BG statistics reigns. Nevertheless some of them can still be handled within statistical mechanical methods, but with a more general entropy, namely [6–8]

$$S_q \equiv k\frac{1 - \sum_{i=1}^{W} p_i^q}{q - 1} = k\sum_{i=1}^{W} p_i \ln_q(1/p_i) = -k\sum_{i=1}^{W} p_i^q \ln_q p_i \quad (S_1 = S_{\mathrm{BG}}), \quad (3)$$

where

$$\ln_q x \equiv \frac{x^{1-q} - 1}{1 - q} \quad (\forall x > 0;\ \ln_1 x = \ln x). \tag{4}$$

It should be clear that, whatever is not yet mathematically justified in BG statistical mechanics, it is even less justified in the present generalization. In addition to this, some of the points that are relatively well understood in the standard theory, can be still unclear in its generalization. In other words, the theory we are presenting here is still in intense evolution. In any case, reviews on the subject can be found in [9–20], and a bibliography can be found in [21].

In the present monograph, we review some important properties related to the extensivity and the entropy production of S_q. We then make some connection with strongly nonlinear dynamical phenomena such as turbulence.

2 Extensivity of the (Nonadditive) Entropy S_q

2.1 Remark on the Thermodynamical Limit

Let us assume a classical mechanical many-body system characterized by the following Hamiltonian:

$$\mathcal{H} = K + V = \sum_{i=1}^{N} \frac{p_i^2}{2m} + \sum_{i \neq j} V(r_{ij}), \tag{5}$$

where $V(r)$ presents no mathematical difficulties at the origin $r = 0$ (e.g. it is either nonsingular, or, if it is singular, it is integrable), and which behaves at long distances like

$$V(r) \sim -\frac{A}{r^\alpha} \quad (A > 0;\ \alpha \geq 0). \tag{6}$$

A typical example would be the $d = 3$ Lennard–Jones gas model, for which $\alpha = 6$. Were it is not the nonintegrable singularity at the origin, another example could be Newtonian $d = 3$ gravitation, for which $\alpha = 1$.

Let us analyse the characteristic average potential energy U_{pot} per particle

$$\frac{U_{\mathrm{pot}}(N)}{N} \propto -A\int_1^\infty dr\, r^{d-1}\, r^{-\alpha}, \tag{7}$$

where we have integrated from a typical distance (taken equal to unity) on. This is the typical energy one would calculate within a BG approach. We see immediately that this integral *converges* for $\alpha/d > 1$ (hereafter referred to as *short-range interactions* for classical systems) but *diverges* for $0 \leq \alpha/d \leq 1$ (hereafter referred to as *long-range interactions*). This already indicates that something anomalous is happening.[2] By the way, the fact is historically fascinating that Gibbs himself was aware of the possibility of such difficulty! Indeed, in his 1902 book [22], he wrote:

In treating of the canonical distribution, we shall always suppose the multiple integral in Eq. (92) *[the partition function, as we call it nowadays]* to have a finite value, as otherwise the coefficient of probability vanishes, and the law of distribution becomes illusory. This will exclude certain cases, but not such apparently, as will affect the value of our results with respect to their bearing on thermodynamics. It will exclude, for instance, cases in which the system or parts of it can be distributed in unlimited space [...]. It also excludes many cases in which the energy can decrease without limit, as when the system contains material points which attract one another inversely as the squares of their distances. [...]. For the purposes of a general discussion, it is sufficient to call attention to the assumption implicitly involved in the formula (92).

On a vein slightly differing from the standard BG recipe, which would demand integration up to infinity in Eq. (7), let us assume that the N-particle system is roughly homogeneously distributed within a limited sphere. Then Eq. (7) has to be replaced by the following one:

$$\frac{U_{\text{pot}}(N)}{N} \propto -A \int_{1}^{N^{1/d}} dr\, r^{d-1}\, r^{-\alpha} = -\frac{A}{d}\tilde{N}, \tag{8}$$

with

$$\tilde{N} \equiv \frac{N^{1-\alpha/d} - 1}{1 - \alpha/d} = \ln_{\alpha/d} N. \tag{9}$$

---

[2] This is essentially the very same reason for which virtually all statistical mechanics textbooks discuss paradigmatic systems like a particle in a square well, the harmonic oscillator, the rigid rotator and a spin $1/2$ in the presence of an external magnetic field, *but not the Hydrogen atom! All* these simple systems, including, of course, the Hydrogen atom, are discussed in the quantum mechanics textbooks. But, in what concerns statistical mechanics, the Hydrogen atom is an illustrious absent. Amazingly enough, with extremely rare exceptions, this highly important system passes with no comments at all in the textbooks on thermal statistics. The—understandable but not justifiable—reason of course is that, since the system involves the long-range Coulombian attraction between electron and proton, the energy spectrum exhibits an accumulation point at the ionization energy (frequently taken to be zero), which makes the BG partition function to diverge.

Therefore, in the $N \to \infty$ limit, $U_{pot}(N)/N$ approaches a constant ($\propto -A/ (\alpha - d)$) if $\alpha/d > 1$, and diverges like $N^{1-\alpha/d}/(1 - \alpha/d)$ if $0 \leq \alpha/d < 1$ (it diverges logarithmically if $\alpha/d = 1$). In other words, the energy is *extensive for short-range interactions* ($\alpha/d > 1$) and *nonextensive for long-range interactions* ($0 \leq \alpha/d \leq 1$).

A totally similar situation occurs if we have, in our Hamiltonian (5), say N rotators localized on a d-dimensional lattice. The coupling constant between site i and site j could then depend on distance like $1/r_{ij}^{\alpha}$, where r_{ij} runs over all sites of the lattice. For example, for a linear chain, we would have $r_{ij} = 1, 2, 3, ...$; for a square lattice we would have $r_{ij} = 1, \sqrt{2}, 2, ...$ In this case, the average potential energy per rotator would be proportional to $\sum_{i \neq j} r_{ij}^{-\alpha}$. This quantity has a behaviour totally similar to $\tilde{N}$ as defined in Eq. (9). This is to say, for $N \gg 1$, it converges to a (positive) constant if $\alpha/d > 1$ and diverges proportionally to $N^{1-\alpha/d}$ if $0 \leq \alpha/d < 1$ (logarithmically if $\alpha/d = 1$).

We are now prepared to make a thermodynamical remark ([23] and references within [24]).

Let us assume an N-sized system characterized by say its *temperature* T, *pressure* p and *external magnetic field* H. Its Gibbs thermodynamical energy will be given by

$$G(N,T,p,H) = U(N,T,p,H) - TS(N,T,p,H)$$
$$+ pV(N,T,p,H) - HM(N,T,p,H), \tag{10}$$

where U, S, V and M are respectively its *internal energy*, *entropy*, *volume* and *magnetization*. It follows that

$$\frac{G(N,T,p,H)}{N\tilde{N}} = \frac{U(N,T,p,H)}{N\tilde{N}} - \frac{T}{\tilde{N}} \frac{S(N,T,p,H)}{N}$$
$$+ \frac{p}{\tilde{N}} \frac{V(N,T,p,H)}{N} - \frac{H}{\tilde{N}} \frac{M(N,T,p,H)}{N}. \tag{11}$$

We can now apply the operation $\lim_{N \to \infty}$ on both members of this equality. We then obtain

$$g\left(\tilde{T}, \tilde{p}, \tilde{H}\right) = u\left(\tilde{T}, \tilde{p}, \tilde{H}\right) - \tilde{T}s\left(\tilde{T}, \tilde{p}, \tilde{H}\right) + \tilde{p}v\left(\tilde{T}, \tilde{p}, \tilde{H}\right) - \tilde{H}m\left(\tilde{T}, \tilde{p}, \tilde{H}\right), \tag{12}$$

where the definitions of the new variables are self-explanatory. For example, $g(\tilde{T}, \tilde{p}, \tilde{H}) \equiv \lim_{N \to \infty} G(N,T,p,H)/(N\tilde{N})$, $s(\tilde{T}, \tilde{p}, \tilde{H}) \equiv \lim_{N \to \infty} S(N,T,p,H) /N$, $\tilde{T} \equiv \lim_{N \to \infty} T/\tilde{N}$ and so on. These scalings have already been verified in various systems (see [24] and references therein). For example the $\alpha/d = 0$ particular case corresponds to the usual mean field approach. Indeed, in this case we have $\tilde{N} = N - 1 \sim N$, which is equivalent to the usual rescaling of the microscopic coupling constant through division by N (see also [25]).

For short-range interactions, $\tilde{N} \to constant$, consequently we recover the usual *extensivity* of Gibbs, Helmholtz and internal thermodynamical energies,

entropy, volume and magnetization, as well as the *intensivity* of temperature, pressure and magnetic field. But for long-range interactions, $\tilde{N}$ diverges
with N, therefore the situation is quite more subtle. Indeed, in order to
have *nontrivial equations of states* we must express the *nonextensive* Gibbs,
Helmholtz and internal thermodynamical energies, as well as the *extensive* entropy, volume and magnetization in terms of the rescaled variables $(\tilde{T}, \tilde{p}, \tilde{H})$.
In general, i.e. $\forall (\alpha/d)$, we see that the variables that are intensive when the
interactions are short-ranged remain a *single class* (although scaling with $\tilde{N}$)
in the presence of long-ranged interactions. But, in what concerns the variables that are extensive when the interactions are short-ranged, the situation
is more complex. Indeed, they split into *two classes*. One of them contains all
types of thermodynamical energies (G, F, U), which scale with $N\tilde{N}$. The other
one contains all those variables (S, V, M) that appear *in pairs* in the thermodynamical energies. These variables remain extensive, in the sense that they
scale with N.

By no means this implies that thermodynamical equilibrium between two
systems occurs in general when they share the same values of say $(\tilde{T}, \tilde{p}, \tilde{H})$.
It only means that, in order to have *finite* mathematical functions for their
equations of states, the variables $(\tilde{T}, \tilde{p}, \tilde{H})$ must be used. Although this has
to be verified, thermodynamical equilibrium might still be directly related to
sharing the usual variables (T, p, H).

It is clear that all this is quite subtle, and easily subject to error. Nevertheless, it constitutes a strong indication that S_{BG} has to be generalized
without violating its extensivity, i.e. as introduced on macroscopic grounds
by Clausius. What we present in the next subsection is perfectly consistent
with this expectation.

2.2 On How Global Correlations Mandate the Generalization of the (Additive) Entropy S_{BG}

Let us consider that the N (identical) elements of a system are independent
in the probabilistic sense, i.e. the probabilities of the states of the N-system
satisfy $p_{i_1,i_2,\ldots,i_N}^{A_1+A_2+\ldots+A_N} = p_{i_1}^{A_1} p_{i_2}^{A_2} \cdots p_{i_N}^{A_N}, \forall (i_1, i_2, \ldots, i_N)$. It can be straightforwardly verified that

$$
\begin{aligned}
S_{\mathrm{BG}}(N) &\equiv -k \sum_{i_1,i_2,\ldots,i_N} p_{i_1,i_2,\ldots,i_N}^{A_1+A_2+\ldots+A_N} \ln p_{i_1,i_2,\ldots,i_N}^{A_1+A_2+\ldots+A_N} \\
&= -k \sum_{i_1,i_2,\ldots,i_N} \left(p_{i_1}^{A_1} p_{i_2}^{A_2} \cdots p_{i_N}^{A_N} \right) \ln \left(p_{i_1}^{A_1} p_{i_2}^{A_2} \cdots p_{i_N}^{A_N} \right) \\
&= -k \sum_{i_1,i_2,\ldots,i_N} \left(p_{i_1}^{A_1} p_{i_2}^{A_2} \cdots p_{i_N}^{A_N} \right) \sum_{i_n} \ln p_{i_n}^{A_n} \\
&= N \left[-k \sum_{i_n} p_{i_n}^{A_n} \ln p_{i_n}^{A_n} \right] = N S_{\mathrm{BG}}(1).
\end{aligned}
\tag{13}
$$

This is to say, the BG entropy is, in this case, not only extensive but even strictly additive. In a more complex system in which the joint probabilities do not exactly, but only asymptotically, factorize, we expect $\lim_{N\to\infty} S_{\mathrm{BG}}(N)/N$ to be *finite*, even if S_{BG} is not strictly proportional to N. This is what typically occurs when the correlations are local, but not global. Such is the case for a nonideal gas (e.g. the Lennard–Jones model), for magnets whose microscopic interactions are nonzero only within some finite neighbourhood (e.g. among first neighbours on some lattice), etc.

When global correlations exist in the system, the situation can drastically change. Indeed, it can happen that S_{BG} is not extensive anymore. And *some other form of entropic functional becomes necessary in order to re-establish Clausius-like extensivity*. We shall now exhibit one such probabilistic system [26, 27]. We consider that the system is constituted by N identical and distinguishable binary variables. The probabilities are indicated in Table 1. The details can be found in [26]. However, the basic idea is that, in the triangle shown in Table 1, only a strip (the "left" strip in fact) whose width is $d + 1$ ($d = 1, 2, 3, ...$) has nonvanishing probabilities. All other probabilities are zero. This is to say, although we have 2^N possible states, only a small fraction of them (increasingly small when N increases) is probabilistically occupied. More precisely, the number of states that have nonzero probability increases with N like N^d. It is the fact that $\lim_{N\to\infty} N^d/2^N = 0$ which makes that S_{BG} cannot be extensive anymore. In contrast, the generalized entropy (3) can be extensive. This occurs if and only if

$$q = 1 - \frac{1}{d}. \tag{14}$$

The extensivity of $S_{1-(1/d)}$ is illustrated in Fig. 1.

Summarizing what we observed in this section: when the probabilistic correlations are either inexistent or only local, S_q is extensive only for $q = 1$; if the correlations occur instead at a global scale, S_{BG} is typically nonextensive, whereas S_q can be extensive for a special value of q. This suggests that, for all types of probabilistic systems, there might exist at least one entropic functional which is extensive, and therefore smoothly matches classical thermodynamics in the $N \to \infty$ limit. Connections with the Central Limit Theorem can be seen in [28–32].

3 Entropy Production Per Unit Time for S_q

In the previous Section we addressed the dependence of the entropy on N. We now address its dependence on time t (discrete or continuous). The property that we shall analyse is the so called *entropy production per unit time*. The reader should however be aware that the same name is used in the literature with meanings that do not necessarily exactly coincide with the present one. We indistinctively address dissipative or conservative dynamical systems

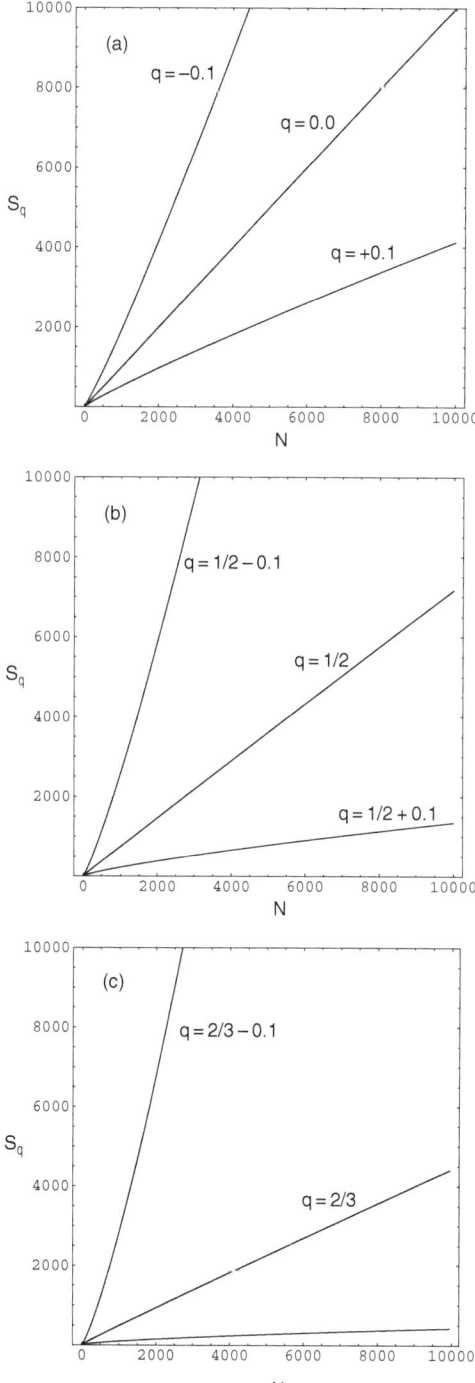

Fig. 1. $S_q(N)$ for anomalous systems: (**a**) $d = 1$, (**b**) $d = 2$, and (**c**) $d = 3$. Only for $q = 1 - (1/d)$ we have a *finite* value for $\lim_{N \to \infty} S_q(N)/N$; it *vanishes* (*diverges*) for $q > 1 - (1/d)$ ($q < 1 - (1/d)$). From [26]

(typically *nonlinear* dynamical systems), with either few or many degrees of freedom (or dimensions, or particles).

Let us partition the phase space in W little cells (typically equal in size), designated by $i = 1, 2, ..., W$. Let us then randomly choose M initial conditions within one of those cells, and follow their time evolution. The occupancies of the cells will be noted $M_i(t)$ ($\sum_{i=1}^{W} M_i(t) = M, \forall t$). We can define a set of probabilities through $p_i(t) \equiv M_i(t)/M$, hence calculate the entropy

$$S_q(t) \equiv \frac{1 - \sum_{i=1}^{W}[p_i(t)]^q}{1 - q} \quad \left[S_1(t) = - \sum_{i=1}^{W} p_i(t) \ln p_i(t) \right], \quad (15)$$

where q can be any desired value. Our definition of *entropy production per unit time* is as follows:

$$K_q \equiv \lim_{t \to \infty} \lim_{W \to \infty} \lim_{M \to \infty} \frac{S_q(t)}{t}. \quad (16)$$

K_1 is a concept analogous to the so called *Kolmogorov–Sinai entropy* (or entropy rate). The latter is based on single trajectories in phase space, whereas the former is based, as we have seen, on an ensemble of initial conditions. They coincide for many systems, but there might well be situations in which they differ. K_q is clearly a generalization of $K_1 \equiv K_{BG}$.

What frequently (perhaps virtually always) occurs is that a special and unique value of q exists (noted q_e, where e stands for *entropy*), such that K_q *vanishes* for $q > q_e$, *diverges* for $q < q_e$ and is *finite* for $q = q_e$. It is of course q_e and K_{q_e} what characterizes the dynamics of the system. If the system is *strongly chaotic*, in the sense that it has at least one *positive* Lyapunov

Table 1. Anomalous probability sets: $d = 1$ (*top*) and $d = 2$ (*bottom*). The left number within parentheses indicates the multiplicity (i.e. Pascal triangle). The right number indicates the corresponding probability. The probabilities, noted $r_{N,n}$, asymptotically satisfy the Leibnitz rule, i.e. $\lim_{N \to \infty}(r_{N,n} + r_{N,n+1})/r_{N-1,n} = 1$ ($\forall n$). In other words, the system is, in this sense, asymptotically scale-invariant. From [26]

$(N = 0)$	$(1, 1)$
$(N = 1)$	$(1, 1/2)(1, 1/2)$
$(N = 2)$	$(1, 1/2)(2, 1/4)(1, 0)$
$(N = 3)$	$(1, 1/2)(3, 1/6)(3, 0)(1, 0)$
$(N = 4)$	$(1, 1/2)(4, 1/8)(6, 0)(4, 0)(1, 0)$
$(N = 0)$	$(1, 1)$
$(N = 1)$	$(1, 1/2)(1, 1/2)$
$(N = 2)$	$(1, 1/3)(2, 1/6)(1, 1/3)$
$(N = 3)$	$(1, 3/8)(3, 5/48)(3, 5/48)(1, 0)$
$(N = 4)$	$(1, 2/5)(4, 3/40)(6, 3/60)(4, 0)(1, 0)$

exponent, then $q_e = 1$ and $K_1 > 0$. If the system is *weakly chaotic* (i.e. its maximum Lyapunov exponent *vanishes*), then typically $q_e < 1$, $K_1 = 0$ and $K_{q_e} > 0$. The quantities q_e and K_{q_e} are intimately related to the *sensitivity to the initial conditions* (see [33–60] and references therein). For example, if the system has a one-dimensional phase space (noted x), the sensitivity to the initial conditions is defined as

$$\xi \equiv \lim_{\Delta x(0) \to 0} \frac{\Delta x(t)}{\Delta x(0)}, \tag{17}$$

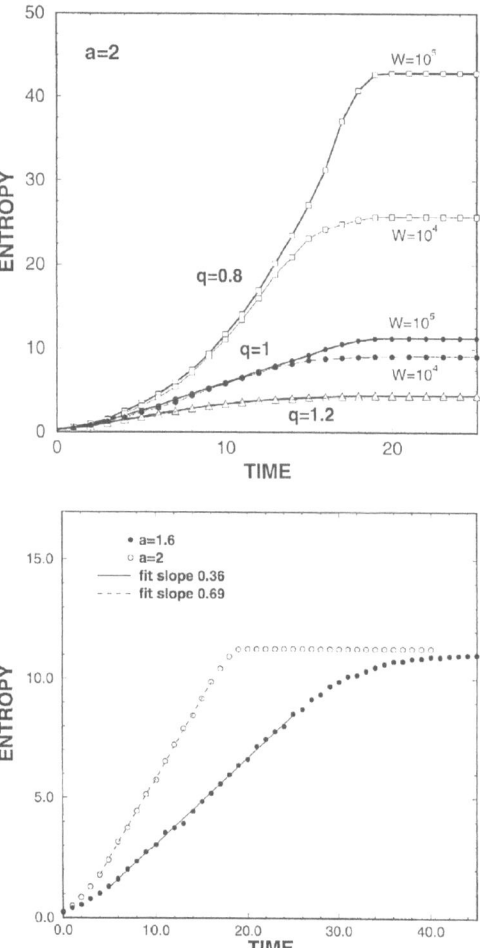

Fig. 2. Time dependence of S_q for strong chaos in the logistic map $x_{t+1} = 1 - ax_t^2$. *Top:* For $a = 2$ we verify $K_1 = \lambda_1 = \ln 2 \simeq 0.69$. The effect of increasing the number W of cells of the partition and that of varying q are exhibited. *Bottom:* For $a = 2$ and $a = 1.6$ we obtain respectively $K_1 = \lambda_1 \simeq 0.69$ and $\simeq 0.36$. K_q vanishes (diverges) if $q > 1$ ($q < 1$). From [61]

where $\Delta x(t)$ denotes the difference between two initially close trajectories.

If the system has a positive (or negative) Lyapunov exponent λ_1, then $d\xi/dt = \lambda_1 \xi$, hence

$$\xi = e^{\lambda_1 t}. \tag{18}$$

If the system has a zero Lyapunov exponent λ_1 (e.g. at the edge of chaos), then $d\xi/dt = \lambda_{q_{sen}} \xi^{q_{sen}}$ (where sen stands for *sensitivity*), hence

$$\xi = e_{q_{sen}}^{\lambda_{q_{sen}} t}, \tag{19}$$

where the q-exponential function is the inverse of the previously defined q-logarithmic function, i.e.

$$e_q^x \equiv [1 + (1-q)\, x]^{1/(1-q)} \quad (e_1^x = e^x) \tag{20}$$

if $[1 + (1-q)\, x] \geq 0$, and zero otherwise.

These concepts are illustrated in Figs. 2 and 3 (for a dissipative one-dimensional unimodal map), as well as in Fig. 4 (for a conservative two-dimensional map [41]).

As a final remark, let us stress the very suggestive fact that the N-dependence and the t-dependence of S_q are strikingly similar. *Strong chaos* eventually leads to a full occupation of phase space. Both aspects mandate S_{BG} as the appropriate entropy. Indeed, it is S_{BG} which is *extensive*, and it is again S_{BG} which has a *finite* entropy production per unit time. *Weak chaos*, although not necessarily, may lead to a partial occupation of phase space. Both aspects typically point onto an entropy like S_q, which, for a special

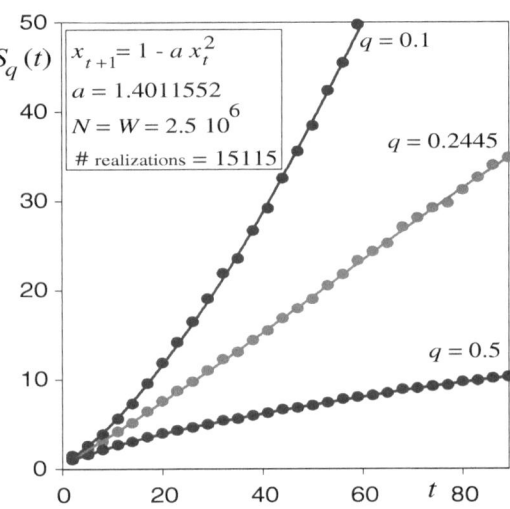

Fig. 3. Time dependence of S_q for the (first) edge of chaos of the logistic map $x_{t+1} = 1 - ax_t^2$ (Courtesy of F. Baldovin)

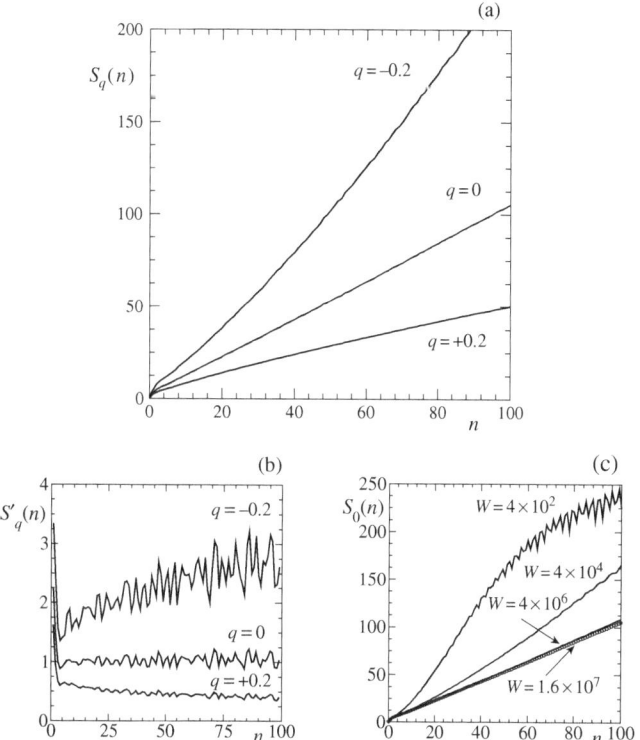

Fig. 4. Time-evolution of the statistical entropy S_q for different values of q. The phase space has been divided into $W = 4,000 \times 4,000$ equal cells of size $l = 5 \times 10^{-4}$ and the initial ensemble is characterized by $N = 10^3$ points randomly distributed inside a partition-square. Curves are the result of an average over 100 different initial squares randomly chosen in phase space. The analysis of the derivative of S_q in (**b**) shows that only for $q = 0$ a linear behaviour is obtained. In fact, a linear regression provides $S_0(n) = 1.029\,n + 1.997$ with a correlation coefficient $R = 0.99993$. (**c**) shows that the linear growth for S_0 is reached from above, in the limit $W \to \infty$. From [41]

value of q (typically below unity), exhibits both *extensivity* and *finite* entropy production. Further comments about this deep and intriguing property can be found in [27, 62].

4 Superstatistics and Connection With Turbulence

A variety of generalizations and extensions of the entropy S_q and/or of its associated statistics are already available in the literature. We briefly review here some of them, namely the *crossover statistics* [63, 64], the *Beck–Cohen superstatistics* [65–74], and finally the quite general *q-spectral statistics* [75].

Crossover statistics contains q-statistics as a particular case. Superstatistics contains crossover statistics as a particular case. Finally q-spectral statistics possibly contains superstatistics as a particular case.

4.1 Crossover Statistics

The optimization of S_q with appropriate constraints yields the following distribution for the stationary state:

$$p_i = \frac{e_q^{-\beta\,(E_i - U_q)}}{\sum_{j=1}^{W} e_q^{-\beta\,(E_j - U_q)}}, \tag{21}$$

where

$$U_q \equiv \frac{\sum_{j=1}^{W} p_j^q E_j}{\sum_{j=1}^{W} p_j^q}, \tag{22}$$

β being an inverse-temperature Lagrange parameter which can be determined as a function of U_q by using relation (22).

Equation (21) can be rewritten as follows:

$$p_i = \frac{e_q^{-\beta_q\,E_i}}{\sum_{j=1}^{W} e_q^{-\beta_q\,E_j}}, \tag{23}$$

where β_q is a simple function of β and U_q. In order to simplify the notation, we shall use $(p_i, E_i) \to (p, E)$. Then, from Eq. (23) it follows that

$$p(E)/p(0) = e_q^{-\beta_q\,E}. \tag{24}$$

This relation is the solution of the following ordinary differential equation:

$$\frac{d\,[p/p(0)]}{dE} = -\beta_q\,[p/p(0)]^q. \tag{25}$$

Clearly, this approach recovers the BG factor $p(E)/p(0) = e^{-\beta_1\,E}$ as the $q = 1$ particular instance.

We can now easily unify the BG factor and the q-factor by postulating

$$\frac{d\,[p/p(0)]}{dE} = -\beta_1\,[p/p(0)] - (\beta_q - \beta_1)\,[p/p(0)]^q. \tag{26}$$

If we have $q = 1$ or $\beta_q = \beta_1$ we recover the BG weight. If we have $\beta_1 = 0$ we recover the q-statistical weight. The general solution of Eq. (26) is given by

$$p/p(0) = \frac{1}{\left[1 + \frac{\beta_q}{\beta_1}\left(e^{(q-1)\,\beta_1\,E} - 1\right)\right]^{\frac{1}{q-1}}}. \tag{27}$$

For $q > 1$ and $0 < \beta_1 \ll \beta_q$ we have an interesting *crossover* (from which this statistics is named). Indeed, for $0 < (q-1)\beta_1 E \ll 1$, we have $p(E)/p(0) \sim e_q^{-\beta_q E}$, whereas, for $(q-1)\beta_1 E \gg 1$, we have $p(E) \propto e^{-\beta_1 E}$. In other words, by increasing the energy E, the stationary state distribution makes a crossover from q-statistics to BG statistics.

As an aside remark, let us notice that, in the limit $\beta_q/\beta_1 \to \infty$ and $p(0)\beta_1/\beta_q \to C$, where C is a constant, Eq. (27) becomes

$$p(E) = \frac{C}{\left[e^{(q-1)\beta_1 E} - 1\right]^{\frac{1}{q-1}}}. \tag{28}$$

If $q = 2$ this distribution becomes

$$p(E) = \frac{C}{e^{\beta_1 E} - 1}. \tag{29}$$

If we multiply this statistical weight by the $d = 3$ photon density of states $g(E) \propto E^2$ and by the energy E, we recover the celebrated Planck law for the black-body radiation

$$u(\nu) \propto \frac{\nu^3}{e^{h\nu/k_B T} - 1}, \tag{30}$$

where we have identified $\beta_1 \to 1/k_B T$ and $E \to h\nu$. In this sense, the distribution (27) can be seen, interestingly enough, as a generalization of Planck statistics.

4.2 Superstatistics

The superstatistical approach was introduced by Beck and Cohen [65], and proceeds along quite different lines. We follow here the version recently published in [74].

We start with a BG factor $e^{-\beta E}$, and focus on the frequent situation where β itself is *not* well defined. Indeed, spatio-temporal fluctuations can make β itself to behave as a random variable. It is from this fact that the word *superstatistics* was coined. Indeed, it is in some sense a *statistics of statistics*. The *BG* factor can be generalized as follows:

$$p(E) = \int_0^\infty d\beta \, f(\beta) \frac{1}{Z(\beta)} g(E) e^{-\beta E}, \tag{31}$$

where $g(E)$ is the density of states and $Z(\beta)$ is the normalization constant associated with $g(E)e^{-\beta E}$ for a given β. If the distribution $f(\beta)$ is a Dirac delta function, then we recover BG statistics. But $f(\beta)$ can be a quite general distribution. It can be, for instance, a χ^2 *distribution* (sometimes also called *Gamma distribution*) with degree n, i.e.

$$f(\beta) = \frac{1}{\Gamma(n/2)} \left(\frac{n}{2\beta_0}\right)^{n/2} \beta^{n/2-1} e^{-n\beta/(2\beta_0)}, \tag{32}$$

where β_0 is the average of β. If we introduce this $f(\beta)$ into Eq. (31), we precisely obtain q-statistics with

$$q = 1 + \frac{2}{n+d}, \qquad (33)$$

d being the space dimension where the system lives. This is to say $p(E)$ decays like a power law with E.

We can in fact define, in general, the parameter

$$q_{BC} \equiv \frac{\langle \beta^2 \rangle}{\langle \beta \rangle^2}, \qquad (34)$$

where $\langle (\ldots) \rangle \equiv \int_0^\infty d\beta\, f(\beta)(\ldots)$, and BC stands for *Beck–Cohen*. If $f(\beta)$ is the χ^2 distribution, then we have $q_{BC} = q$.

It might happen that it is not β, but $1/\beta$, which follows a χ^2 distribution. In this case, we have

$$f(\beta) = \frac{\beta_0}{\Gamma(n/2)} \left(\frac{n\beta_0}{2} \right)^{n/2} \beta^{-n/2-2}\, e^{-n\beta_0/(2\beta)}, \qquad (35)$$

If we introduce this $f(\beta)$ into Eq. (31), we obtain a $p(E)$ which exponentially decays with E.

A third interesting situation refers to β being log-normally distributed, i.e.

$$f(\beta) = \frac{1}{\sqrt{2\pi}s\beta}\, e^{-\frac{[\ln(\beta/\mu)]^2}{2s^2}}, \qquad (36)$$

where μ and s^2 are convenient mean and variance parameters. The distribution $p(E)$ has no analytical expression in this case, but it can of course be numerically calculated without any particular difficulty. As we shall mention later on, it is precisely this case which appears to play a special role for both Lagrangian and Eulerian turbulence.

Generally speaking, we see from its definition (31) that superstatistics $p(E)$ is mathematically well defined if $f(\beta)$ has a Laplace or Laplace-like transform. We have not checked this for the crossover distribution (27). This is to say we have not attempted to calculate the inverse Laplace or Laplace-like transform of distribution (27). However, (27) being a smooth unification of BG statistics and q-statistics, it seems quite plausible that the corresponding $f(\beta)$ distribution would be well defined. It is for this reason that we believe that crossover statistics is a particular case of superstatistics.

It is clear that distributions $p(E)$ might exist which have no $f(\beta)$ distribution to be generated from. Essentially because the inverse Laplace or Laplace-like transform does not exist. Such cases cannot be seen as superstatistical. This leads us to the Sect. 4.3, where we present a procedure which does not seem to require the existence of Laplace-like transforms, being thus more general.

4.3 q-Spectral Statistics

Within this approach, we do not assume, like in superstatistics, that β is a random variable. We rather assume that q is a random variable. In other words, instead of having the distribution $f(\beta)$, we are now going to have a *spectral distribution* $\kappa(q)$, such that

$$\left| \int_{-\infty}^{\infty} dq\, \kappa(q) \right| < \infty. \tag{37}$$

Since q-statistics is itself a superstatistics—the χ^2 one—, and since the present approach considers q itself to be some kind of random variable, spectral statistics appears to be, amusingly enough, some sort of "super-superstatistics"!

The procedure consists in a natural generalization of the differential equation (26). In that crossover statistics, we have two q-indices that are involved, namely $q = 1$ and an arbitrary value of q. We can then generalize (26) as follows:

$$\frac{d\,[p/p(0)]}{dE} = \int_{-\infty}^{\infty} d\tau\, \kappa(\tau)[p/p(0)]^{\tau}. \tag{38}$$

If $\kappa(\tau) \propto \delta(\tau - q)$ ($\delta(x)$ being Dirac Delta distribution), $p(E)$ is just q-statistics. If $\kappa(\tau) \propto a\,\delta(\tau - 1) + b\,\delta(\tau - q)$, we basically have the crossover statistics generated by Eq. (26). We may have a more general situation, namely that corresponding to $\kappa(\tau) \propto a\,\delta(\tau - r) + b\,\delta(\tau - q)$, whose solution can be analytically solved in terms of hypergeometric functions (see [63, 64]). It is clear that we can have many classes of spectral functions $\kappa(q)$, and some of them are discussed in [75], to where we refer the reader for further details.

Superstatistics are well defined for $f(\beta)$ that have a Laplace-like transform. The q-spectral statistics are well defined when $\kappa(q)$ is such that Eq. (38) admits a physically acceptable solution. Which one is more general? The answer is not trivial, and would deserve special analysis. However, since the trivial case $\kappa(\tau) \propto \delta(\tau - q)$ already is, as mentioned above, a superstatistics, it might well be that the present approach is more general than the superstatistical one. Rigorously speaking, this technical point remains, at the present stage, as an open question.

4.4 Application to Turbulence

Many quantities are of interest in order to physically understand a hard problem such as turbulence. A central such quantity is the distribution of differences of velocities. These differences can refer to the same location at two different times, or at two different locations at the same time, or even to more complex situations. The velocity difference distributions are, over several decades, very well fitted by q-Gaussians [76–90] (see examples in Figs. 5 and 8). However, when further decades are accessible (i.e. for large differences of velocities), a gradual departure from χ^2 superstatistics might be observed.

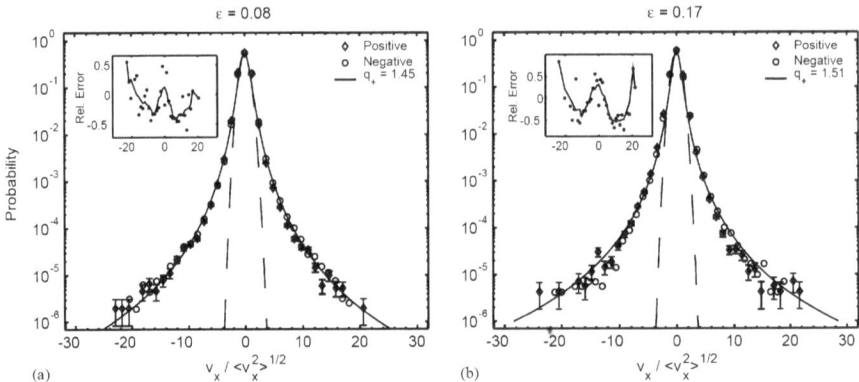

Fig. 5. Transverse velocity distributions at (**a**) $\epsilon = 0.08$ and (**b**) $\epsilon = 0.17$. Details in [84]

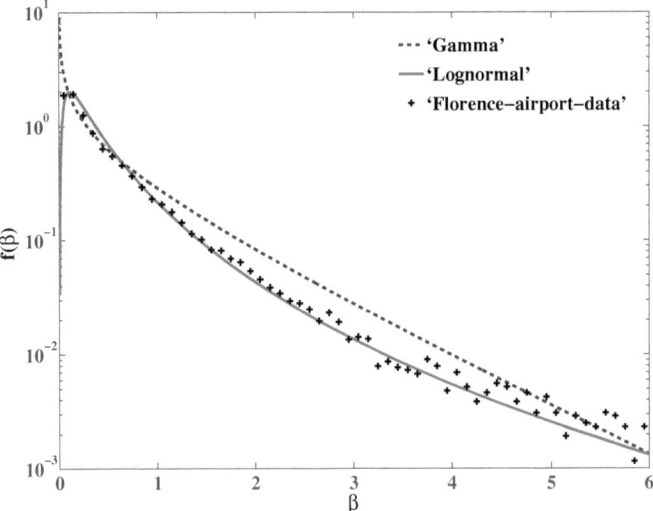

Fig. 6. Rescaled probability density of the fluctuating parameter β, as obtained for the Florence airport data. Also shown is a Gamma distribution (*dashed line*) and a lognormal distribution (*solid line*) sharing the same mean and variance as the data. The data are reasonably well fitted by the log-normal distribution. From [91]

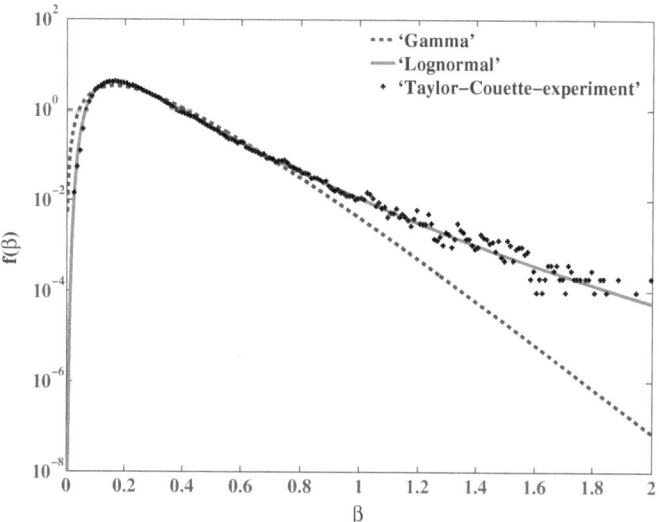

Fig. 7. Probability density of the local variance parameter β as extracted from a time series of longitudinal velocity differences measured in a turbulent Taylor–Couette flow at Reynolds number Re = 540000. A log-normal distribution yields a good fit. From [91]

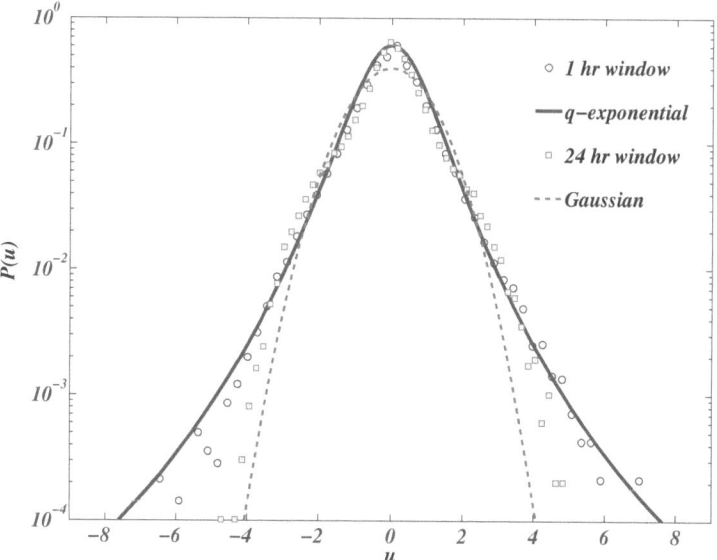

Fig. 8. Probability density $p(u)$ of temporal wind velocity differences as observed at the airport. From [91]

It seems then that, in some cases, the log-normal superstatistics does a better job [74, 91] (see examples in Figs. 6, 7 and 8).

There is no surprise that q-statistics and log-normal superstatistics coincide for relatively small velocity differences. This is in fact expected for *all* superstatistics on quite general grounds [65]. Now, why log-normal superstatistics describes so well the velocity difference distributions of a variety of turbulent phenomena remains, to the best of our knowledge, an open question.

5 Conclusions

We have shown that the value of the index q for which the entropy S_q is extensive is intimately related to the nature of the probabilistic correlations between the N elements of the system. If the correlations are inexistent or only local, then $q = 1$. If they are global, then $q < 1$. This exhibits a fundamental point: if we wish *to preserve Clausius-like extensivity for the entropy in order to adequately match classical thermodynamics, we may need to consider entropies S_q different from the* BG *one* (see also [92–96]). How these properties are entangled with the range of the interactions has also been discussed.

We have also shown that the value of the index q for which the entropy S_q yields a *finite* entropy production per unit time is intimately related to the sensitivity to the initial conditions. If the system is strongly chaotic, i.e. if the maximal Lyapunov exponent is positive, then $q = 1$. If the system is weakly chaotic, i.e. if the maximal Lyapunov exponent vanishes, then $q < 1$. This exhibits another fundamental point: if we wish *to produce, while exploring phase space, a finite amount of entropy per unit time, we may need to consider entropies S_q different from the* BG *one.*

These two facts are so amazingly similar that there must be some deep relation between the N-dependence and the t-dependence of the entropy. In fact, it seems reasonable to conjecture (as done in [62]) that, under some circumstances, it is one and the same dependence. In other words, the entropy would depend on Nt, and not separately on N and on t.

Finally, we have briefly reviewed three successive generalizations of q-statistics, namely *crossover statistics*, which is contained in *superstatistics*, which in turn is possibly contained in *q-spectral statistics*. How q-statistics and superstatistics (especially the log-normal superstatistics) are connected with turbulence has also been exhibited. The basic ingredient for the applicability of nonextensive statistics is to have a power-law, and *not* exponential, dependence on time of the sensitivity to the initial conditions. This fact, together with the adequacy of such statistics for describing the velocity distributions in very turbulent systems, suggests that the physically most enlightening standpoint might be to focus on relatively long-ranged interactions between structures such as vortices or others.

Acknowledgements

I acknowledge warm hospitality of W. Hillebrandt and F. Kupka at both the Garching Max-Planck-Institute for Astrophysics and Schloss Ringberg, as well as financial support by SI International and AFRL. I thank F. Kupka also for his critical reading of the manuscript. Finally, I thank the authors of Figs. 5, 6, 7 and 8, who gratiously authorized me to reproduce them here.

References

1. Rapisarda, A., Pluchino, A.: Nonextensive thermodynamics and glassy behavior, in Nonextensive Statistical Mechanics: New Trends, New perspectives. In: Boon, J.P., Tsallis, C. (eds.) Europhysics News **36**, 202 (2005)
2. Tamarit, F.A., Anteneodo, C.: Relaxation and aging in long-range interacting systems, in Nonextensive Statistical Mechanics: New Trends, New perspectives. In: Boon, J.P., Tsallis, C. (eds.) Europhysics News **36**, 194 (2005)
3. Thurner, S.: Nonextensive statistical mechanics and complex networks, in Nonex-tensive Statistical Mechanics: New Trends, New perspectives. In: Boon, J.P., Tsallis, C. (eds.) Europhysics News **36**, 218 (2005)
4. Carati, A.: Thermodynamics and time averages. Physica A **348**, 110 (2005)
5. Carati, A.: Time-averages and the heat theorem. In: Beck, C., Benedek, G., Rapisarda, A., Tsallis, C. (eds.) Complexity, Metastability and Nonextensivity, World Scientific, Singapore **42**, 55 (2005)
6. Tsallis, C.: Possible generalization of Boltzmann-Gibbs statistics. J. Stat. Phys. **52**, 479 (1988)
7. Curado, E.M.F., Tsallis, C.: Generalized statistical mechanics: connection with thermodynamics. J. Phys. A **24**, L69 (1991) [Corrigenda: **24**, 3187 (1991) and **25**, 1019 (1992)]
8. Tsallis, C., Mendes, R.S., Plastino, A.R.: The role of constraints within gener-alized nonextensive statistics. Physica A **261**, 534 (1998)
9. Salinas, S.R.A., Tsallis, C. (eds.): Nonextensive Statistical Mechanics and Ther-modynamics. Braz. J. Phys. **29**, Number 1 (1999)
10. Abe, S., Okamoto, Y. (eds.): Nonextensive Statistical Mechanics and Its Ap-plications, Series Lecture Notes in Physics. Springer, Berlin (2001)
11. Grigolini, P., Tsallis, C., West, B.J. (eds.) Classical and Quantum Complexity and Nonextensive Thermodynamics, Chaos, Solitons and Fractals **13**, Issue 3 (2002)
12. Kaniadakis, G., Lissia, M., Rapisarda, A. (eds.): Non Extensive Thermody-namics and Physical Applications. Physica A **305**, 1/2 (2002)
13. Sugiyama, M. (ed.): Nonadditive Entropy and Nonextensive Statistical Mechan-ics, Continuum Mechanics and Thermodynamics **16**. Springer-Verlag, Heidel-berg (2004)
14. Swinney, H.L., Tsallis, C. (eds.): Anomalous Distributions, Nonlinear Dynam-ics and Nonextensivity. Physica D **193** (2004)
15. Gell-Mann, M, Tsallis, C. (eds.): Nonextensive Entropy – InterdisciplinaryAp-plications. Oxford University Press, New York (2004)
16. Kaniadakis, G., Lissia, M. (eds.): News and Expectations in Thermostatistics. Physica A **340** (2004)

17. Herrmann, H.J., Barbosa, M., Curado, E.M.F. (eds.) Trends and Perspec-tives in Extensive and Non-extensive Statistical Mechanics. Physica A **344**, 3/4 (2004)
18. Beck, C. Benedek, G., Rapisarda, A., Tsallis, C. (eds.) Complexity, Metasta-bility and Nonextensivity. World Scientific, Singapore (2005)
19. Boon, J.P., Tsallis, C. (eds.): Nonextensive Statistical Mechanics: New Trends, New Perspectives. Europhysics News **36**, Number 6 (EDP Sciences, Les Ulis-France, 2005)
20. Abe, S., Sakagami, M., Suzuki, N. (eds.): Complexity and Nonextensivity: New Trends in Statistical Mechanics. Prog. Theor. Phys. Suppl. **162**, (2006)
21. http://tsallis.cat.cbpf.br/biblio.htm
22. Gibbs, J.W.: Elementary Principles in Statistical Mechanics – Developed with Especial Reference to the Rational Foundation of Thermodynamics (C. Scrib-ner's Sons, New York, 1902; Yale University Press, New Haven, 1948; OX Bow Press, Woodbridge, Connecticut, 1981)
23. Tsallis, C.: Extensive versus nonextensive physics, in New trends in magnetism, magnetic materials and their applications. In: Morán-López, J.L.,Sanchez, J.M. (eds.) p. 451. Plenum Press, New York (1994)
24. Andrade, R.F.S., Pinho, S.T.R.: Tsallis scaling and the long-range Ising chain: A transfer matrix approach. Phys. Rev. E **71**, 026126 (2005)
25. Anteneodo, C., Tsallis, C.: Breakdown of the exponential sensitivity to the initial conditions: Role of the range of the interaction. Phys. Rev. Lett. **80**, 5313 (1998)
26. Tsallis, C., Gell-Mann, M., Sato, Y.: Asymptotically scale-invariant occupancy of phase space makes the entropy Sq extensive. Proc. Natl. Acad. Sc. USA **102**, 15377 (2005)
27. Caruso, F., Tsallis, C.: Extensive q-entropy in quantum magnetic systems, preprint (2006) [cond-mat/0612032]
28. Moyano, L.G., Tsallis, C., Gell-Mann, M.: Numerical indications of a q-generalised central limit theorem. Europhys. Lett. **73**, 813 (2006)
29. Umarov, S., Tsallis, C., Steinberg, S.: A generalization of the central limit the-orem consistent with nonextensive statistical mechanics, preprint (2006) [cond-mat/0603593]
30. Umarov, S., Tsallis, C., Gell-Mann, M., Steinberg, S.: q-generalization of sym-metric α-stable distributions. Part I, preprint (2006) [cond-mat/0606038]
31. Umarov, S., Tsallis, C., Gell-Mann, M., Steinberg, S.: q-generalization of sym-metric α-stable distributions. Part II: preprint (2006) [cond-mat/0606040]
32. Umarov, S., Tsallis, C.: Multivariate generalizations of the q–central limit the-orem, preprint (2007) [cond-mat/0703533]
33. Tsallis, C., Plastino, A.R., Zheng, W.-M.: Power-law sensitivity to initial con-ditions – New entropic representation, Chaos, Solitons and Fractals **8**, 885 (1997)
34. Costa, U.M.S., Lyra, M.L., Plastino, A.R., Tsallis, C.: Power-law sensitivity to initial conditions within a logistic-like family of maps: Fractality and nonex-tensivity. Phys. Rev. E **56**, 245 (1997)
35. Lyra, M.L., Tsallis, C.: Nonextensivity and multifractality in low-dimensional dissipative systems. Phys. Rev. Lett. **80**, 53 (1998)
36. Tirnakli, U., Ananos, G.F.J., Tsallis, C.: Generalization of the Kolmogorov-Sinai entropy: Logistic -like and generalized cosine maps at the chaos threshold. Phys. Lett. A **289**, 51 (2001)

37. Borges, E.P., Tsallis, C., Ananos, G.F.J., Oliveira, P.M.C.: Nonequilibrium probabilistic dynamics at the logistic map edge of chaos. Phys. Rev. Lett. **89**, 254103 (2002)
38. Borges, E.P., Tirnakli, U.. Mixing and relaxation dynamics of the Henon map at the edge of chaos, in Anomalous Distributions, Nonlinear Dynamics and Nonextensivity. In: Swinney, H.L., Tsallis, C. (eds.). Physica D **193**, 148 (2004)
39. Borges, E.P., Tirnakli, U.: Two-dimensional dissipative maps at chaos threshold: Sensitivity to initial conditions and relaxation dynamics. Physica A **340**, 227 (2004)
40. Ananos, G.F.J., Baldovin, F., Tsallis, C.: Anomalous sensitivity to initial conditions and entropy production in standard maps: Nonextensive approach. Eur. Phys. J. B **46**, 409 (2005)
41. Casati, G., Tsallis, C., Baldovin, F.: Linear instability and statistical laws of physics. Europhys. Lett. **72**, 355 (2005)
42. Baldovin, F., Robledo, A.: Sensitivity to initial conditions at bifurcations in one-dimensional nonlinear maps: Rigorous nonextensive solutions. Europhys. Lett. **60**, 518 (2002)
43. Baldovin, F., Robledo, A.: Universal renormalization-group dynamics at the onset of chaos in logistic maps and nonextensive statistical mechanics. Phys. Rev. E **66**, R045104 (2002)
44. Baldovin, F., Robledo, A.: Nonextensive Pesin identity. Exact renormalization group analytical results for the dynamics at the edge of chaos of the logistic map. Phys. Rev. E **69**, 045202(R) (2004)
45. Mayoral, E., Robledo, A.: Multifractality and nonextensivity at the edge of chaos of unimodal maps. Physica A **340**, 219 (2004)
46. Robledo, A.: Criticality in nonlinear one-dimensional maps: RG universal map and nonextensive entropy, in Anomalous Distributions, Nonlinear Dynamics and Nonextensivity. In: Swinney, H.L., Tsallis, C. (eds.) Physica D **193**, 153 (2004)
47. Robledo, A.: Critical fluctuations, intermittent dynamics and Tsallis statistics. Physica A **344**, 631 (2004)
48. Robledo, A.: Aging at the edge of chaos: Glassy dynamics and nonextensive statistics. Physica A **342**, 104 (2004)
49. Robledo, A.:Unifying laws in multi-disciplinary power-law phenomena: Fixed-point universality and non-extensive entropy, in Nonextensive Entropy – Inter-disciplinary Applications. In: Gell-Mann, M., Tsallis, C. (eds.) Oxford UniversityPress, New York (2004)
50. Robledo, A.: Intermittency at critical transitions and aging dynamics at edge of chaos, Proceedings of Statphys-Bangalore (2004). Pramana-J. Phys. **64**, 947 (2005)
51. Robledo, A.: Critical attractors and q-statistics, in Nonextensive Statistical Mechanics: New Trends, new perspectives. In: Boon, J.P., Tsallis, C. (eds.) Europhysics News **36**, 214 (2005)
52. Mayoral, E., Robledo, A.: Tsallis' q index and Mori's q phase transitions at edge of chaos. Phys. Rev. E **72**, 026209 (2005)
53. Robledo, A., Baldovin, F., Mayoral, E.: Two stories outside Boltzmann-Gibbs statistics: Mori's q-phase transitions and glassy dynamics at the onset of chaos, in Complexity, Metastability and Nonextensivity. Proc. 31st Workshop of the International School of Solid State Physics (20–26 July 2004, Erice-Italy). In:

Beck, C., Benedek, G., Rapisarda, A., Tsallis, C. (eds.) p. 43. World Scientific, Singapore (2005)

54. Robledo, A.: Unorthodox properties of critical clusters. Mol. Phys. **103**, 3025 (2005)

55. Baldovin, F., Robledo, A.: Parallels between the dynamics at the noise-perturbed onset of chaos in logistic maps and the dynamics of glass formation. Phys. Rev. E **72**, 066213 (2005)

56. Hernandez-Saldana, H., Robledo, A.: Fluctuating dynamics at the quasiperiodic onset of chaos, Tsallis q-statistics and Mori's q-phase thermodynamics. Physica A **370**, 286–300 (2006)

57. Grassberger, P.: Temporal scaling at Feigenbaum points and nonextensive thermodynamics. Phys. Rev. Lett. **95**, 140601 (2005)

58. Robledo, A.: Comment on "Temporal scaling at Feigenbaum points and nonextensive thermodynamics" by P. Grassberger, cond-mat/0510293 (2005)

59. Robledo, A.: Incidence of nonextensive thermodynamics in temporal scaling at Feigenbaum points. Physica A **370**, 449–460 (2006)

60. Tsallis, C.: Comment on "Temporal scaling at Feigenbaum points and nonextensive thermodynamics" by P. Grassberger, cond-mat/0511213 (2005)

61. Latora, V., Baranger, M., Rapisarda, A., Tsallis, C.: The rate of entropy increase at the edge of chaos. Phys. Lett. A **273**, 97 (2000)

62. Tsallis, C., Gell-Mann, M., Sato, Y.: Extensivity and entropy production, in Nonextensive Statistical Mechanics: New Trends, New Perspectives. In: Boon, J.P., Tsallis, C. (eds.), Europhysics News **36**, 186 (2005)

63. Tsallis, C., Bemski, G., Mendes, R.S.: Is re-association in folded proteins a case of nonextensivity? Phys. Lett. A **257**, 93 (1999)

64. Tsallis, C., Anjos, J.C., Borges, E.P.: Fluxes of cosmic rays: A delicately balanced stationary state. Phys. Lett. A **310**, 372 (2003)

65. Beck, C., Cohen, E.G.D.: Superstatistics. Physica A **322**, 267 (2003)

66. Tsallis, C., Souza, A.M.C.: Constructing a statistical mechanics for Beck-Cohen superstatistics. Phys. Rev. E **67**, 026106 (2003)

67. Souza, A.M.C., Tsallis, C.: Stability of the entropy for superstatistics. Phys. Lett. A **319**, 273 (2003)

68. Souza, A.M.C., Tsallis, C.: Stability analysis of the entropy for superstatistics. Physica A **342**, 132 (2004)

69. Beck, C.: Superstatistics: Theory and Applications, in Nonadditive entropy and nonextensive statistical mechanics. In: Sugiyama, M. (ed.) ContinuumMechanics and Thermodynamics **16**, 293. Springer-Verlag, Heidelberg (2004)

70. Cohen, E.G.D.: Superstatistics, in Anomalous Distributions, Nonlinear Dynamics and Nonextensivity. In: Swinney, H.L., Tsallis, C. (eds.) Physica D **193**, 35 (2004)

71. Cohen, E.G.D.: Boltzmann and Einstein: Statistics and dynamics – An unsolved problem, Boltzmann Award Lecture at Statphys-Bangalore-2004. Pramana **64**, 635 (2005)

72. Beck, C.: Superstatistics: Recent developments and applications, in Complexity, Metastability and Nonextensivity, In: Beck, C., Benedek, G., Rapisarda, A., Tsallis, C. (eds.) p. 33. World Scientific, Singapore (2005)

73. Souza, A.M.C., Tsallis, C.: Generalizing the Planck distribution, in Complexity, Metastability and Nonextensivity.In: Beck, C., Benedek, G., Rapisarda, A.,Tsallis, C. (eds.) p. 66 World Scientific, Singapore (2005)

74. Beck, C., Cohen, E.G.D., Swinney, H.L.: From time series to superstatistics. Phys. Rev. E **72**, 056133 (2005)
75. Tsekouras, G.A., Tsallis, C.: Generalized entropy arising from a distribution of q-indices. Phys. Rev. E **71**, 046144 (2005)
76. Beck, C.: Application of generalized thermostatistics to fully developed turbulence, Physica A **277**, 115 (2000)
77. Beck, C.: Scaling exponents in fully developed turbulence from nonextensive statistical mechanics, Physica A **295**, 195 (2001)
78. C. Beck, On the small-scale statistics of Lagrangian turbulence. Phys. Lett. A **287**, 240 (2001)
79. Beck, C., Lewis, G.S., Swinney, H.L.: Measuring non-extensivity parameters in a turbulent Couette-Taylor flow. Phys. Rev. E **63**, 035303 (2001)
80. Beck, C.: Generalized statistical mechanics and fully developed turbulence, Physica A **306**, 189 (2002)
81. Baroud, C.N., Swinney, H.L.: Nonextensivity in turbulence in rotating two-dimensional and three-dimensional flows. Physica D **184**, 21 (2003)
82. Beck, C.: Superstatistics in hydrodynamic turbulence, in Anomalous Distributions, Nonlinear Dynamics and Nonextensivity. In: Swinney, H.L., Tsallis C. (eds.) Physica D **193**, 195 (2004)
83. Jung, S., Storey, B.D., Aubert, J., Swinney, H.L.: Nonextensive statistical mechanics for rotating quasi-two-dimensional turbulence, in Anomalous Distributions, Nonlinear Dynamics and Nonextensivity. In: Swinney, H.L.,Tsallis, C. (eds.) Physica D **193**, 252 (2004)
84. Daniels, K.E., Beck, C., Bodenschatz, E.: Defect turbulence and generalized statistical mechanics, in Anomalous Distributions, Nonlinear Dynamics and Nonextensivity. In: Swinney, H.L., Tsallis, C. (eds.) Physica D **193**, 208 (2004)
85. Reynolds, A.M., Veneziani, M.: Rotational dynamics of turbulence and Tsallis Statistics. Phys. Lett. A **327**, 9 (2004)
86. Arimitsu, T., Arimitsu, N.: Multifractal analysis of fluid particle accelerations in turbulence, in Anomalous Distributions, Nonlinear Dynamics and Nonextensivity. In: Swinney, H.L., Tsallis, C. (eds.) Physica D **193**, 218 (2004)
87. Arimitsu, T., Arimitsu, N.: Harmonious representation of PDF's reflecting large deviations. Physica A **340**, 347 (2004)
88. Arimitsu, T., Arimitsu, N.: Multifractal analysis of turbulence and granular flow, in Complexity, Metastability and Nonextensivity. In: Beck, C., Benedek, G., Rapisarda, A., Tsallis, C. (eds.) p. 236. World Scientific, Singapore (2005)
89. Rizzo, S., Rapisarda, A.: Environmental atmospheric turbulence at Florence airport, 8th Experimental Chaos Conference (14–17 June 2004, Florence, Italy), American Institute of Physics Conference Proceedings **742**, 176 (2004).
90. Rizzo, S., Rapisarda, A.: Application of superstatistics to atmospheric turbulence, in Complexity, Metastability and Nonextensivity. In: Beck, C., Benedek, G., Rapisarda A., Tsallis, C. (eds.) p. 246. World Scientific, Singapore (2005)
91. Beck, C., Cohen, E.G.D., Rizzo, S.: Atmospheric turbulence and superstatistics, in Nonextensive Statistical Mechanics: New Trends, New perspectives. In: Boon, J.P., Tsallis C. (eds.) Europhysics News **36**, 189 (2005)
92. Rajagopal, A.K., Rendell, R.W.: Classical statistics inherent in a quantum density matrix, Phys. Rev. A **72**, 022322 (2005)
93. Marsh, J., Earl, S.: New solutions to scale-invariant phase-space occupancy for the generalized entropy S_q. Phys. Lett. A **349**, 146 (2005)

94. Abe, S.: Temperature of nonextensive systems: Tsallis entropy as Clausius entropy, Physica A **368**, 430–434 (2006)
95. Zander, C., Plastino, A.R.: Composite systems with extensive S_q (power law) entropies. Physica A **364**, 145–156 (2006)
96. Vignat, C., Plastino, A., Plastino, A.R.: Correlated Gaussian systems exhibiting additive power-law entropies. Phys. Lett. A **354**, 27–30 (2006)

Turbulent Convection and Numerical Simulations in Solar and Stellar Astrophysics

F. Kupka

Max-Planck-Institut für Astrophysik, Garching, Germany
fk@mpa-garching.mpg.de

Astrophysical fluid flows and particularly stellar convection are considered to be *turbulent* rather than *laminar*. This judgement is based on the Reynolds number Re := $(UL)/\nu$ which compares the relative importance of non-linear inertial to viscous forces. In astrophysical flows Re is usually in the range of $10^{10} \ldots 10^{14}$. U and L are typical velocities and length scales for the flow. L can either characterize the size of the entire system, or the size of the largest structures found in the flow, or the length scale for which the kinetic energy of the flow has its maximum. The velocity U is associated with this length scale. Hence, the Reynolds number is a dimensionless, but scale-dependent quantity. Even if the kinematic viscosity ν were several orders of magnitudes larger than in a terrestrial flow with similar U, ν would easily be overpowered by UL due to the enormous size of the astrophysical systems, which results in huge values of L and thus huge values of Re. Various definitions have been suggested for the meaning of the term *turbulence*. A practical one was given in Chap. I of [85]: a turbulent flow must be unpredictable (small errors in initial conditions amplify so much that deterministic predictions become impossible), mix a fluid more efficiently than molecular processes, and involve a wide range of spatial scales. This does not exclude *coherent structures* which are defined as spatial regions that at a given time show some organization with respect to any quantity related to the flow. The term *fully developed* is used to describe turbulence which can evolve without imposed constraints due to spatial boundaries, external forces and viscosity. Thus, independently of how large Re might be, no real turbulent flow can be fully developed at the largest scales or at the scales of energy exchange with its environment. In line with this notion the concept that *coherent structures emerge from chaos under the action of external constraints* has been suggested (Chap. IV in [85]) to understand many of the results found in the analysis of real turbulent flows.

Keeping these definitions in mind, we now take a closer look on *turbulent convection in stars* and especially in our Sun. Most of the dynamical processes on our Sun are related to its turbulent convection zone ([128], e.g.). A better understanding of these processes clearly has a direct impact on matters far

Kupka, F.: *Turbulent Convection and Numerical Simulations in Solar and Stellar Astrophysics*. Lect. Notes Phys. **756**, 49–105 (2009)
DOI 10.1007/978-3-540-78961-1_3   © Springer-Verlag Berlin Heidelberg 2009

beyond physics. But the very same processes are also interesting on their own: stellar turbulent convection is a multi-scale phenomenon, both in space and time, requiring a detailed understanding of hydrodynamics and its interaction with external or self-generated magnetic fields, with radiation, thermodynamics, nuclear processes, diffusion phenomena and self-gravitation.

Here, we discuss some basic and also surprising properties of the high Re number flows in stellar convection zones. Convection near the visible surface of stars is chosen as the main subject, since the most accurate observational (experimental) data are available for that case. In Sect. 1 we introduce the background for the hydrodynamical description of stars, the origin of convective instabilities in stars, and the main effects of convection accounted for in stellar models. In Sect. 2 we review the observational evidence for turbulence and turbulent convection in the Sun and other stars. Since modelling of turbulent convection is treated in detail in the chapter "Turbulence in Astrophysical and Geophysical Flows" and in the chapter "Turbulence in the Lower Troposphere: Second-Order Closure and Mass-Flux Modelling Frameworks" in this book, we only provide a brief overview on that subject in Sect. 3 and then focus on how to perform numerical simulations to probe the available models. The consequences of finite numerical resolution and the different concepts for dealing with unresolved length scales are presented together with the constraints put by accurate time integrations during such simulations. In Sect. 4 we discuss how realistic simulations have to be set up to minimize the limitations introduced by the available computational resources. Resolving shear driven turbulence in convection simulations provides a particularly demanding example. We continue with a discussion of the requirements that are put on time integrations when computing ensemble-averaged quantities. In Sect. 5 the influence of boundary conditions on simulations of solar surface convection is shown. Ensemble-averaged higher order moments are provided for comparisons of solar convection with convection in geophysical systems. In Sect. 6 we turn to the problem of convective turbulent mixing and scaling relations in layers around stellar convection zones, still one of the most urging questions in stellar astrophysics pushing both models and simulations of turbulent convection to their limits. Section 7 provides a summary and conclusions.

1 Convective Heat Transport and Mixing in Stars

The theory of stellar structure predicts matter in a star to be in a gaseous phase. Because of high temperatures the gas exists mostly in ionized form. A detailed treatise on the theory of stellar structure can be found, for instance, in [137]. In stars the gas is stratified because of self-gravitation created by the huge amount of mass, some $2 \cdot 10^{32}$ to 10^{35} g, in a comparatively small volume. During a long period of their evolution, when they convert hydrogen in their central region (*core*) into helium by nuclear fusion, stars have quite "terrestrial" average densities $\bar{\rho}$ between about 0.05 and 50 g cm^{-3}. Their central densities ρ_c are typically two orders of magnitudes larger than $\bar{\rho}$ while densities at the *surface*, the visible layers of stars, are some six orders of magnitudes smaller. During the later stages of evolution much lower values of $\bar{\rho}$ occur, in the range of 10^{-7} to 10^{-4} g cm^{-3}, as a result of huge stellar *envelopes*, which have expanded by two to three orders of magnitude in comparison with earlier phases during which hydrogen fusion takes place in the stellar core. At the same time central densities become more and more extreme. In the end this evolution process creates compact objects with $\bar{\rho}$ of 10^6 g cm^{-3} (*white dwarfs*) and even up to about 10^{14} g cm^{-3} (*neutron stars*) [117, 137], after much of the extended envelope has been transferred back into the interstellar environment. Initially, temperatures T range from about 2000 to 50,000 K at the surface to several 10^7 K in the centre. Core temperatures can exceed 10^9 K during late phases of stellar evolution [71, 137]. This distribution of temperatures and densities is the reason why matter in stars can be described as a fluid. Consequently, the equations of hydrodynamics are applicable to the study of the structure and evolution of stars.[1]

1.1 Hydrodynamical Description of Stars

The equations of hydrodynamics describe the evolution of mass density ρ, momentum density ρv and internal energy density $\rho\varepsilon$ of the gas. They read

$$\frac{\partial \rho}{\partial t} + \text{div}\,(\rho v) = 0, \tag{1}$$

$$\frac{\partial}{\partial t}(\rho v) + \text{div}\,[\rho(v \otimes v)] + \text{div}\,\boldsymbol{\Pi} = -\rho\,\text{grad}\,\Phi, \tag{2}$$

$$\frac{\partial}{\partial t}(\rho E) + \text{div}\,[(\rho E + p)v] + \text{div}\,h - \text{div}(\pi v) = -\rho v\,\text{grad}\,\Phi. \tag{3}$$

Equation (1) is the well-known continuity equation, while Eq. (2) is the Navier–Stokes equation. The expression $v \otimes v$ is the dyadic product of v with itself. The pressure tensor $\boldsymbol{\Pi}$ appearing in (2) is commonly rewritten as

$$\boldsymbol{\Pi} = p\boldsymbol{I} - \boldsymbol{\pi}, \tag{4}$$

with $\boldsymbol{I}$ being the unit tensor, p the isotropic pressure from velocity fluctuations of the particles of the fluid and $\boldsymbol{\pi}$ the tensor viscosity with its components

$$\pi_{ik} = \eta\left(\frac{\partial v_i}{\partial x_k} + \frac{\partial v_k}{\partial x_i} - \frac{2}{3}\delta_{ik}\,\text{div}\,v\right) + \zeta\delta_{ik}\text{div}\,v. \tag{5}$$

---

[1] In the outermost layers of cool stars, dust particles can form. During late stages of white dwarf and neutron star evolution, solid state and more exotic phases can occur. We exclude these objects and stages from discussions in this review.

As usual, here δ_{ik} denotes the Kronecker symbol, while η and ζ are the molecular and the bulk viscosity coefficients respectively. The kinematic viscosity ν depends on the molecular viscosity η via $\nu = \eta/\rho$. The form of (5) is derived in [83], e.g., and is an approximation known as *Newtonian fluid*. Φ denotes the sum of all external potentials (including gravitation). In the energy equation (3) the specific internal energy ε is related to the total specific energy E (in units of energy per mass) through $E \equiv \frac{1}{2}|v|^2 + \varepsilon$ whereas h denotes the energy flux by heat conduction. The chapter "An Introduction to Turbulence" provides a derivation of (1), (2), (3), (4) and (5) from statistical physics. A continuum mechanics approach can be found, for example, in [83]. A possible choice for h is the phenomenological ansatz

$$h = -K_h \operatorname{grad} T. \tag{6}$$

In this relation, K_h is the heat conduction coefficient. Equation (6) assumes that conduction is a purely diffusive process. In stars, radiation is much more efficient for energy transport than heat conduction, with the exception of conditions of very high densities as found in the interior of compact stars (see [137]). If the mean free path of photons l_{ph} is very small, radiative transfer can also be described as a diffusion process:

$$h = -K_{rad} \operatorname{grad} T. \tag{7}$$

Here, K_{rad} is the *radiative conductivity*. For typical conditions in stellar interiors, l_{ph} is of the order of 2 cm along which temperature changes in radial direction are $\sim 3 \cdot 10^{-4}$ K (see [71]), which justifies (7). Thus, a very accurate approximation for K_{rad}, which becomes exact in the limit of small l_{ph}, is

$$K_{rad} = \frac{4acT^3}{3\kappa\rho} = \frac{16\sigma T^3}{3\kappa\rho}, \tag{8}$$

where a, c and σ are the radiation constant, vacuum speed of light and Stefan–Boltzmann constant, respectively, and κ is the *Rosseland mean opacity* (see [71]). The latter is the specific cross-section of a gas for photons emitted and absorbed at local thermodynamical conditions and averaged over all frequencies (thus, $[\kappa] = cm^2\,g^{-1}$ and $\kappa = \kappa(T, \rho)$ for a given chemical composition).

Close to the surface of a star, l_{ph} increases and in fact, once photons are emitted into the interstellar environment, it becomes almost arbitrarily large. In that case, equations (1), (2) and (3) have to be coupled to the equations of radiative transfer. The resulting equations of *radiation hydrodynamics (RHD)* augment (1), (2) and (3) by equations for energy and momentum density of photons [91]. Moreover, coupling terms have to be added to (2) and (3). For Eq. (2) this is essentially the contribution of radiative pressure to the momentum density balance. It is common practice to include this term into the definition of Π in (4) by writing $p = p_{gas} + p_{rad}$. In this form it also couples to Eq. (3). In addition, the term h can no longer be approximated through

(7), but has to be obtained from solving a transport equation for the radiation field. It is therefore more convenient to rewrite (3) into the general form

$$\frac{\partial}{\partial t}\left(\rho E\right)+\text{div}\left[(\rho E+p)\boldsymbol{v}\right]-Q_{\text{rad}}-Q_{\text{cond}}-Q_{\text{nuc}}-\text{div}(\boldsymbol{\pi}\boldsymbol{v})=-\rho\boldsymbol{v}\,\text{grad}\,\Phi, \quad (9)$$

where $Q_{\text{rad}}=\text{div}\,\boldsymbol{F}_{\text{rad}}$ is the radiative heating rate with $\boldsymbol{F}_{\text{rad}}$ as the radiative energy flux. $Q_{\text{cond}}=\text{div}(K_{\text{h}}\,\text{grad}\,T)$ is the conductive heating rate and Q_{nuc} is the heating rate due to local nuclear energy generation. Near the surface of stars, $Q_{\text{nuc}}=0$ and $Q_{\text{cond}}\approx0$. For this case it is sufficient to consider

$$\frac{\partial}{\partial t}\left(\rho E\right)+\text{div}\left[(\rho E+p)\boldsymbol{v}\right]-Q_{\text{rad}}-\text{div}(\boldsymbol{\pi}\boldsymbol{v})=-\rho\boldsymbol{v}\,\text{grad}\,\Phi, \quad (10)$$

instead of (3). In [91] the interested reader can find different approximations and generalizations of (10) useful for the study of RHD problems.

For some problems in stellar astrophysics one cannot neglect that the stellar plasma generates magnetic fields (dynamo effect) and interacts with both external and self-generated fields. To investigate them it is essential to generalize Eqs. (1), (2) and (3) to the equations of *magneto-hydrodynamics (MHD)*. Derivations and discussions of the MHD equations can be found, for instance, in [16, 39, 97] (see also the chapter "Magnetohydrodynamic Turbulence"). A derivation of the MHD equations from equations governing microscopic particle motions can be found in [20]. For the study of sun spots and other magnetically active regions at the solar surface and of solar activity in general it is necessary to couple the MHD equations to the equations of radiative transfer. In the following, we do not discuss this case of *radiation magneto-hydrodynamics (RMHD)*, but rather focus on the non-magnetic case. This is sufficient to describe the mean structure of the Sun, the generation of oscillations ("p-modes") and the bulk properties of magnetically quiet regions, which dominate over most of the solar surface.

In Eqs. (2) and (10) the gravitational field appears as an external source term through the potential Φ. Whenever a relativistic treatment is not absolutely necessary, the latter can be calculated from the Poisson equation

$$\triangle\,\Phi=4\pi\rho\,G, \quad (11)$$

the Newtonian limit of Einstein's gravitational field equations (see, for instance, Chap. 12 in [82]). This is a fully consistent treatment, since in the relativistic case the hydrodynamical equations have to be modified as well to account for Lorentz invariance or full covariance (see [82, 83]). In stellar astrophysics relativistic corrections are mainly of interest for the study of dynamical stability of white dwarfs, the structure of neutron stars and core collapse supernovae (see [117]). If rotation is not important, spherical symmetry is assumed to hold. For this case the analytical solution of (11) is known (see the discussion of Eq. (13)). If we study convection within the surface region of such a star and the vertical domain considered is small compared to the stellar radius, we can approximate the solution of (11) by a constant acceleration $g = -\mathrm{grad}\,\Phi$.

Instead of describing a star as a single-component fluid, it may also be modelled as a multi-component fluid. For the stellar case the most interesting components are hydrogen and helium. In later stages of stellar evolution, heavier elements may have to be modelled as well. Trace elements can be studied in a postprocessing step, if they do not change the temporal evolution of the flow. In general, Eq. (1) has to be replaced by a set of equations for the concentration of each species. Equations (2) and (3) are modified accordingly. The motivation behind a multi-component treatment is the study of gradients in chemical composition in the star which build up because of nuclear processes or diffusion [71].

The system of dynamical equations for the fluid is closed by an equation of state, $p = p(\rho, \varepsilon)$ (or more commonly, $p = p(\rho, T)$, since $\varepsilon = \varepsilon(\rho, T)$) and its associated thermodynamical derivatives. They are usually given in tabular form. The same is done for the opacity $\kappa = \kappa(\rho, T)$ and the rate of nuclear energy generation $Q_{\mathrm{nuc}}(\rho, T) = \rho(\epsilon_{\mathrm{n}}(\rho, T) - \epsilon_{\nu}(\rho, T))$. Here, ϵ_{n} is the nuclear energy production per unit of mass and time. ϵ_{ν} denotes the energy loss due to neutrinos produced in the star which are not already accounted for by the contributions to ϵ_{n}. Details can be found in [71, 117, 137]. The same procedure could also be used for viscosity – we return to this in Sect. 4.1. Each of these functions of *microphysical input quantities* is usually computed for a fixed chemical composition. Interpolation is often used to avoid recomputing the tables for each new mixture, although this is not always a viable option, since the dependencies on the mixture can be highly non-linear.

1.2 Ensemble Averages, Convective Instability and Overshooting

In most cases, numerical solutions of the full set of hydrodynamical equations are too complex and computationally too expensive to allow the construction of complete stellar structure models. The complexity increases even further, if the evolution of a star is to be followed over billions of years of stellar time, since hydrodynamical processes near the surface of a star operate on a typical time scale of a few seconds to a couple of minutes (see Sect. 4.3). Hence, the theory of stellar structure and evolution is based on *ensemble averages* of the full hydrodynamical equations under the assumption of quasi-ergodicity of their solutions. In most cases, the physical quantities are also averaged horizontally (over plane parallel layers or spheres). For the remainder of this section we consider such *averages as a function of depth*, unless stated otherwise. In the case of purely radiative energy transfer and $v \approx 0$ (and thus perfect spherical symmetry of the star) these averages read ([71, 137]):

$$\frac{\partial M_r}{\partial r} = 4\pi r^2 \rho, \tag{12}$$

$$\frac{\partial p}{\partial r} = \frac{-G M_r \rho}{r^2}, \tag{13}$$

$$\frac{\partial T}{\partial r} = \frac{-3\,\kappa(T,\rho)\,\rho\,L_r}{16\pi r^2 ac\,T^3},\tag{14}$$

$$\frac{\partial L_r}{\partial r} - 4\pi r^2 \rho(\epsilon_n - \epsilon_\nu - Q_g).\tag{15}$$

Here, M_r is the mass contained inside a radius r and L_r is the luminosity (outwards directed energy loss at r per unit of time due to radiation). The term Q_g denotes energy produced or lost per unit of time by expansion or contraction of the star, which takes place on the very long Kelvin–Helmholtz time scale (28), whence consistently $v \approx 0$. Equations (12) and (13) are the velocity independent part of (1) and (2) for the case of spherical symmetry. The latter allows an analytical solution of (11) and is given by the radial component of $\boldsymbol{g} = (-g, 0, 0)$, the well-known result that $g = GM_r/r^2$. Equations (14) and (15) are the velocity independent part of (9) under the assumption that Eqs. (7) and (8) hold. Equations for the evolution of the chemical composition are added to this system. Together they yield the (spherically and ensemble averaged) *stellar structure* at time t.

What is the role of turbulent convection in this context? A problem of (12), (13), (14) and (15) is that in general this system describes a physical solution which becomes unstable to small vertical perturbations whenever $\partial T/\partial r$ is sufficiently large. As first noted by K. Schwarzschild [115], a fluid stratified along a temperature gradient $(\partial T/\partial r)$ is unstable to buoyancy, if that gradient is larger than the adiabatic one, $|\partial T/\partial r| > |(\partial T/\partial r)_{\mathrm{ad}}|$, $\partial T/\partial r < 0$. The adiabatic temperature gradient is given by the equation of state of the fluid. Since (12), (13), (14) and (15) depend only on r, Eq. (13) is used in the astrophysical literature to rewrite this condition into a relation between the dimensionless logarithmic gradients,

$$\nabla := \frac{\partial \ln T}{\partial \ln p} = \frac{p}{T}\frac{\partial T}{\partial r}\frac{\partial r}{\partial p},\tag{16}$$

$$\nabla_{\mathrm{ad}} := \left(\frac{\partial \ln T}{\partial \ln p}\right)_{\mathrm{ad}} = \frac{p}{T}\left(\frac{\partial T}{\partial r}\right)_{\mathrm{ad}}\frac{\partial r}{\partial p}.\tag{17}$$

Thus, $\nabla - \nabla_{\mathrm{ad}}$ can be used to characterize the stability of the gas against buoyancy at a given location r as follows:

$$\text{if}\quad \nabla - \nabla_{\mathrm{ad}} > 0 \quad\Rightarrow\quad \text{unstable layer at } r,\tag{18}$$
$$\text{if}\quad \nabla - \nabla_{\mathrm{ad}} \leqslant 0 \quad\Rightarrow\quad \text{stable layer at } r.$$

This criterion assumes that the mean molecular weight remains constant. In the same way we can rewrite the radiative temperature gradient, Eq. (14):

$$\nabla_{\mathrm{rad}} = \frac{+3\,\kappa(T,\rho)\,p\,L_r}{16\pi ac\,GT^4 M_r}.\tag{19}$$

An equivalent version of (18) is based on the entropy gradient $\partial S/\partial r$: if $\partial S/\partial r \geqslant 0$, the layer is stable against buoyancy, if $\partial S/\partial r < 0$, the layer is unstable (cf. [137]). For $\nabla_{\mathrm{rad}} \leqslant \nabla_{\mathrm{ad}}$, the fluid is *stable* against convection, if $\nabla = \nabla_{\mathrm{rad}}$. Such layers of a star are called *radiative*. If $\nabla_{\mathrm{rad}} > \nabla_{\mathrm{ad}}$, then $\nabla = \nabla_{\mathrm{rad}}$ is *unstable* according to (18). Because of radiative losses the convective energy transport is not entirely adiabatic and, hence, within a *convective layer* we expect [137]:

$$\nabla_{\mathrm{rad}} > \nabla > \nabla_{\mathrm{ad}} > 0. \tag{20}$$

Equations (16), (17), (18), (19) and (20) assume some simplifications. A real fluid has a finite viscosity which creates some threshold for the convective instability. The same is true for radiative exchange of the fluid with its environment, because it prevents fully adiabatic heating and cooling. In stellar astrophysics the stabilization due to molecular viscosity is always very small because of the small Prandtl number in stars (Sect. 4.1). In a numerical simulation of convection this effect can nevertheless lead to an erroneous stability against small perturbations, since *numerical viscosity* is generally much larger (Sect. 3.2). Stability due to large radiative losses is more important: in the atmospheres of late B type stars with surface temperatures around 12,000 K the gas should be unstable according to (18), but no observational indications exist for such an instability [84]. Magnetic fields change the stability in a more complex way. A general treatise is given in [39]. For stellar astrophysics, there is another, important generalization of (18) which takes the effects of mean shear flows and gradients in chemical composition into account. Both can either enhance or decrease the stability of the fluid against buoyancy (see [30] and the chapter "Turbulence in Astrophysical and Geophysical Flows").

There are several reasons to have $\nabla_{\mathrm{rad}} > \nabla_{\mathrm{ad}}$ in stars as follows from analyzing (19) (cf. [71]). First of all, the opacity κ can be large. In that case the radiative conductivity (8) becomes small and so does the convective heat flux (7). In such layers radiation cannot sustain a large energy flux. Our Sun is a good example for this scenario and the main reason for a large κ in the upper part of its convection zone is the ionization of hydrogen. The condition $\nabla_{\mathrm{rad}} > \nabla_{\mathrm{ad}}$ can also be fulfilled, if ∇_{ad} is lowered. This happens, e.g. when atoms or molecules obtain additional internal degrees of freedom, for example, when hydrogen is partially ionized [137]. As a consequence, buoyancy is increased even further. That mechanism was in fact the first explanation suggested for the convective instability of the solar surface layers [132] and operates jointly with a large κ. The third reason for $\nabla_{\mathrm{rad}} > \nabla_{\mathrm{ad}}$, which is independent of opacity and equation of state properties, is an intrinsically high luminosity L_r. As an analysis of (12), (15) and (19) for $r \to 0$ reveals, this occurs in the core of a star whenever its nuclear energy production ϵ_{n} is high. Such conditions exist, e.g. for massive stars during their hydrogen burning phase, as opposed to ionization, which is responsible for convective envelopes in cool stars [71].

It is instructive to compare the local stability criterion (18) with the situation found in a 3D numerical simulation of convection. Figure 1 shows horizontal time averages of $\nabla - \nabla_{\mathrm{ad}}$ for a convection zone embedded between

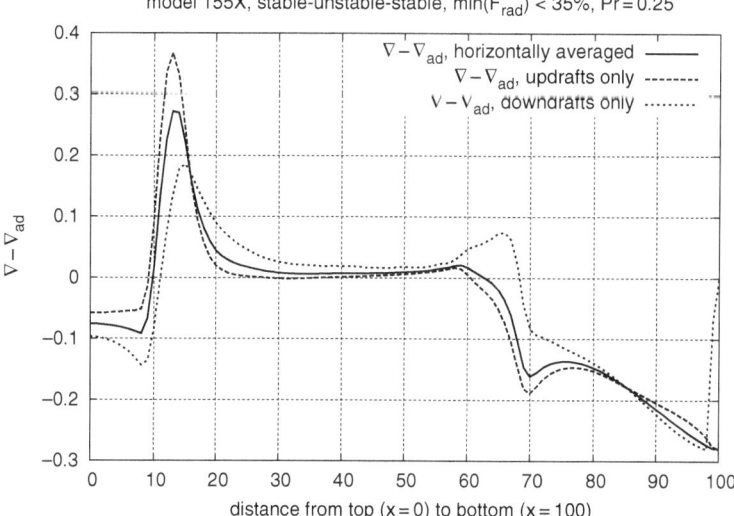

Fig. 1. Super-adiabaticity of temperature gradients ($\nabla - \nabla_{\mathrm{ad}}$) in a numerical simulation of convection. The ensemble average for each horizontal layer is compared to averages taken only over up- and downflow areas respectively

two stable layers. The fluid obeys a perfect gas law with a constant ratio $\gamma = c_p/c_v$ of specific heats (at constant pressure and volume respectively). A constant g is assumed which defines the vertical direction. K_{rad} is a prescribed function of depth held constant in time, as is the fixed Prandtl number $\mathrm{Pr} = \nu/\chi = c_p \rho \nu / K_{\mathrm{rad}}$. Lateral boundary conditions are periodic, vertical ones are impenetrable, but stress-free (thus no shear stress is exerted on the wall, see Sect. 5), with a fixed T on top and a constant $\partial T/\partial x$ (input heat flux) at the bottom. In the interior of the convection zone (between x-values of 20 and 60) convection is rather efficient, since the radiative flux drops to $< 35\%$. This simulation is described in more detail in [79–81]. Note that the average taken only over areas for which the vertical velocity is pointing upwards (updrafts only) yields *marginally stable* values for $\nabla - \nabla_{\mathrm{ad}}$ inside the *convectively unstable* region. In contrast, the averages taken only over areas with downwards pointing velocity (downdrafts only) clearly remain slightly *super-adiabatic* (positive) throughout the interior of the convection zone.[2] Obviously, here the downwards flowing material is responsible for the *on average unstable* stratification of the convection zone (see also Sect. 2), which is defined by the points where the horizontal average of $\nabla - \nabla_{\mathrm{ad}}$ changes sign (18). The *super-adiabatic peak* near the top of the convection zone is typical also for convection zones that reach the surface of stars. It is caused by the radiative heat exchange of the fluid with the uppermost layers which have a

---

[2] The increase in $\nabla - \nabla_{\mathrm{ad}}$ for the downflow at the bottom is due to the lower boundary conditions, which does not influence $\nabla - \nabla_{\mathrm{ad}}$ further above.

high radiative conductivity (for the solar case cf. Fig. 5 and the comparisons
in [109]). The downflow regions show a similar peak at the bottom of the
convection zone. This is caused by *plumes* penetrating into the *on average
stable* stratification. They are not in thermal equilibrium with their environment until they have reached a certain depth. Note that the peaks of $\nabla - \nabla_{\rm ad}$
are shifted in opposite ways near top and bottom and that they are twice as
large for the updrafts than for the downdrafts for the peak near the top.[3] In
conclusion, convective mixing cannot be confined to regions where on average
$\nabla - \nabla_{\rm ad} > 0$. There is *overshooting (OV)* into adjacent layers, which the local,
linear stability criterion (18) neglects, since it does not account for advection
and thus the inertia of fluid in motion. Figure 1 also demonstrates that the
physical conditions in up- and downflows can appreciably differ (cf. [123] for
the solar case), which complicates the definition of suitably averaged physical
quantities in turbulent convective flows. We shall return to the question of
the nature of turbulence in stellar convection zones in Sect. 2.

1.3 Effects of Convection on Stellar Structure and Mixing

Model calculations for the structure and evolution of stars must account for
convection [71, 137]. The most important physical process neglected in (12),
(13), (14) and (15) is the energy flux produced by the convective instability.
Convection models used in stellar astrophysics such as the *mixing length theory* [15, 17, 137] compute the heat flux generated by the advective transport
in layers where $\nabla - \nabla_{\rm ad} > 0$. In the astrophysical literature this quantity
is known as the *convective flux*, $F_{\rm conv}$, though it should more accurately be
called the enthalpy flux (the interested reader may find the relations derived
in [30] useful in this context). The physical reason behind this generalization
of (12), (13), (14) and (15) is that even though the ensemble average of the
momentum density ρv (or v) might be (nearly) zero, ensemble averages of
products of ρ, v, ρv, E, T, etc. with each other can be nonzero. Some of them
are quite large even after a horizontal mean value has been subtracted. The
products originate from the non-linearities of (1), (2) and (3) respectively (10).
Dynamical equations constructed for such quantities from the basic equations
(1), (2) and (3), as is done in the Reynolds stress approach, contain products
of even higher order (see [26–28] and the chapter "Turbulence in Astrophysical
and Geophysical Flows"). If fluctuations in specific heat and density are ne-
glected, $F_{\rm conv} = c_p \rho \overline{w\theta}$ [26]. Here, as before, c_p and ρ are horizontal ensemble
averages, while w and θ denote the differences between the instantaneous val-
ues and the horizontal ensemble average of vertical velocity and temperature,
respectively. The ensemble average $\overline{w\theta}$ is the main quantity to be retrieved
from a stellar convection model. The role $F_{\rm conv}$ has to play in the energy flux
balance of a star can be traced back to div $[(\rho E + p)v]$ in (3), resp. (10), which

---

[3] The cusps within the stable zones are partially caused by the form of $K_{\rm rad}$ as a
function of depth in that simulation.

due to $E = \frac{1}{2}|\boldsymbol{v}|^2 + \varepsilon$ and (7) implies that both $(\rho\varepsilon + p)\boldsymbol{v}$ and $\boldsymbol{F}_{\mathrm{rad}}$ should appear in a flux balance once $\boldsymbol{v}$ differs from zero. From the same equations it follows that there is an additional term related to $\frac{1}{2}\rho|\boldsymbol{v}|^2\boldsymbol{v}$ in such a balance. As for F_{conv} the horizontal ensemble average of this quantity yields a nonzero vertical component, the turbulent kinetic energy flux F_{kin}. Standard models of stellar convection neglect this term [17, 71, 137]. This can hold only for special symmetries, such as up- and downflow covering identical horizontal areas, which is usually not the case (cf. solar observations as shown in Fig. 2 and numerical simulations of solar granulation [109, 124]). Modelling F_{kin} is the domain of Reynolds stress models of stellar convection [27, 28, 35, 75, 138]. In the end, both mixing length and Reynolds stress models replace (14) with a new equation for T or $\partial T/\partial r$ which is coupled to further, non-linear algebraic or differential equations. Stellar models with convection zones predict more compact objects than their purely radiative counterparts. The reason for this is that the much smaller temperature gradients of convection zones allow for a much larger mass to be contained in hydrostatic balance within the same volume (see the comparisons of radiative and convective stellar envelopes in [71]). The radiative counterpart of the simulation shown in Fig. 1 has twice the temperature of the relaxed convective model at the bottom, despite the temperature at the top and the mass contained in the box are identical.

An important consequence of convective energy transport in stars is *turbulent pressure*. It originates from the $\mathrm{div}\,[\rho(\boldsymbol{v}\otimes\boldsymbol{v})]$ term in (2), which after subtracting mean values and ensemble averaging yields a nonvanishing contribution. Its vertical component can conveniently be written as $\mathrm{grad}\,p_{\mathrm{turb}}$ with $p_{\mathrm{turb}} \approx \rho\overline{w^2}$. If we account for p_{turb} in a stellar model, the structure of (13) can be retained even in the fully compressible case after appropriately renormalizing p and g [27, 28], thus $p = p_{\mathrm{gas}} + p_{\mathrm{rad}} + p_{\mathrm{turb}}$. Knowing p_{turb} is necessary for accurate models of the surface layers of stars which have convective envelopes

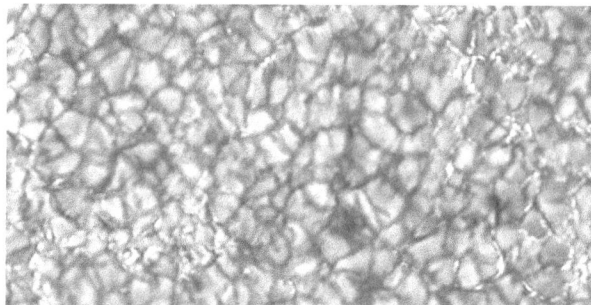

Fig. 2. Solar granulation in a magnetically more quiet region during activity maximum, as observable at wavelengths of the *G-band* ($\sim$ 430 nm) caused by the CH molecule. This figure is part of a much larger image taken with the KIS/VTT instrument at Obs. del Teide, Tenerife, by Oskar von der Lühe, Kiepenheuer-Institute for Solar Physics, and reproduced here with his kind permission

and whenever dynamical phenomena are of interest for which pressure acts as a restoring force. This includes the p-mode oscillations occurring in our Sun (see [5, 68] and references in [45]). The quantity $\overline{w^2}$ can be computed within most convection models, but physically well-defined and numerically robust generalizations of (13) are more difficult to derive than models which only account for F_{conv}. Thus, p_{turb} is not included in the standard mixing length model [17, 71, 137] despite p_{turb}/p reaches maximum values of $\sim 14\%$ in numerical simulations of solar granulation [109, 110] and even $\sim 35\%$ in the same type of simulations for hotter stars with surface temperatures $\sim 7300\,\mathrm{K}$ instead of just $\sim 5800\,\mathrm{K}$ [131]. Fortunately for stellar evolution theory, these large deviations are confined to a rather small layer near the stellar surface.

In addition to F_{conv}, F_{kin} and p_{turb}, which change the thermal structure of the star, there is another by-product of convection which influences stellar evolution: the mixing of stably stratified layers next to convection zones, called overshooting in stellar physics, but known as *entrainment* in the geophysical sciences (cf. Fig. 1 and Sect. 1.2). Heuristic models of this process have found limited success based on tuning model parameters to reproduce a set of test data [30, 42]. Unfortunately, those parametrizations are not very robust and a more detailed modelling is required (see Sect. 6 and the chapter "Turbulence in Astrophysical and Geophysical Flows").

Each of the physical processes introduced here can be studied for the solar convection zone, which provides a good example for explaining the strategies used and the difficulties encountered in modelling stellar turbulent convection.

2 Turbulence in the Solar and in Stellar Convection Zones

The solar convection zone is a rotating, approximately spherically symmetric shell. Its mean density changes from about $0.2\,\mathrm{g\,cm^{-3}}$ at the bottom to some $3.2 \cdot 10^{-7}\,\mathrm{g\,cm^{-3}}$ near the surface. The mean temperature changes from about $2.15 \cdot 10^6\,\mathrm{K}$ near the bottom to about $6200\,\mathrm{K}$ near the top (see Tables 2.4 and 6.1 in [128] for a typical solar model and Chap. 18 in [137] for experimental tests of this structure based on helioseismology). This implies a contrast in ρ of 625,000:1 and a contrast in T of 350:1 in a domain extending over 30% of the solar radius. Convective velocities increase from a few dozen $\mathrm{m\,s^{-1}}$ to about $2 \ldots 3\,\mathrm{km\,s^{-1}}$, which corresponds to an increase in Mach numbers from $\sim 10^{-4}$ to about 0.3 (cf. Table 6.1 in [128] and Fig. 18.11 in [137]). Figure 2 shows an image of the solar surface in visual light. Bright elements, the solar granules, are seen embedded in a network of less bright (and due to the finite image contrast actually dark) intergranular lanes. Following approximately Stefan's law (emissivity equals σT^4) the bright elements correspond to high temperature regions. Dark areas indicate regions of lower temperature. The white features seen in parts of the intergranular network are regions of enhanced magnetic activity (*plages*). There are also a few very dark regions (*pores*) which are again related to magnetic activity. Still, most of the image is covered by magnetically quiet, ordinary granulation.

2.1 Surface Granulation and Turbulence

We turn to the notion that astrophysical flows are "turbulent almost by definition", as the non-linear inertial forces resulting from div $[\rho(\boldsymbol{v} \otimes \boldsymbol{v})]$ in (2) completely dominate over their viscous counterparts. Do huge values of Re really mean that astrophysical flows always look disordered and chaotic with many small scale structures which we know from laboratory experiments on the flow behind a grid? Nothing like that is visible in Fig. 2. How should we interpret this picture? Clearly, with U of the order of $2 \ldots 3 \,\mathrm{km\,s^{-1}}$, L some $1200 \ldots 1300 \,\mathrm{km}$, and ν taken to be about $1740 \,\mathrm{cm^2\,s^{-1}}$, we obtain that $\mathrm{Re} \approx 1.3 \cdot 10^{10}$. Here, the average convective velocities U at the solar surface are obtained from the analysis of spectral line broadening in the visual light observed from the Sun or from numerical simulations [4, 128]. L is the average size of solar granules (Chap. 1.3 and 6.3 in [128]) and the length scale for which the kinetic energy of the flow is close to its maximum. The value for ν is taken from [48] (Tables 1 and 2 for a layer at the surface with $T = 5660 \,\mathrm{K}$). If instead we take L to be the depth of the solar convection zone ($\sim 180{,}000 \,\mathrm{km}$), U as the average velocity found in solar structure models ($\sim 100 \,\mathrm{m\,s^{-1}}$), and ν from Chapman's [40] relation $\nu[\mathrm{cm^2\,s^{-1}}] = 1.2 \cdot 10^{-16} \, T^{5/2} \rho^{-1}$ for fully ionized gases for T and ρ at the bottom of the solar convection zone, we obtain $\mathrm{Re} \approx 10^{14}$, though this value is more appropriate for the lower part of the convection zone. Could an increase in resolving power of solar telescopes change the laminar appearance of the granules? A comparison of Figs. 1 in [116, 122, 127] based on observations with telescope apertures of 30, 50 and 100 cm, respectively, corroborates the impression we gain from Fig. 2. Could it be that the physics accounted for in the definition of Re is not sufficient to describe the observed properties of solar granules?

We first note that solar granulation is a *surface phenomenon*. The granules are located on top of a convection zone most of which is nearly perfectly adiabatic and where photons can travel distances of only a few cm. The top of the granules behaves differently, since the average stratification becomes stable $(\nabla - \nabla_{\mathrm{ad}} \sim 0)$ and photons are emitted into space. The statistical properties of the flow at the solar surface are certainly not invariant under translation and rotation in space at the scale L. Hence, there is no physical reason why Kolmogorov scaling $v(\ell) = v(L)\,(\ell/L)^{1/3}$ should hold for velocity fluctuations $v(\ell)$ at $\ell \sim L$, since some of its premises are not fulfilled (cf. the chapter "An Introduction to Turbulence"). Indeed, L is within the length scale range of *kinetic energy injection* into the system. The latter results from the negative buoyancy of the gas cooled at the surface [124]. Thus, L is not contained in the *inertial range* described by Kolmogorov scaling and solar granules should not reveal the signatures of fully developed turbulence. Rather, they are an excellent example for coherent structures. The convective instability as defined in (18) is not directly related to turbulence, but to a special kind of Rayleigh Taylor instability of the fluid which drives the flow because light (buoyant) fluid underneath heavier one is unstable to small perturbations (see the chapter "An Introduction to Turbulence"). Turbulence comes in only as a secondary instability in case of large enough driving of the flow. This driving is quantified by the Rayleigh number, which relates the product of viscous and conductive time scales to the square of the buoyancy time scale: $Ra \equiv \tau_{visc}\tau_{cond}/\tau_b^2$ (cf. Chap. III in [85]). Near the surface of the granules $Ra \sim 10^{10}$. Ra rises rapidly beyond 10^{20} for the solar interior. This illustrates the dominance of buoyancy over viscous and conductive effects in solar convection as well as the extremely strong driving experienced by the fluid. In the end this results in huge values of Re which can tempt researchers into the fallacy that the solar granules themselves are an indicator of fully developed turbulence. The boundary layer phenomenon of solar granules has sometimes been studied with the picture of a Kolmogorov cascade in mind. Indeed, one can formally take the Fourier spectrum of kinetic energy, as derived from observations, and try to find a scaling $E(k) \sim k^{-5/3}$ as an indicator for a Kolmogorov spectrum $E(k) = C_K \epsilon^{2/3} k^{-5/3}$ (C_K is the Kolmogorov constant, ϵ the dissipation rate of turbulent kinetic energy, $k = 2\pi/l$ the wave number – see also the chapter "An Introduction to Turbulence"). However, current solar observations only resolve scales $l \gtrsim 0.05\ldots 0.1\,L$. Thus, any detection of Kolmogorov scaling at such coarse resolution should be considered as possibly spurious or accidental. Similar holds for analyses of numerical simulations of solar granulation at such resolution (we return to this point in Sect. 4.2). One has to keep these limitations in mind when comparing the following independent explanations that have been suggested for the 'laminar' appearance of solar granules.

For ordinary granulation one can neglect rotation and magnetic fields.[4] But there are two important physical facts which cannot be neglected. First of all, the fluid near the surface is extremely stratified. Density changes by a factor of 1000 from layers which are transparent in the visual solar spectrum down to layers which are opaque (cf. Table 4.1 in [128]). Since that occurs within less than 1000 km (and thus $< L$) which can be traversed by the fluid within two or three multiples of the time required for a sound wave to do so, a dramatic expansion of upwards rising fluid and an equally dramatic compression of downwards sinking fluid is expected to occur. Compressibility is not represented in the comparison of inertial to viscous forces, $Re = (UL)/\nu$. Secondly, fluid reaching the observable solar surface is subject to strong (radiative) cooling. Fluid rising up is thus expanded, smoothed out, cooled at the surface and then advected downwards again, with turbulence produced mostly in the downdrafts. The latter has little chance to reappear at the surface again. In astrophysics, this picture of downdrafts driving convection was developed in [101, 123] and further refined in [102, 124, 125]. It applies to

---

[4] Rotational forces are small compared to buoyancy at granulation length scales L and magnetic fields sufficiently weak in magnetically quiet regions (cf. [128]).

the simulation discussed in Sect. 1.2. In meteorology the role of heating vs. cooling at the boundary was discussed in [92] (see also for further references).

Note there is a practical problem when probing small scales in stellar surface flows. Most photons we observe at a given frequency f originate from layers which are just about to become optically thick and provide the *optical surface at frequency f*. Layers further below are optically thick and hence shrouded by the optical surface. Layers above it are transparent (*optically thin*). There is little information from photons on layers transparent to them, since by definition there is hardly any interaction between the gas and such photons.[5] The length scale $l_f \sim (\kappa_f \rho)^{-1}$ along which a layer becomes optically thick at f sets a lower limit for scales l which can be probed by simple observational means. For $l < l_f$ at any layer above the optical surface variations in intensity are difficult to detect because of the transparency of the gas. Temperature differences at scales $l < l_f$, which might have been advected from layers further below, are subject to strong smoothing due to radiative cooling [121]. Thus, small-scale temperature fluctuations have a very short cooling time scale (~ 0.1 s for structures of size $l \sim 10$ km). In turn, Doppler broadening due to local velocity fluctuations is difficult to trace, if the latter are smaller than 1 km s^{-1}, because the thermal Doppler broadening at solar photospheric temperatures is much larger. Hence, velocity fluctuations present on scales $l \ll l_f$ may remain undetectable.[6] Note that one can get partially around this problem and attempt to increase vertical resolution by considering various f for which photons predominantly originate from different geometrical depths (cf. also [4]) and compare them differentially. However, since for any f the observations stem from an entire depth range $\sim l_f$, probing small-scale velocity fields this way is difficult, too.

In an earlier proposition [116] it was argued that turbulence at unresolved scales acts on the larger, observed scales by giving rise to a *turbulent viscosity* $\nu_t(l)$. This enhances the molecular viscosity ν such that the Reynolds number at L is actually Re $= (UL)/(\nu + \nu_t(l)) \sim$ O(1). While compressibility and (radiative) cooling have been used to argue that even at very high Re the flow does not have to be fiercely turbulent at the top of the granules [124, 125], the concept of turbulent viscosity in [116] aimed at reconciling observations with the expectation of a highly turbulent flow at very high Re. This argument was completed in [31] by demonstrating that the non-linear interactions in a turbulent flow, which originate from the advection term div $[\rho(\boldsymbol{v} \otimes \boldsymbol{v})]$ in (2), indeed lead to a renormalization $\nu \to \nu + \nu_t$ for length scales close to

---

[5] In the present discussion we neglect collective effects such as radiative pumping, excitation of atoms by photons originating from hotter, underneath lying layers. Those are difficult to use as tracers for the velocity field.

[6] When flying with an aeroplane, we can experience this phenomenon, too. Clouds are readily identified as regions of high turbulence (as confirmed by the bumpy feeling when flying through them), but the consequences of turbulence are also encountered in "clear" sky (see [21] for the underlying physical mechanism): visual light does not warn us about disturbances which are transparent to its photons!

the maximum of turbulent kinetic energy in the flow. Since in the turbulence model used in that work the up- and downflow motions in a convection zone are just a part of the turbulent flow field, the maximum of kinetic energy within the model is also at the scale of granulation. This motivated the conclusion that an effective renormalization of ν takes place at the required scales $\sim L$. The bottom line of this argument is that while the absence of non-linear interactions implies the presence of laminar structures, the latter are expected to exist at scales L around the maximum of turbulent kinetic energy also in a turbulent flow. This is in agreement with the definitions of turbulence and coherent structures introduced on p. 49. Although the argument leaves out the influence of compressibility and radiative cooling, it also demands a modification of the meaning of $\mathrm{Re} = (UL)/\nu$ for the observed scales $\sim L$. But while compressibility and radiative cooling are directly accessible to observations, it has to rely on the applicability of its underlying ideas to solar granulation, as scales $l \ll l_f$ escape direct detection. Note that the picture of a turbulent flow behaving like a laminar flow at scales $\sim L$ is not identical to the picture, which claims granules to be part of a cascade initiated at the bottom of the solar convection zone that neglects the role of cooling at the surface and expansion (contradicted by [123]).

There may be another process which might enhance the laminar-like properties of solar granulation which so far has found little attention. In [48] it was first pointed out that the second or bulk viscosity term, $\zeta\delta_{ik}\mathrm{div}\,\boldsymbol{v}$, which appears in (2) via (5), should not be neglected for the surface region of the solar convection zone, which is generally done in stellar convection modelling. As is discussed in [83] (Chap. VIII, § 81), ζ can become large compared to the usual molecular viscosity η, if slow relaxation processes, for instance various chemical reactions, interact with the flow, which in turn leads to a large dissipation of kinetic energy. In [48] such a possible source was identified to be the finite relaxation time characterizing the exchange of the translational energy of electrons and the internal binding energy of electrons in hydrogen atoms. Calculations for a pure hydrogen gas show that ζ/η can become as large as 10^7 near the optical surface, though this ratio rapidly drops to less than 100 already 1500 km further below (Table 2 in [48], H is the dominant source of e⁻ for the top of the solar convection zone). If shear (the first two terms in (5)) and compressibility effects (the terms in (5) depending on div $\boldsymbol{v}$) have comparable size, then the Reynolds number could alternatively be written as $\mathrm{Re}_{\mathrm{comp}} = (UL)/(\eta/\rho + \zeta/\rho)$. $\mathrm{Re}_{\mathrm{comp}}$ is 3000 instead of $1.3 \cdot 10^{10}$ for the surface of solar granules. This is still a value usually characterizing a turbulent flow, but together with the aforementioned effects it may enhance the solar granules to behave more like laminar flow as far as compressibility effects are concerned. Thus far, no studies appear to have been published which further investigate the possibility of a large bulk viscosity at the solar surface.

2.2 Turbulence Inside the Solar Convection Zone

Of course, there is no reason why the situation *inside* the Sun, below the photosphere, should be the same as on its surface: the optically thin scenario no longer holds and the stratification effects become smaller. In addition, ζ strongly decreases [48]. Figure 3 shows how the entropy changes from the photosphere to the quasi-adiabatic stratification inside the convection zone in a 2D simulation [100] performed at very high resolution (an order of magnitude higher than for the observations in Fig. 2 even when averaging over several grid points of the simulation). Note the very smooth and sharp surface within the top layers, which indicates the entropy jump, the super-adiabatic peak just below the observable photosphere (similar to Fig. 1). Visible light originates from the smooth layers above that surface. In turn, the layers underneath are fiercely turbulent and show all the shear instabilities expected from a (2D) flow at high Re number. Shearing inevitably occurs between strong downflows (indicated in Fig. 3 by low-entropy material advected from the surface) and the surrounding regions of more gentle upflows. To catch these flow features in a granulation simulation requires sophisticated numerical procedures to which we return in Sect. 4.2. One should keep in mind that 2D simulations can only provide some first guidance on what to look for in more detail in 3D: the properties of (1), (2) and (3), in particular with respect to the redistribution of turbulent kinetic energy, change in a fundamental manner by adding a third spatial dimension (see [85] and Sect. 4.1). Nevertheless, a

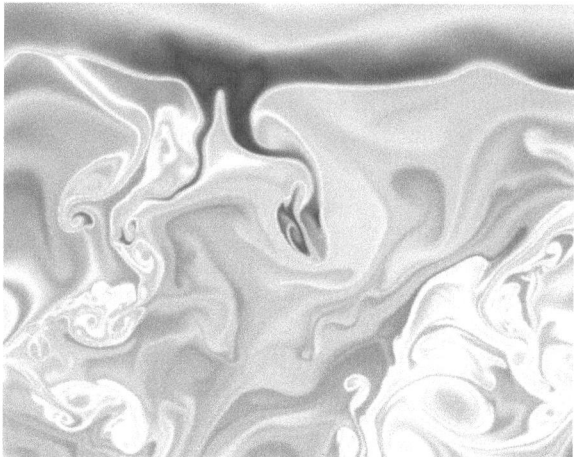

Fig. 3. Snapshot of entropy in a simulation of solar granulation in 2D at a very high resolution of 1.8 km vertically and 2.8 km horizontally. The domain is 2 Mm deep and 2.6 Mm wide and embedded in a larger region 3 Mm deep and 11.2 Mm wide simulated at four times lower resolution [100]. Low values of entropy are indicated in dark grey, intermediate values in very light grey and high values in medium grey, a scale chosen to enhance grey scale image contrast

similar difference between smooth surface layers and small scale turbulence in the interior is found in (see also http://www.univie.ac.at/acore/gallery.html) high-resolution (7×10^2 km^3) 3D simulations confirming a view voiced at the dawn of stellar convection modelling: the appearance of convection at the surface and inside the Sun does not have to be alike [15].

2.3 Turbulence at Different Scales: Micro- and Macroturbulence

Long before numerical simulations of convection became feasible the phenomenological concepts of *microturbulence* and *macroturbulence* were developed in observational stellar spectroscopy [84]. To explain extra broadening observed for spectral lines of "intermediate strength" in cool stars (including our Sun) the concept of microturbulence was proposed. It postulated an enhanced Doppler broadening of such spectral lines due to turbulence on length scales $l \ll l_f$. Typical values derived from solar spectra are $\xi_t \sim 0.85$ km s^{-1} [4] and assume a simple, Maxwell–Boltzmann distribution of "fluid elements" of size l independent of depth. Whether this effect is a good tracer for turbulence at unresolved scales is rather uncertain. Synthetic spectra calculated on the basis of 3D numerical simulations of solar granulation allow very accurate matches of spectral line strengths and line profiles [4] without invoking a non-zero ξ_t. Moreover, in stellar spectroscopy ξ_t is commonly used as a fudge parameter also for other physical mechanisms held responsible for extra line broadening. However, the necessity to introduce a non-zero ξ_t in an analysis clearly indicates the presence of an extra broadening mechanism and if the lower parts of a stellar photosphere are convectively unstable, convection is an obvious suspect to be held responsible for that broadening.

High resolution spectra of cool stars with sufficiently small projected rotational velocity, including those obtained from our Sun, also allow the detection of spectral line profile asymmetries. These are commonly related to the asymmetry between up- and downflows, between hot and cold flows. For the Sun such flows are directly observed and result in the solar granulation (Fig. 2). The flows lead to line profile distortions also in spectra averaged over the entire disk of an object, i.e. the usual type of data we collect for stars other than the Sun. Since the line profile is affected here by processes on scales $l > l_f$, and even $l > L$, the designation macroturbulence was proposed for it in stellar spectroscopy. One of the main tests for numerical simulations of solar granulation is indeed the recovery of line profiles without having to invoke an ad hoc broadening mechanism operating on such scales. Recent simulations are able to do so [4]. As discussed above, this cannot rule out completely velocity fluctuations on a scale $l \ll l_f$, but the successful results with $\xi_t = 0$ set some upper limits on their size and relevance for spectral line formation. We note here that $\xi_t > 0$ increases the effective opacity κ in the photosphere by increasing the strength of some of the absorption lines [84]. Hence, the thermal structure of a convective photosphere differs from a radiative one also because of this feedback mechanism between the radiation and the flow.

Modelling a stellar photosphere with a non-zero ξ_t within a stellar structure calculation attempts to account for some of this feedback in the framework of simplified, horizontally averaged models. In contrast, stellar spectroscopy treats macroturbulence as a purely geometrical effect that does not change the thermal structure of the photosphere. As a means of probing turbulence in stellar atmospheres, both concepts have to be used very cautiously.

2.4 Probing the Interior of Turbulent Convection Zones in Stars

As we have seen, convective stellar photospheres are rather peculiar systems. One should bear in mind that their governing equations are those of RHD (and even RMHD) and not just the hydrodynamical equations (1), (2) and (3). Numerical experiments as shown in Fig. 3 imply that conclusions drawn from the properties of visible surface layers may not be applicable to the layers underneath. How can we probe stellar turbulent convection zones inside stars? A common requirement for models of entire stars is to successfully predict integral quantities. For instance, an evolution model of a star of one solar mass with the chemical composition of our Sun should recover the radius and luminosity of the Sun at its present age. Unfortunately, it is rather easy to reproduce these quantities by adjusting parameters in convection models and insufficiently known input quantities such as the initial helium content [62, 71, 137]. Another option is to consider samples of stars such as binary pairs [59], open clusters [2], or even entire classes of stars [42]. For all those methods the systematic effects restrict the constraints which can be put on stellar models. Ultimately, all those tests are limited by the fact that different internal structures can recover the same global property which in the end allows a model to predict the correct global quantity for the wrong reasons.

The main alternatives to this type of approach are helioseismology and, more recently, asteroseismology [43]. Both use the frequencies, amplitudes and life times of oscillations of the Sun and other stars, respectively, to conclude on the internal *structure* of the pulsating object and on its global properties. The oscillations are measured by the usual observational methods of astrophysics (flux variations from optical photometry and Doppler shifts and line profile variations from spectroscopy, see [72] for some examples and [55, 56] for a review on the physics of stellar oscillations and their role in astrophysical research). Stars suitable for seismology, such as our Sun, oscillate simultaneously in many different *modes*. If the restoring force of the oscillation is dominated by *pressure* and the mode is akin to a standing acoustic wave, the designation p-mode is used [45] to distinguish them, for instance, from standing gravity waves with buoyancy as the main restoring force (g-modes). The key distinction to the aforementioned global methods is based on the possibility to compare the properties of different p-modes. Each of them has its own acoustic cavity, hence there are regions inside the star which differ in the excitation, damping and propagation properties for different modes. By comparing the results for two different modes we can learn about the differences between the associated resonant cavities and in particular about the region probed by one mode but not by the other [43, 72]. A large set of knowledge has been assembled this way about our Sun [74] including evidence of subsurface flows, measurements of the depth of the solar convection zone, overshooting and mixing underneath it, and the differential rotation pattern in its interior. We note that although seismology essentially measures sound travel times (and through comparing different modes also local sound speeds), the thermal structure, the mean values of $T(r)$, $P(r)$ and $\rho(r)$, can be constrained fairly well, too, at least in the solar case: physically plausible uncertainties in the equation of state [3, 10] (and opacities [10, 86]) leave only limited room for changes in the model structure to match an observed oscillation pattern. Thus, helioseismology can be used to approximately recover the solar structure [43, 74] and set tight constraints on solar models. Asteroseismology is more limited in capabilities since it has to rely on modes which are also detectable from averages over the entire stellar disk, while helioseismology can be performed both 'globally' and 'locally' (using resolved images of the solar surface). Nevertheless, recent advances in observations have allowed putting some constraints on the interior of stars other than the Sun [72] and in cases such as oscillating close binary stars asteroseismology can also be combined with more traditional methods [1].

The mechanisms actually driving stellar oscillations can be probed by the means of helio- and asteroseismology, too. Stochastic excitation by turbulent convection [61] has been suggested to drive the pulsations observed in our Sun. This mechanism is now commonly accepted as the cause of solar oscillations (cf. the extensions of this idea in [5, 6] and the summaries in [44, 45, 69]).

Since solar oscillations are so tightly related to turbulent convection, they offer possibilities for probing the predictions for F_{conv}, F_{kin}, p_{turb}, and overshooting from models and simulations of convection. The temperature gradient can be probed, since a change in speed of sound modifies sound travel times (or from a different point of view the size of the resonant cavity for the oscillation modes) and hence the mode frequencies change, too. Because the actual temperature gradient in a convective layer is a consequence of both radiative transfer (F_{rad}) and convective energy transport (the sum of F_{conv} and F_{kin}), variations in ∇ correspond to variations in F_{conv} (cf. also the discussion of Eq. (20) and the derivation in [27, 28]). In [12] it was demonstrated that the new convection model suggested in [32, 33] yields improved predictions of solar oscillation frequencies. That model differs from the standard mixing-length model [15, 17, 137] of stellar evolution theory by taking into account that the convective energy transport has contributions from velocity fluctuations distributed over many different length scales rather than from just a single mixing length ℓ. As a consequence, the model predicts a very rapid transition between regions of inefficient convection (where $F_{\mathrm{rad}} \gg F_{\mathrm{conv}}$) and efficient, adiabatic convection (with $F_{\mathrm{rad}} \ll F_{\mathrm{conv}}$). This rapid transition in energy transport capability yields very steep temperature gradients (large values of $\nabla - \nabla_{\mathrm{ad}}$) near the solar surface [32, 33, 95] compared to the standard model [32, 33, 95] and numerical simulations of solar granulation [109, 110]. Remarkably, numerical simulations also yield an improvement [109, 110] similar to the model of [32, 33] in spite of much flatter *average* temperature gradients. It is interesting to check Figs. 1 and 3 in this context: note the much steeper gradient in the updraft region already discussed in Sect. 1.2 and the large fluctuations of the temperature gradient (averages over up- and downflows) and entropy along the horizontal layers which define the solar surface. This horizontal inhomogeneity averages out locally steep gradients (cf. Figs. 13 and 14 in [124]). The improved agreement with helioseismology in case of [109, 110] compared to the standard mixing length model originates mostly from the fact that the simulations naturally include p_{turb} and inhomogeneous cooling at the surface. Flux blocking by spectral lines in the solar photosphere, which is only included in [110], appears to play a more subtle role, since both [110] and [109] find comparably small differences from observations. The role of p_{turb} is to support a lower gas pressure and mass density at a given T. This pushes the photosphere further upwards [110], increases the size of the acoustic cavity and lowers the frequencies compared to the standard model. In the simulations p_{turb} is much more efficient in doing so, because a significant p_{turb}/p_{tot} is found at the point where the stratification becomes locally stable ($\sim 5\%$ as from Fig. 3 of [110]), while *local* convection models including [17, 32] force both F_{conv} and p_{turb} to be zero at that point.

Can helioseismology tell the difference? Since p_{turb} allows hydrostatic balance for a given temperature profile with a smaller amount of mass (both ρ and p_{gas} are lower than if $p_{turb} = 0$), the mass contained near the surface is reduced. This changes the pulsation amplitudes (as well as excitation and damping rates) of the fluid (cf. [69] and references therein) and is hence accessible to helioseismology. In [113] it was found that the underestimation of excitation rates by the standard mixing length model [17, 137] is considerably reduced for the new convection model suggested in [34], which is a refined version of the model of [32, 33]. This improvement comes in addition to the much smaller difference with observed pulsation frequencies compared to the standard model. However, a clear discrepancy remained for pulsation modes which predominantly originate from layers where p_{turb} is large. On the other hand, numerical simulations such as those in [110] have been found to agree with the data to an acceptable level also for the excitation rates of p-modes mostly driven near the solar surface [14, 112]. Hence, helioseismology is indeed capable to probe predictions for p_{turb} and ∇ simultaneously. Is it possible to separate F_{conv} from F_{kin}, as only the sum of both is related to the temperature gradient? Because F_{kin} crucially depends on the flux of vertical kinetic energy $\overline{\rho w^3}/\overline{\rho}$ and thus also on the skewness $S_w = \overline{w^3}/\overline{w^2}^{3/2}$, this relation may be probed by investigating the dependence of pulsation frequencies and amplitudes on the skewness S_w (see Sect. 1.3 for definitions). Indeed, it has recently been found [13, 14] that S_w is a key ingredient for accurate predictions of observed excitation rates of solar p-modes.

Overshooting below the solar convection zone has been probed with helioseismology, too (see [7, 94, 111], e.g.). Constraints can be set on both the extent of the region where $\nabla \neq \nabla_{rad}$ as well as on the change in chemical composition in the entrainment region and also inside the convective zone itself [8, 9]. In all those tests the fact used is that changes in temperature gradients and chemical composition (or in a more crude sense, mean molecular weight) also modify the travel times and other propagation properties of sound waves. Helio- and asteroseismology thus have become the most important techniques to probe our models and simulations of convection inside stars.

3 Modelling and Simulating Stellar Turbulent Convection

3.1 Convection Models for Stellar Astrophysics

Convection models predict ensemble-averaged quantities which define the mean structure of a star (such as $T(r)$) and describe the convective transport (F_{conv}, etc.). As there is no ab initio procedure known how to derive dynamical or stationary limit equations for ensemble averages from just the hydrodynamical equations (1), (2) and (3), any model used in astrophysical calculations has to rely on additional assumptions which each have to be tested. The general validity of such tests is an important issue, since a stellar evolution model which predicts the correct radius of the present Sun may fail to predict the radius of the Sun on the red giant branch ~ 5 billion years from now or on its pre-main sequence track ~ 4.6 billion years ago (cf. [93]). Some widely used models of turbulent convection have already been mentioned in Sect. 2.4. The main drawback of this class of models is their dependence on the assumption of locality, which implies among others that injection of energy into and dissipation of energy from the flow is a local process. The existence of overshooting (cf. Fig. 1) demonstrates that in general this cannot be valid.

A more advanced methodology to compute ensemble-averaged models of turbulent convection is the Reynolds-stress approach. Its underlying idea is to construct (dynamical or stationary limit) equations for the higher order moments of the independent variables of (1), (2) and (3) from the basic equations, ensemble average them, and close the resulting equations with additional physical assumptions. One advantage of these models is their capability to account for nonlocality. An introduction into this methodology is given in the chapter "Turbulence in Astrophysical and Geophysical Flows" (see also the chapter "Turbulence in the Lower Troposphere: Second-Order Closure and Mass-Flux Modelling Frameworks"). Pioneered in astrophysics by [138] it has found rapid development only more recently [26–28, 35]. In this area astrophysics owes a lot to the progress made in the atmospheric sciences (cf. [36, 41, 65, 66]). Independent developments are less frequent (the most sophisticated ones are probably those of [63, 64, 139]). This is not only because the atmospheric sciences have a longer tradition in the field (compare [144] with [138]). A different parameter space is encountered in stellar convection, while only a limited degree of complexity is acceptable for applications such as stellar evolution calculations (cf. [137]). Suitable data for challenging the models with helioseismology and sufficiently realistic numerical simulations have become available only more recently. Direct tests with such observational data [140–142] and numerical simulations [76, 77, 79–81] are now being performed in a step-by-step process to identify which parts of these models work and which ones do not. Because of their important role in providing a complementary set of data for probing turbulent convection models and for studying stellar convection by numerical experiments, we now turn to hydrodynamical simulations of turbulent convection. We discuss the basic concepts underlying this approach in detail in the remainder of this section and in Sect. 4.

3.2 Hydrodynamical Simulations of Turbulent Convection

Numerical simulations of convection solve some variant of the hydrodynamical equations, say (1), (2) and (10), for a chosen domain, the *simulation box* or *volume*. Its extent is characterized by its linear dimensions H_i in each *spatial direction i*. In astrophysics the simulation box often has a simple geometrical shape which is chosen with respect to a coordinate system appropriate for the physical problem. The simulation volume is split into a finite number of discrete volumes which in turn may be represented by grid points, finite elements, orthogonal basis functions or other mathematical means [25, 105, 107, 129]. If these *smallest units* are geometrically regular, they can easily be characterized by a grid resolution h_i along the direction i (the volume taken by such a *resolution element* may be used as a suitable measure, too). The h_i may all be of the same size in which case $N_i = H_i/h_i$ is the number of grid points or cells per direction. The h_i define the minimum length scale explicitly accounted for by the simulation. *Boundary conditions* for the hydrodynamical equations are specified with respect to the physical problem and the finite simulation volume. One or just a few *initial conditions* are selected and numerical solutions are studied for those choices. Initial and boundary conditions have to be *discretized consistently* together with the hydrodynamical equations and the simulation volume to obtain a finite system of (algebraic) equations with a well-defined solution for a finite time t [107]. Depending on the physical problem it may be necessary to neglect the initial part of the simulation run in subsequent analyses to avoid artefacts (transient solutions) introduced by a peculiar choice of initial conditions. Such *relaxation* ensures that the physical system has essentially "forgotten" its initial conditions. Statistical evaluations of time averages of the simulation can be made from that point in time, t_{rel}, onwards to compute ensemble averages under a quasi-ergodic assumption (all regions in phase space are visited according to their realization probability). Note that *typical solutions* are not known beforehand as this would require an exact statistical theory for the solutions of the hydrodynamical equations. Quite often, experimental data or simplified models are used to provide some guidance for shortening the initial relaxation time scales. Unless stated otherwise we assume that $H_i = H_j = H$, $h_i = h_j = h$ and $N = N_i = N_j$ (i and j represent each of either two or three possible directions). Hence, both the box and the discretization (grids, cells, etc.) are regular. As a benefit we can link the *resolution of the simulation*, h, the *resolved length scales*, ℓ and the number of grid points per direction, N, in a simple way avoiding technical difficulties.

If we use hydrodynamical simulations to study stellar convection zones, we have to face a number of restrictions. The most basic one is introduced by the huge spatial extent of stellar convection zones and the fact that the non-linear terms in (1), (2) and (9) always dominate over those containing the tensor viscosity $\boldsymbol{\pi}$, (4) and (5), which results in the huge values of Re quoted in Sect. 2. It is instructive to consider the simple advective–diffusive equation, also known as convection–diffusion equation [129], to understand the consequences of a high Re for turbulent flow simulations. This equation reads

$$u_t + a u_x = b u_{xx}, \tag{21}$$

where a and b are positive constants (advection velocity and diffusion coefficient), u is the velocity, x the spatial coordinate and t denotes time. As a one-dimensional, linear problem it already demonstrates the necessity of trading enhanced (numerical) viscosity against (insufficient) resolution. Equation (21) can be approximated numerically by finite differences. Taking equidistant steps Δt in time enumerated by an index n and a grid of points which have equidistant spacing Δx and are enumerated by an index m one can solve

$$\frac{v_m^{n+1} - v_m^n}{\Delta t} + a \frac{v_{m+1}^n - v_{m-1}^n}{2\,\Delta x} = b \frac{v_{m+1}^n - 2\,v_m^n + v_{m-1}^n}{(\Delta x)^2} \tag{22}$$

instead of (21) provided that $(b\Delta t)/(\Delta x)^2 \leqslant 1/2$. In the limit of $\Delta t \to 0$ and $\Delta x \to 0$ this yields a consistent and stable and, hence, convergent approximation for (21) on a grid [129]. If b is very small, this *stability constraint* imposed by the size of b is not very restrictive, because Δt can still be chosen quite large. That is exactly the situation we find in a high Re flow. However, the solutions of (21) are oscillatory [129] unless in addition $\Delta x \leqslant 2\,b/a$, i.e. the Reynolds number at the scale Δx (the *cell Reynolds number*) is less than 2. The combination of both conditions is in fact very restrictive, if the advective term $a u_x$ dominates over the diffusive one, $b u_{xx}$. It demands nothing less than resolving all spatial and temporal scales down to the scale of viscous dissipation processes (at which $b u_{xx}$ dominates over $a u_x$). In Eq. (22) the coupling between different scales $\ell \geqslant \Delta x$ occurs through a statistical, linear model ($b u_{xx}$) that is not represented properly for $\Delta x > 2\,b/a$. In Eq. (2) both the non-linear term $\mathrm{div}\,[\rho(\boldsymbol{v} \otimes \boldsymbol{v})]$ and the term $\mathrm{div}\,\boldsymbol{\Pi}$ introduce similar constraints on resolution. Solutions of the full hydrodynamical equations (1), (2) and (3) fulfilling all the resolution requirements are called *direct numerical simulations*. They require to resolve the Kolmogorov (dissipation) length scale $l_{\mathrm{d}} = L\,\mathrm{Re}^{-3/4}$, as obtained from the Kolmogorov scaling relation

$l/l_{\rm d} \sim {\rm Re}(l)^{3/4}$ for turbulent flows in three spatial dimensions (see the chapter "An Introduction to Turbulence" and the chapter "Turbulence in Astrophysical and Geophysical Flows"). Values for $l_{\rm d}$ between ~ 1 cm and ~ 4 cm are obtained for the solar convection zone with L and Re as quoted in Sect. 2.1. Pure ignorance towards this constraint does not help: contrary to the case of (21) numerical solutions of (1), (2) and (3) not resolving $l_{\rm d}$ do not just oscillate, but blow up because of the non-linearity of the advective terms in (1), (2) and (3). The number of grid points required by such a constraint makes this approach useless for simulations of stellar convection with realistic microphysical parameters, i.e. realistic values of $K_{\rm rad}$ and ν.

The way out of this restriction are *large eddy simulations* (LES) which have an affordable number of N grid points or resolution elements along each direction. They rely on a sub-grid-scale model, numerical viscosity, or hyperviscosity to account for the interaction of scales $l < h$ with scales $\ell \geqslant h$ that are resolved in the simulation. Note that the designation LES is often used in a more restrictive context. We turn to some of the underlying issues further below. In any case, cell Reynolds numbers of order unity are achieved by these methods which limit or avoid oscillations (or even blowups) at the resolution scale h. One can understand this principle by a different numerical method for solving (21), the upwind differencing scheme

$$\frac{v_m^{n+1} - v_m^n}{\Delta t} + a\frac{v_m^n - v_{m-1}^n}{\Delta x} = b\frac{v_{m+1}^n - 2\,v_m^n + v_{m-1}^n}{(\Delta x)^2} \qquad (23)$$

which is algebraically equivalent to

$$\frac{v_m^{n+1} - v_m^n}{\Delta t} + a\frac{v_{m+1}^n - v_{m-1}^n}{2\,\Delta x} = \left(b + \frac{a\Delta x}{2}\right)\frac{v_{m+1}^n - 2\,v_m^n + v_{m-1}^n}{(\Delta x)^2}. \qquad (24)$$

This scheme is consistent and stable, and hence convergent, if $2(b\Delta t)/(\Delta x)^2 + a\Delta t/\Delta x \leqslant 1$ and there is no additional condition on the resolution Δx to avoid oscillatory solutions [129]. This is much less restrictive, if au_x is large compared to bu_{xx}. The trade-off is an enhanced *artificial viscosity*, $b + (a\Delta x)/2$, which of course converges to b for $\Delta x \to 0$. As pointed out in [129], there is no simple mathematical criterion to decide whether (24) or an oscillatory solution of (22) is a better approximation to (21) for a coarse grid spacing Δx. Indeed, we need *physical criteria* to make our choice! For Eqs. (1), (2) and (3) the situation is more clear-cut: a numerical solution which is blowing up is evidently useless, while one that is more viscous may still be useful. The physical questions we have to ask about such approximations with $h \gg l_{\rm d}$ are: how much are the properties of the solution modified at the energy carrying scales $\sim L$? What is the influence of the different methods on observable quantities? Are the mixing properties notably changed? Are flows behaving approximately like volume averages of simulations that have been performed at higher resolution? With these questions in mind we turn to the different methods for performing simulations of turbulent convection with $\ell \geqslant h \gg l_{\rm d}$.

Hyperviscosity and artificial diffusivity are concepts based on the pseudo-viscosity methods invented by von Neumann and Richtmyer to stabilize classical finite difference schemes in the presence of shocks (Chap. 12 in [107]). Hyperviscosity and most artificial diffusivity methods are obtained by adding a term 'similar to viscosity' to (2) and (3). That term is based on an even order spatial derivative of the v and front factors which depend on the resolution $h = \Delta x$ [47]. Second order derivatives are most common (e.g. [124, 135]), but higher order derivatives have been used as well [19]. The extra damping provided by such a term, e.g. $\nu^{\mathrm{hyp}} \sim (\nabla \overline{S})$ with $\overline{S} = \sqrt{2S_{ij}S_{ij}}$ for the strain rate tensor $S_{ij} = (\partial v_i/\partial x_j + \partial v_j/\partial x_i)/2$, affects mostly length scales of a few multiples of Δx: in wave number space the damping is $\sim k^4$ due to the two extra derivatives which appear in the term actually added to Eq. (2), $\mathrm{div}\,((\nu_k^{\mathrm{hyp}}\partial v_i/\partial x_k + \nu_i^{\mathrm{hyp}}\partial v_k/\partial x_i)\rho/2)$. A distinction is sometimes made between damping terms dealing with shocks (ν^{shk}) and with effects due to unresolved scales (ν^{hyp}). This leads to adding a combined term $\nu^{\mathrm{ad}} = \nu^{\mathrm{shk}} + \nu^{\mathrm{hyp}}$ with components that are not necessarily of the same order [124, 135]. Contrary to the simplistic approach in (24) these methods do not reduce the convergence order of the method (and thus accuracy) away from shocks and scales $\ell \gg \Delta x$ are much less affected (see Chap. 5 in [129]). Their goal is not to mimic sub-grid-scale ($h > l \geqslant l_{\mathrm{d}}$) turbulence, but to stabilize the simulation based on an analysis of physical processes which can cause instabilities in insufficiently resolved calculations. The methods per se have no way of "knowing" whether we deal with a turbulent or a laminar flow: gradients which are steep on a scale $\lesssim 4\,\Delta x$ are simply smoothed out.

Non-linear numerical viscosity as used in modern shock-capturing schemes such as the PPM (piece-wise parabolic) method [46] or, more recently, the essentially non-oscillatory (ENO) methods [87, 88, 118, 119] originate from a very different mathematical framework. They combine higher order accuracy with the capabilities of Godunov type methods (Chap. 12 in [107]) to reproduce discontinuous solutions of the hydrodynamical equations. A trade-off compared to first order methods such as Godunov's [60] or the simple upwind scheme in (24) is that small oscillations can occur. However, they remain finite and (in the case of ENO schemes) of the order of the truncation error of the scheme, thus maintaining the high approximation order of the scheme even across discontinuities (see Chap. 14 in [105]). The potential of these schemes is demonstrated in Fig. 4: artificial diffusivity with standard coefficients of order unity (see [135] for further details and references) is added to the scheme. This degrades the resolution by about a factor of 4. In turn the simulation run without artificial diffusivity is qualitatively similar to a calculation at eight times higher resolution [100] which had added the same type of artificial diffusivity with the same parameters as in the calculation for the right panel of Fig. 4 (and also for Fig. 3). This demonstrates on the one hand that simulations with artificial diffusivity also converge to a similar flow structure at high enough resolution. On the other hand, it illustrates how pseudo-viscosity models can degrade the resolution at small multiples of the grid scale.

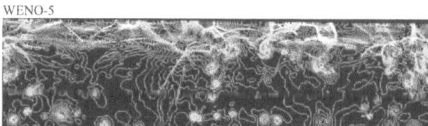 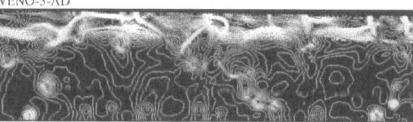

Fig. 4. Weighted ENO 5-th order (WENO-5) scheme of [87] without (*left panel*) and with artificial diffusivity (*right panel*). The resolution of this two-dimensional simulation of solar granulation is 15 km vertically and 23 km horizontally (H.J. Muthsam and C. Obertscheider, personal communication). Isolines of constant pressure are shown in a plot of relative density fluctuations

What are the disadvantages of these methods apart from increased computational costs? First of all, physical processes operating on scales $l < h$ are of course neglected unless accounted for through a sub-grid-scale model (for instance, turbulent combustion, see the chapter "Turbulent Combustion in Thermonuclear Supernovae"). A second potential danger is that the nonlinear numerical viscosity can create its own artefacts on the grid scale h, e.g. small but unphysical oscillations. Instead of adding hyperviscosity to avoid such problems one can add a very small physical viscosity (scaling up ν) or use a sub-grid–scale model together with the shock-capturing scheme. There are different opinions on this point in the literature. One extreme position is that the intrinsic numerical dissipation of shock-capturing schemes already mimics the effects of a sub-grid-scale model [18]. This leaves scaling up ν or decreasing the grid spacing, if a simulation at a given resolution h is unsatisfactory. The capability of the PPM scheme to mimic small scale turbulence has been studied with high-resolution simulations in [104] while [54] claim that such schemes are too viscous compared to sub-grid-scale models. That notion is not surprising, since the design goal of shock-capturing has not been the simulation of turbulence on unresolved scales, but the representation of steep gradients or discontinuities with affordable grid spacings h. Hence, they avoid having to introduce an artificial diffusivity term ν^{shk} in the calculations, but not necessarily also ν^{hyp}. One should be cautious about straightforward comparisons of different tests of turbulence simulations, if they are based on calculations with different resolution. For instance, [104] use 512^3 grid points while [54] use up to 128^3. To resolve shear driven turbulence on an equi-spaced grid requires typically $\sim 400^3$ grid points (see Sect. 4.2). Hence, the performance of the different methods should be expected to be *strongly scale dependent!* Consequently, even for the same physical problem the answer to the question which method is most reliable (or most efficient) depends on the affordable resolution and the required accuracy. Since physical processes, which change the dynamics of the flow, may be accounted for with a given number of grid points N in one physical system, but not so in another, intercomparisons of methods for different systems have to be considered with even more caution.

Another approach to LES, which is guided by turbulence modelling, is to use a scheme with little or no numerical dissipation and combine it with a physical sub-grid-scale model (originally, the term LES was used only for this class of simulations). The classical model by [120] is still the most popular for this purpose. It is based on identifying the dissipation processes not accounted for due to the grid resolution h with an *eddy viscosity* which enhances the shear-stresses of the velocity field as resolved on the grid. This has some similarity to the concept of turbulent viscosity mentioned in Sect. 2.1. In simulations of flows with shocks, an artificial diffusivity (ν^{shk}) has to be added or the coefficient of the model [120] has to be boosted accordingly [109]. The application of eddy viscosity models to flows with shock fronts has been criticized, since underlying assumptions such as isotropy at scales $l < h$ and the existence of a Kolmogorov cascade at those scales do not hold. A more refined version of this approach is the dynamical sub-grid-scale model of [58]. It uses information from the flow at different scales $\ell \geqslant h$ to determine the (empirically defined) parameter of the eddy viscosity model [120] (see the chapter "Turbulent Combustion in Thermonuclear Supernovae"). We note here that some researchers consider the designation 'LES' only appropriate for simulations which can account for shear driven turbulence or show a Kolmogorov spectrum (see Sect. 2.1) or another scaling relation for the turbulent kinetic energy in Fourier space for some of the length scales resolved on the grid. Because such simulations require massive computational resources, other researchers prefer to call LES also more affordable simulations which only contain the scales around the maximum of turbulent kinetic energy, even if the simulation results in this case are more sensitive to the model used for the unresolved scales. The choice between these two points of view may have to be left to comparisons with data and simulations with higher resolution.

3.3 Time Integration in Hydrodynamical Simulations

Hydrodynamical simulations of turbulent convection also require numerical approximations to calculate the evolution of the system as a function of time. Very few numerical schemes, such as leapfrog combined with symmetric discretizations in space (centred finite differences, Fourier collocation), do not alter the amplitudes of purely advective or oscillatory solutions of differential equations [25, 107, 129]. Such numerical schemes might appear attractive for the integration of the non-linear advection terms in (1), (2) and (3), because they only introduce phase errors into the solution. Hence, they are *dispersive*, but not *dissipative*. Dispersivity of a numerical integration method implies that solution properties at different scales ℓ are propagated with different phase speed $\boldsymbol{v}(\ell)$ on the numerical grid. Consequently, waves of frequency f in the solution of the hydrodynamical equations are propagated with individual speeds $v(f)$, which gives rise to a phase error and an associated dispersion relation [129]. Dispersive errors affect all solution methods for *hyperbolic* terms in partial differential equations except for special cases with no practical relevance. Numerical dissipation of a time integration scheme also provides a different amount of damping for amplitudes of the numerical solution at various scales ℓ. This, too, originates from the *truncation error* of the numerical integration scheme and can be minimized by higher order schemes [129]. It must not be confused with physical dissipation due to viscosity. The latter converts macroscopic motions into microscopic ones (heat), but does not violate the conservation of (total) energy. A numerical time integration error may easily violate exact conservation laws, as the latter are integrated only with a finite accuracy. For higher order methods the dissipative errors may be smaller than phase errors of lower order methods [107], unless we consider isolated systems or extremely long integration times. Since purely dispersive schemes fail for solutions with shocks and also often for non-linear advection with very steep gradients, some small amount of numerical damping during time integration is usually accepted in simulations of turbulent flows. From a physical point of view, this limits the phase space that can be visited by the numerical simulation and hence the accuracy of ensemble averages for a given Δt.

Most time integration methods used for the non-linear advection terms in (1), (2) and (3) are explicit. This allows the construction of the next time step by extrapolations from existing information (the solution given at the current time or during recent time steps). Moreover, in a compressible flow sound waves are easily generated. They constrain Δt to follow the time evolution of a sound wave which travels relative to the flow speed $|v|$ at speed c_s between grid points with a minimum distance of Δx. This is the famous Courant–Friedrichs–Lewy (CFL) condition [107, 129]. If we are not interested in the fate of sound waves (or other waves travelling at high speed) and the flow velocity itself is subsonic ($|v| \ll c_s$), we have to resort to some form of implicit time integration. This couples all the grid points of the simulation domain during the construction of the solution at the new time $t + \Delta t$. The coupling is expressed through large (and in the case of the equations (1), (2) and (3) also non-linear) systems of algebraic equations. Since their solution is computationally expensive, there is a threshold which has to be exceeded to benefit from the less restrictive stability conditions of implicit methods.

Implicit time integration also trades its less restrictive stability conditions for an increased dispersivity. Close to the stability limit traditional implicit methods have a much large dispersion error than comparable explicit methods (cf. the examples given in [129]). This error increases rapidly when increasing the time steps even further. Hence, care has to be taken, if accurate information on waves is to be extracted from a hydrodynamical simulation with large Δt (see also [129]). The same holds, if Mach numbers $|v|/c_s$ of the flow are close to 1. The mean velocities are close to sound speed in this case and, in addition, acoustic events such as generation of sound waves by the flow can easily occur. For such cases implicit integration methods may still be used for relaxation of the simulation towards a statistically steady state, but explicit methods are preferred to study the relaxed system [109]. Note that the source and diffusion terms in (2) and (9) often require implicit time integration.

4 Realistic Simulations of Stellar Convection

4.1 Choosing Scales: Effective Reynolds and Prandtl Number

Numerical simulations of stellar convection with realistic microphysics (equation of state, K_{rad}, ν) and a box size H require to include scales ℓ such that both $H > L > h$ and $H > \ell > h$ while $h \gg l_d$. How should we define the characteristic scale of the flow, L, ahead of a simulation? Let us consider the spectrum of kinetic energy E_{kin} after Fourier transformation at this point, for which $E_{kin} = \int E(k) dk$. In principle, L should be related to the length scale $\ell_{max} = 2\pi/k$ at which the maximum of (turbulent) kinetic energy occurs in Fourier space, since most of the dynamics of the system is expected to take place around that scale. The actual horizontal size of the computational box, H_{hor}, is not imposing L. It is the other way round, since L results from physical constraints such as mass conservation, cooling time scale and finite flow speed (see [124]). Indeed, if $L = H_{hor}$, one has to worry seriously about the influence of horizontal boundary conditions. What about using the vertical depth of the simulation box, H_{vert}, or the fraction of the vertical extent which contains the *convective layers* (see Sect. 1.2)? The depth of the entire simulation box has limited meaning, too, since it could be extended arbitrarily into the radiative layers. This leaves the depth z of the convective layer as a more useful choice. The depth of the entire convection zone has some well-defined physical meaning, because it defines the maximum vertical scale on which flow structures are unstable to buoyancy. But this is not an accessible quantity for most simulations of solar or stellar convection, since $H = \min(H_{hor}, H_{vert}) \ll z$ may be imposed by limited computational resources. Moreover, the total size of z does not necessarily correspond to the scale ℓ_{max} (certainly not for convection at the surface of solar-like stars). From that point of view the horizontal size of the main coherent flow structures, the width of upflows (resp. downflows or the sum of both) appears to be the most preferable choice for L. It has already been used in Sect. 2.1. Note that ℓ_{max} of the convective flow becomes larger for layers further inside the Sun. Hence, the upper layers within a convection zone set the strongest constraints on the extent of the range of scales $L > \ell > h$, while the lowermost convective layers define the minimum size for H_{hor}. As we can expect from the discussion in Sect. 3.2, a stable hydrodynamical simulation will have a cell Reynolds number Uh/ν_{eff} of order unity. The *effective viscosity* ν_{eff} includes contributions from the numerical integration scheme, a sub-grid-scale model (if present) and the kinematic viscosity originating from molecular processes. Numerical simulations of stellar convection with realistic microphysics currently all have $\nu_{eff} \gg \nu$ due to $L \gg l_d$. An LES is thus characterized by an *effective Reynolds number* $Re_{eff} = UL/\nu_{eff}$ (also called numerical Reynolds number, but the exact meaning varies). The amount of numerical viscosity is difficult to quantify. Since ν_{eff} is scale dependent, the naive estimate $Re_{eff} = UL/\nu_{eff} = (L/h)(Uh/\nu_{eff}) \gtrsim (L/h)$ is far too pessimistic, because $\nu_{eff} \to \nu$ if $Re_{eff} \to Re$. For simulations in 3D (three spatial dimensions) one can redefine $\mathrm{Re}_{\mathrm{eff}}$ to account for this scale dependency via

$$\mathrm{Re}_{\mathrm{eff}} \approx (l/h)^{4/3} \qquad (25)$$

assuming a Kolmogorov-like scaling $l/l_{\mathrm{d}} \sim \mathrm{Re}^{3/4}$ for $L \geqslant l \geqslant l_{\mathrm{d}}$ (cf. Chap. VI.6 in [85]). Because this scaling has its own region of applicability (see the chapter "An Introduction to Turbulence"), (25) should only be considered as an estimate. If a flow is restricted to two spatial dimensions, vorticity is conserved in a reference frame following the fluid motion [11]. In this case the dynamics is quite different: turbulent kinetic energy is preferentially transferred to large scales rather than small ones as in 3D, while enstrophy, the ensemble averaged variance of fluctuations in vorticity, $\mathrm{curl}\,\boldsymbol{v}$, is piled up at small scales. This eventually leads to the estimate (cf. Chap. VIII in [85])

$$\mathrm{Re}_{\mathrm{eff},2D} \approx (L/h)^2, \qquad (26)$$

which explains why it is so much easier to achieve high values of Re or $\mathrm{Re}_{\mathrm{eff}}$ in simulations of flows in 2D. The different dynamics, however, limits the usefulness of this simplification (for a real flow, even L and l_{d} might change).

The efficiency of radiative transfer and the inefficiency of molecular (or actually atomic) diffusion in stars yields Prandtl numbers Pr in the range of 10^{-6} to 10^{-10} for both the stellar interior and the stellar surface layers (see Sect. 1.2 for definitions). However, both molecular processes and radiation can contribute to viscosity. Following Chap. 43.1 of [71] the computations of Pr are often based on the *radiative viscosity* $\nu_{\mathrm{rad}} = a\,T^4/(c\,\kappa\,\rho^2) = 3\,c_p T\chi/(4\,c^2)$ alone. The resulting physical quantity should rather be called $\mathrm{Pr}_{\mathrm{rad}}$. A detailed computation with more accurate tables for ν (see [48]) demonstrates that both ν_{rad} and ν have very similar values at the solar surface. But in the underneath lying layer of hydrogen ionization ν_{rad} drops to only 0.1 to 0.01 of ν. It becomes larger than ν only further inside the convective region. For the Sun, surface values of Pr are close to $\sim 10^{-9}$ and gradually increase to $\sim 10^{-7}$ at the bottom of the convection zone. The extremely low Prandtl number is one of the most important differences between stellar convection and any terrestrial flow. It also explains why the detailed treatment of radiative processes receives much more attention in numerical simulations of stellar convection, while viscous processes are left to rather simple modelling [100, 109, 124, 135, 136]. For such purposes ν_{rad} provides a sufficient estimate for ν which can easily be calculated, if T, ρ, chemical composition and opacity are known. As the viscosity in numerical simulations is equal to ν_{eff} on the resolved scales $\ell \geqslant h$, LES of stellar convection are characterized by an *effective Prandtl number* $\mathrm{Pr}_{\mathrm{eff}} = \nu_{\mathrm{eff}}/\chi$. If ν_{eff} just results from a sub grid-scale model, it is also called *turbulent viscosity*, ν_{t}. The energy transport by scales $l < h$ gives rise to a *turbulent diffusivity* χ_{t} which is separately accounted for in such simulations. The ratio of these two is known as the turbulent Prandtl number Pr_{t}. An example for applying this modelling approach to solar surface convection can be found in [109]. Note that these turbulence quantities describe processes operating on scales $l < h$ interacting with scales $\ell \geqslant h$. They should not be confused with their identically named siblings in Reynolds-stress models of convection which describe the kinetic and thermal energy transport on *all* scales (cf. the chapter "Turbulence in Astrophysical and Geophysical Flows").

4.2 Shear Driven Turbulence and Grid Refinement

What value of $\mathrm{Re}_{\mathrm{eff}}$ do we have to achieve, if we want to observe turbulent shear instabilities in an LES on scales $\ell \geqslant h$, which are created on the scales resolved in the simulations? The critical Reynolds number for the transition from laminar to turbulent flow most often quoted in the literature refers to circular pipe flow and implies that $\mathrm{Re}_{\mathrm{eff}} \gtrsim 2300$. The shear–stress between an up- and a downflow in a convection zone is perhaps more similar to plane Couette flow (occurring in fluid embedded between two plates moved in opposite directions). This would imply $\mathrm{Re}_{\mathrm{eff}} \gtrsim 1000$ (cf. [85]). Since there is a considerable spread in the numerical value for different flows which depends on details of the actual experiment, we shall use here the condition that $\mathrm{Re}_{\mathrm{eff}} \gtrsim 2300$ as a rough estimate. This number should not be confused with the much lower critical Reynolds number for the onset of turbulence in stratified shear flow as discussed, e.g. in [114]. How can we fulfill this requirement in a 3D numerical simulation of convection? From (25) we have that $\mathrm{Re}_{\mathrm{eff}} \approx (L/h)^{4/3}$. Because dissipation occurs within a simulation at scales $\sim h$, a simulation with $H \geqslant L > h$ requires $N = L/h \gtrsim 330$ grid points per spatial direction to reproduce fully developed shear driven turbulence. The latter is expected to be caused by the Kelvin–Helmholtz instability which in turn results from the shearing forces between up- and downflows in a convective zone. L, the length scale of the maximum of kinetic energy in the flow, should be contained safely within the simulation box. We therefore require that $H > L$, whence $N \gtrsim 330$ is a lower limit. It might seem sufficient to perform a simulation at a much lower $\mathrm{Re}_{\mathrm{eff}}$. In that case the sub-grid-scale model or some kind of viscosity has to act exactly like the non-linear forces on the scale h for the cell average over h to behave like the volume average of a fully resolved calculation. This is unfortunately not within the mathematical nature of the methods used for modelling unresolved scales: if we really want to find out how shear driven turbulence influences the convective flow in detail, we cannot avoid resolving the shear instability itself. Otherwise the simulation requires careful comparisons with experimental data to ensure that the properties of the simulated flow at scales $\ell \geqslant h$ are within an acceptable range of uncertainty introduced by the sub-grid-scale modelling. Moreover, the lower limit of $N \gtrsim 330$ is optimistic, because the effective resolution h of an LES is usually less than the actual grid spacing Δx, especially, if numerical or artificial viscosity are large.

In solar granulation shear-driven turbulence can be excited as a secondary instability [100] by the up- and downflows of convection resulting from the primary (convective) instability. For $L \sim 1300$ km relation (25) implies $h \sim 4$ km, if we want to fully resolve the shear driven instability within a 3D simulation.

For 2D simulations this is already achieved at $h \sim 27$ km because of (26). This might explain why the results shown in Fig. 4 at $h \lesssim 23$ km reveal many features of the much higher resolution simulations discussed in [100], where $h \lesssim 3$ km. Artificial diffusivity was used in most of the latter.

As just mentioned, $H = L$ is too small for a realistic simulation of solar granulation, even when using periodic boundary conditions. Granules interact with each other both horizontally and vertically. In simulations of Boussinesq convection the aspect ratio $A = H_{\mathrm{hor}}/H_{\mathrm{vert}}$ is used to relate the horizontal width to the vertical extent of the simulation. Aspect ratios of 10 and 20 have been used [37] to study the possible origin of solar mesogranulation, large scale structures with an average size of ~ 6000 km. The consequences of choosing H_{hor} between $\sim L$ and $\sim 4\,L$ on simulations of solar granulation were studied in [109]. If $H_{\mathrm{hor}} \gtrsim 4\,L$, the influence of horizontal periodicity on p_{turb}, $\nabla - \nabla_{\mathrm{ad}}$, and other ensemble-averaged quantities describing the flow and the stratification of the convection zone becomes acceptably small. That size is just enough to hold a single mesogranular structure as studied in the MHD simulations of [37]. Hence, $H_{\mathrm{hor}} \gtrsim 4\,L \ldots 6\,L$ is recommended for applying horizontal periodic boundary conditions in solar granulation simulations. If we also want to study shear driven turbulence in such simulations, we end up with $N \sim 1000 \ldots 2000$ as a lower limit for computations which are done with a constant resolution throughout the simulation volume. In 3D that requirement implies using $10^9 \ldots 10^{10}$ grid points which brings the currently most powerful supercomputers to their capability limits for just a single simulation. Would it help to use a larger h in regions of larger L, i.e. in layers further inside the Sun, to keep the ratio L/h constant? This can save some resources, but the total dynamical range of the flow is larger anyway for a deep simulation, since ν and l_{d} do not change very much, and the increase in L means pushing H. As a measure on its own a constant L/h is insufficient.

An increasing number of numerical simulations of turbulent flows nowadays uses *adaptive mesh refinement* to increase numerical resolution within the simulation volume where it is most needed (see [103] for reviews and applications). The basic idea behind this approach is to place more grid points into regions of rapid local variation of the numerical solution. The problem with this approach for simulating large scale convective zones is that the refinement strategy works well essentially, if resolution is needed along surfaces in 3D or lines in 2D, i.e. if the structures which require enhanced resolution are not volume filling. Since up- and downflows cover large volume fractions, such a grid refinement strategy does not really help (a much more suitable problem for this approach would be an isolated thunderstorm cloud in the atmosphere of the Earth where the region separating the cloud form its environment justifies enhanced numerical resolution). An older variant of this approach, however, is quite useful for studying secondary turbulent instabilities in solar and stellar convection zones: grid nesting. It provides high resolution in a small interior domain and lower resolution in a much larger volume which contains the high resolution domain. This approach was used in [100] (in 2D). It allows resolving shear-driven turbulence in 3D without sacrificing a realistic aspect ratio: $H \gtrsim 6\,L$ in the low-resolution part while $L/h \gtrsim 330$ in the central, high-resolution part is feasible with a total of just $\sim 10^8$ grid points. Such simulations have to be evaluated carefully, as features originating from the lower resolution part of the simulation may enter the high-resolution domain and create transient features near the boundaries of the high-resolution part.

How robust are our estimates based on (25)? For a direct numerical simulation study of the typical laboratory setting of convection between two plates, it has recently been found [108] that this type of flow shows notable deviations from a Kolmogorov scaling $l/l_{\rm d} \sim {\rm Re}^{3/4}$. Other setups such as those of [134] with higher Rayleigh number but much lower aspect ratio found an $l/l_{\rm d} \sim {\rm Re}^{3/4}$ scaling at least for some cases. An inspection of numerical resolution reveals that all these simulations had less than $N^3 = 300^3$ grid points for the simulation volume, though $N > 300$ was sometimes used for a preferred direction. That may not suffice to resolve turbulent shear (${\rm Re}_{\rm eff} \lesssim 2000$). Hence, while the up- and downflow patterns of convective flows (see Fig. 2) are a robust feature readily found in direct numerical simulations with ${\rm Re} \leqslant 1000$ and $l_d \geqslant h$ [99, 108], in LES of solar granulation with ${\rm Re}_{\rm eff} < 1000$ (see Sect. 5), in geophysical simulations (see the chapter "Turbulence in the Lower Troposphere: Second-Order Closure and Mass-Flux Modelling Frameworks"), and in many others, such calculations cannot be used straightforwardly to extrapolate *all* the scaling properties of higher resolution simulations. This holds even more so for 2D simulations which at similar resolution as their 3D counterparts can represent a different domain in terms of dynamics and non-linear interactions.

4.3 Time Scales, Relaxation and Computing Statistics

For how long do we have to perform a numerical simulation of convection to obtain useful statistical information? To answer this question we first introduce a few time scales. An important quantity for a convective flow is the turn over time scale $\tau_{\rm c} = H_{\rm vert}/U = H/U$, or more generally,

$$\tau_{\rm c} = \int_{\rm bottom}^{\rm top} u^{-1}(x)\mathrm{d}x = \int_0^H u^{-1}(x)\mathrm{d}x, \qquad (27)$$

where $u(x)$ is the averaged root mean square velocity along the vertical coordinate x. Close to the surface of a star $\tau_{\rm c}$ is usually just a small multiple of the acoustic time scale $\tau_{\rm ac} = \int_0^H c_{\rm s}^{-1}(x)\mathrm{d}x \approx H/\overline{c_{\rm s}}$. Here, $c_{\rm s}$ is the local average sound speed (the alternative definition of $\tau_{\rm ac} = 2\,H_p/\overline{c_{\rm s}}$, with $H_p = P/(\rho g)$ as the local pressure scale height, is less useful in this context). In the interior of a star, $\tau_{\rm c} \approx 10^2 \ldots 10^4\,\tau_{\rm ac}$ (see Sect. 4.4). $\tau_{\rm c}$ provides a crude estimate how long it would take a test particle to travel from one boundary of the convection zone to the other. This is to be understood in a statistical sense: only few test particles will complete this journey, because they are also advected sidewards and even carried back in the opposite direction. τ_c is thus an optimistic lower boundary for τ_{mix}, the time it takes to mix a convective zone. For imponetrable vertical boundaries the definition (27) has to be refined to avoid singularities due to $u = 0$. A frequent substitute for τ_c is the buoyancy time scale τ_b or inverse Brunt–Väisälä frequency, the time scale on which density perturbations grow inside a convective zone [39]. In the interior of a convection zone $\tau_b \approx \tau_c$. Close to stable layers τ_b becomes arbitrarily large, since perturbations do not grow at all for a stable layer, where τ_b is related to wave propagation [45] instead of advective motions. In simulations with extended radiative layers the interval $[0, H]$ in (27) is often restricted to the convectively unstable layers or to layers where $u(x)$ is larger than some lower bound. Local time scales τ_c, τ_{ac} and τ_b can be computed for a vertical subdomain $[x_1, x_2]$.

In addition to τ_c, τ_{ac} and τ_b, there are several important time scales which are introduced by the sources and sinks of energy appearing in Eq. (9): gravity, radiation and nuclear energy production. The latter governs most phases of stellar evolution, but becomes comparable to the other two only during late, dynamical phases such as thermonuclear supernovae (see the chapter "Turbulent Combustion in Thermonuclear Supernovae"). In other cases it is much longer [71, 137] and can be neglected for hydrodynamical problems. So what about gravity? The Kelvin–Helmholtz time scale estimates for how long a star can sustain its present luminosity from its current gravitational potential energy:

$$t_{\mathrm{KH}} = |\Omega|/\mathcal{L} \approx G\mathcal{M}^2/(\mathcal{R}\mathcal{L}). \tag{28}$$

Here, Ω is the gravitational binding energy and $\mathcal{M}$, $\mathcal{R}$ and $\mathcal{L}$ are mass, radius and luminosity of the star. More accurate estimates introduce factors of order unity [137]. For a spherically symmetric star Ω can also be computed from $|\Omega| = 3 \int_{M_s(r_1)}^{M_s(r_2)} P/\rho \, dM_s$ by means of the virial theorem [71], where $M_s(r_{\mathrm{bottom}}) - M_s(r_{\mathrm{top}})$ is the total mass inside the simulation box and $M_s(r) = \mathcal{M} - M_r$ is the mass integrated downwards from the surface. This way of computing t_{KH} is also useful for surface convection simulations in a rectangular box geometry as long as differences to spherical symmetry are small. If the mass distribution of a star as a function of depth changes, t_{KH} also provides the time scale required for a stellar model to readjust itself into an equilibrium state [137]. The radiative time scale t_{rad} on the other hand, describes how long it takes for a perturbation in temperature to disappear due to radiative transfer. In the optically thick case it is equal to the radiative diffusion time scale, which for a perturbation of size l implies that $t_{\mathrm{rad}} \approx l^2/\chi$. As long as convection is efficient, t_{rad} is not important. In that case, $t_{\mathrm{rad}} \gg \tau_c$, if we take $l = z$ as the size of the convective zone and compute τ_c from integration over the same domain. Whenever radiative losses of the fluid are large and convection is inefficient, t_{rad} becomes important. It may even constrain the time step of a numerical simulation of surface convection [53] (the latter scenario requires a more general computation of radiative cooling valid for the optically thin case, see [121]). Apart from the layers of partial ionization of hydrogen and helium in a star, which contain very little of its total mass, the radiative temperature gradient ∇_{rad} agrees with the actual temperature gradient ∇ to within an order of magnitude. It is thus not surprising that t_{rad} and t_{KH} often agree with each other to within a factor of 10 for layers inside a star and t_{rad} is used interchangeably with t_{KH} as the thermal adjustment time of stars [71]. These interpretations are of interest for numerical simulations, because they imply that t_{KH} describes how long a simulation requires to adjust itself to new model parameters. But do we really have to perform simulations for such a long time scale?

For a convection zone filling the entire computational domain, a simulation time t_{s} of several multiples of τ_{c} is sufficient to establish the mean structure of the physical system, since it is rapidly mixed by the velocity field and brought into a new dynamical equilibrium.[7] For numerical simulations of surface convection in the Sun and in cool stars, each of the local time scales $\tau_{\mathrm{c}}, \tau_{\mathrm{ac}}, t_{\mathrm{rad}}$ and t_{KH} is very short for the stably stratified layers on top, where radiative losses occur (typically fractions of a second to at most a few minutes, except for stars with very large radii). The time scale for relaxing the simulation through advection is thus governed by the value of τ_{c} for the entire simulation domain. As a result, the mean structure is established within a few multiples of τ_{c} [90, 109]. Is such a short t_{s} sufficient to compute higher order correlations statistics and consider the system as thermally relaxed?

If the starting configuration for the simulation is far from hydrostatic balance, a much more vigorous flow can be driven. This accelerates relaxation towards a state close to equilibrium. Thus, t_{KH} will be initially much shorter and change during the simulation towards its equilibrium value. If the starting configuration is close to hydrostatic balance, much smaller changes occur due the flow. In both cases the relevant physical process for advancing the physical state of the simulation domain, advection of mass, occurs on short time scales of a few multiples of τ_{c}. For a 3-Mm-deep box of a typical solar granulation simulation, t_{KH} is roughly 1 day or $\sim 10^2 \, \tau_{\mathrm{c}}$. Reported relaxation times are a few hours for the Sun [109] and other cool stars [90]. This is clearly shorter than t_{KH}, in agreement with the observation that equilibrium is established by advection of mass. Figure 5 shows the ensemble averages of higher order moments of the horizontal velocity field for the solar granulation simulations discussed in Sect. 5. The variables u and v designate the deviations from the horizontal mean of the two horizontal flow components. Their ensemble averages have been obtained from first horizontally averaging the quantities of interest which in turn are averaged in time. The resulting functions depend on depth only. The quantity $\nabla - \nabla_{\mathrm{ad}}$ has been computed from averaging the local gradients $(\nabla, \nabla_{\mathrm{ad}})$. The agreement for the root mean square velocities is remarkable. However, skewness of the horizontal flow is less well converged (it should be zero for

---

[7] We exclude here simulations starting from very small velocities: if they are too small, they may not even destabilize the fluid to create a sustainable flow.

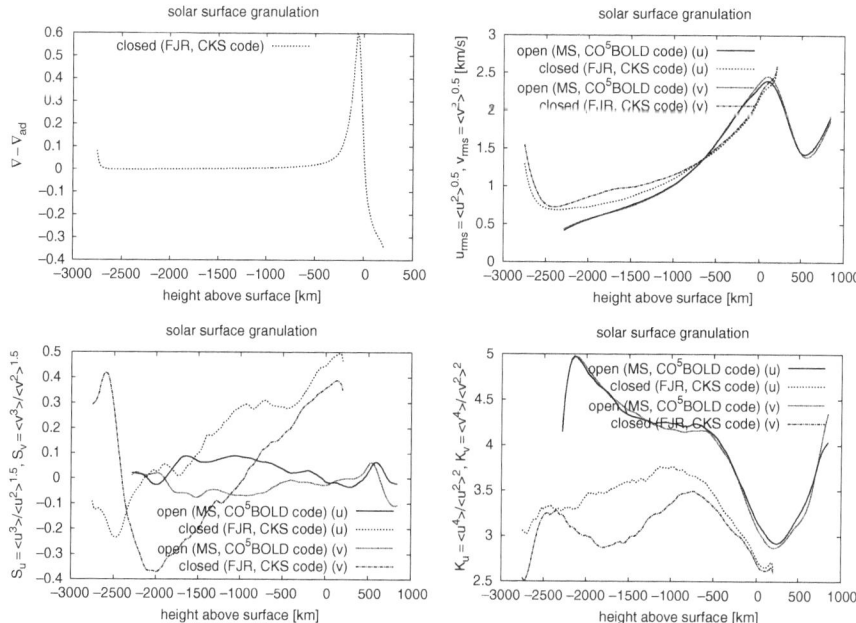

Fig. 5. Moments of horizontal velocities in solar granulation simulations with open and closed vertical boundary conditions. The *upper right* panel shows the root mean square averages of both components, u and v. The simulations are well relaxed, since $(\overline{u^2})^{1/2} \approx (\overline{v^2})^{1/2}$. *Lower left* and *right* panels show skewness and kurtosis. Ideally, $S_u = S_v = 0$, which is clearly the most challenging quantity. The superadiabatic gradient $\nabla - \nabla_{\mathrm{ad}}$ is shown as a reference. Just underneath the surface (i.e. below 0 km) is the superadiabatic peak. The feature to the far left is caused by the closed boundary at the bottom. Data kindly provided by M. Steffen and F.J. Robinson, based on the CO^{5}BOLD code [52] and the Chan–Kim–Sofia (CKS) code [38, 70]

simulations of infinite horizontal extent in the absence of any external forcing breaking horizontal symmetry). Part of this discrepancy is related to the small aspect ratio (just a few granules per box) of the simulation run for a closed box. However, while u_{rms} and v_{rms} barely differ between averages over 52 min (shown) and 26 min of averaging time t_{av} for the simulation with open vertical boundaries, S_u and S_v continue to improve. It appears that a t_{av} close to t_{KH} is necessary for a similar agreement as for u_{rms} and v_{rms}, unless $H_{\mathrm{hor}} \gg H_{\mathrm{vert}}$. The latter implies that each snapshot of the simulation at a given time is already a good representation of the ensemble average, which is not the case here. Not all the ensemble averaged quantities are equally sensitive to intermittent events, which can take place on time scales much longer than τ_c and contribute significantly to their statistics. Therefore, convergence of ensemble averages and hence the necessary t_{av} are checked in a case-dependent way [78] by comparing averages over smaller subintervals in time to exclude long-term trends (e.g. comparing $[0, t_{\mathrm{av}}/2]$

vs. $[t_{av}/2, t_{av}]$). Since initial conditions may be atypical, the first, *relaxation* part t_{rel} of t_s has to be neglected, whence $t_s = t_{rel} + t_{av}$. The averaging time t_{av} and the simulation time t_s thus depend on the physical question of interest.

But is there also a case for long relaxation? Possible issues include oscillations, long-term drifts, and inefficient mass flux combined with poorly known initial conditions. The first case occurs in simulations of surface convection in solar-like stars. In general, the initial conditions of such simulations are slightly out of hydrostatic equilibrium, since a stellar structure model, a simulation in 2D, or a coarse grid simulation in 3D will have a different p_{turb}. This triggers artificial p-modes as a result of which the centre of gravity within the simulation box will oscillate vertically. The life-time τ_{damp} of such p-modes is similar to those naturally triggered by the convective flow itself: of the order of a few days for a simulation 3-Mm deep (cf. [57], see [45] for a review on p-mode physics). Consequently, $\tau_{damp} > t_{KH}$. Though this may not matter for computing spectral line profiles, it is highly unwanted when probing simulations through helioseismology. Fortunately, the oscillations can be removed more rapidly by damping them numerically over $t \sim 5\,\tau_c$, which brings the relaxation time scale t_{rel} into the previously mentioned range.

Another issue are slow drifts over t_{KH}: statistical averages will slowly change, if there is a remaining excess heat or gravitational energy which has not yet been released through radiation at the upper boundary of the simulation box. In simulations with no internal source of energy and a constant heat flux entering just the bottom of the simulation domain, this can be noticed by a nonconstant flux of a few per cent of the total energy flux through the domain (cf. [99]). For probing closures in Reynolds stress models that accuracy is often sufficient, but for a comparison with a fully self-consistent stellar structure calculation based on a Reynolds stress model of convection such a slow drift of ensemble averages introduces systematic errors. Stellar models are relaxed on the time scale t_{KH} for the entire star or at least the domain considered for comparison: such stationary limit configurations are very useful to accelerate the time integration of the very long phases of nearly perfect thermal equilibrium over the "life time" of a star (cf. [71, 137]). In such comparisons discrepancies caused by insufficient thermal relaxation can be misleading, if high accuracy (better than 5–10% relative error) is expected for sensitive quantities (say, S_w or F_{kin}).

Poorly known initial conditions can be a problem in simulations with a low mass flow in the lower parts of the domain. To understand this point let us first consider the Sun. The kinetic energy of the solar photosphere due to convection is roughly comparable to its gravitational potential energy. If we also consider that $t_{KH} \lesssim 30$ s and $t_{KH} < \tau_c$ for the photospheric layers, it is evident that even poor initial conditions for the surface do not influence t_{rel} for a simulation extending say 3 Mm below the photosphere, because τ_c for the entire domain is of the order of 10 min. However, for the entire solar convection zone t_{KH} is about 10^5 years! From the solar model tabulated in

[128] one can estimate that τ_c is about 1 month for that region. The energy transported to the surface is just a tiny fraction of the gravitational potential energy of the entire convection zone. Since the solar convection zone in most layers is nearly adiabatic (hence the mean structure is well known) and mixing through advection only requires a few rotation periods (roughly equal to τ_c), it is feasible to perform simulations of the lower part of the solar convection zone (e.g. [24]) with realistic values of $\mathcal{M}$, $\mathcal{R}$ and $\mathcal{L}$. But what about the stable layers underneath? The resolution requirements introduced by those layers are so challenging that presently artificially enhanced values of ν and χ have to be used [24]. They help in a more rapid mixing of those layers for which good initial conditions are not known. Reasonable ensemble averages can again be computed after t_{rel} of a few τ_c, although a very small and slow drift of the ensemble averages in time has to be expected once more long-term computations become feasible (say with t_s of a few decades to study the solar cycle). Simulations of overshooting below a stellar convection zone can also be performed for A-stars, which have about twice the mass of the Sun with very shallow surface convective layers. Their comparatively small extent is caused by high radiative losses due to higher temperatures and lower densities. Hence, the entire vertical extent of the convective layers and their surroundings can be considered in a single numerical simulation in 3D with present computational resources [53]. In this case, the stratification remains close to radiative so a good initial stratification is possible and high convective velocities lead to rapid mixing. The situation is different for arbitrary convection zones deep inside stars. An example of what happens when the mass distribution cannot be guessed in advance and a large amount of excess energy has to be transported off is given in [99]. In this case strong overshooting changes the initial mass distribution beneath a convection zone sandwiched in between two deep, optically thick and convectively stable layers. Potential energy is required to build up the flow itself, since a good guess for $\nabla - \nabla_{ad}$ is not known in advance. Hence, it took $t_s \sim 0.5 \, t_{KH} \sim 5000 \, \tau_{ac}$ (or more than $10^3 \tau_c$) until converged statistical averages could be computed from that simulation (cf. the comments in [79] and compare the results shown in [99] with [76]). However, even if a convection model had been used to guess the initial mass distribution more accurately, $t_s \sim 10^3 \tau_{ac}$ would have been inevitable due to the mass redistribution below the convection zone where advection is no longer efficient. A variant of this problem is the case of open vertical lower boundary conditions for which a constant input flux should be imposed at the bottom: to simulate solar surface convection for the correct value of $\mathcal{L}$ requires an iterative change of the input entropy on $t_{rel} \sim t_{KH}$ (see [135] and Sect. 5). Starting from a 'more similar input model' is just the same procedure distributed over more than one simulation run. Whether one can avoid relaxation on a scale $t_{rel} \sim t_{KH}$ or not hence depends on the absence or presence of a strong feedback between the (not necessarily in advance known) average thermal structure and the convection zone.

4.4 Sound Speed and Convection

Is stellar convection really subsonic? A large range of Mach numbers Ma $=$ u/c_s is found in models and simulations of stellar convection zones of different stars or even inside the same star. Convection zones deeply inside stars usually have Ma of $O(10^{-4})$ to $O(10^{-2})$. Examples are the bottom of the convection zone of our Sun and convective core regions of massive stars. On the contrary, surface convection zones can have Ma of $O(1)$. The root mean square velocities quoted for the Sun in Sect. 1 imply Ma $\sim 0.3 \ldots 0.4$. For hotter stars Ma ~ 0.7 has been found in numerical simulations [131]. As a result, it is common to find even Ma > 1.0 at least locally, for instance, inside fast downdrafts. Such regions are prone to excess heating caused by shocks. The latter form when local flow velocities exceed c_s which in turn reduces the efficiency of convective energy transport. But that is neither an obstacle for convection to occur, nor for turbulence, nor for the combination of both. Shock fronts are also encountered at the solar surface (see, for instance, [67]). From a spectroscopic point of view shocks can be difficult to distinguish from other processes, say, if only disk-averaged line profiles in a limited wavelength range are available. Indeed, observed "macroturbulent line broadening" in a stellar spectrum (Sect. 2.3) may be caused by shock fronts, a convective up- and downflow pattern, or other types of velocity fields (an example for this problem are the line profiles found for supergiant stars, see [84]). Note that sound waves can easily be generated in regions where the average flow speed is close to c_s. This adds another possible sink of energy to the convective flow. Thus, convection, turbulence, local shock fronts and sound waves can appear within the same simulation domain. For that reason, numerical simulation codes used to study stellar surface convection must be able to tackle shock phenomena, independently of how they deal with turbulence on unresolved scales. At this point it is important to recall the difference between the (near) discontinuity of some physical variables at a given location (the shock front) and the velocity the discontinuity is moving with. It is completely false that a shock wave always propagates at supersonic velocities. Shocks may even stall, as is the case at the surface of neutron stars during their formation in a core-collapse supernova [98]. Nor does there have to be a flow across the shock interface: in *contact discontinuities* the associated mass flow across the interface is zero.

5 Boundary Conditions and Reynolds Averages

One important difference between the convective planetary boundary layer of the atmosphere of the Earth or laboratory experiments on convection and stellar convection zones is the absence of a 'wall boundary layer' in stars. A solid boundary provides a distinct source of turbulence through shear stresses between the flow in the interior of the convection zone and the boundary itself. Another consequence is the formation of a boundary layer near the 'wall'. Local instabilities in this wall boundary layer are an important source of intermittency (see also Sect. IV.3 in [85]). Solid boundaries set a strict upper limit to the length scales in a flow at a given location. In stars including our Sun the penetrable (vertical) boundaries of a convection zone do not set strict upper limits to the size of flow structures. The stably stratified layers enclosing convective layers can act both as a source or a sink of turbulence (see Sect. 2.1 and the chapter "Turbulence in Astrophysical and Geophysical Flows"). Just like the role of shear between convective up- and downflows the influence of closed (solid) boundaries as compared to open (penetrable) ones is still debated for the case of stellar convection. In the following, we discuss some results hopefully motivating further research. To proceed we first explain the implementation of boundary conditions.

In lateral directions, the boundaries of simulation boxes are commonly assumed to be open and periodic. This assumption relies either on spherical symmetry of the simulation domain or on considering identical copies of rectangular boxes. As discussed in Sect. 4.2, periodic copies of a simulation box require a sufficiently large aspect ratio to provide space for a minimum number of distinct up- and downflow structures. The intention behind this is to reduce self-interaction between the copies of the physical system and thus provide a more realistic representation of ensemble states. To this end it is sufficient that $H_{\mathrm{hor}} \gtrsim 3 \ldots 6 \, L$, unless we are explicitly interested in the long-range interactions (extended tests can be found in [109]). This idea is the basis of the "box-in-a-star" approach which simulates just a limited volume within the star. Spherical symmetry is used in simulations of entire shells within a star or the core of a star as a variant of the "star-in-a-box" approach, in which the simulation volume is assumed to contain the entire star [52]. The latter are known as *global* simulations which are much more limited in spatial resolution. The box-in-a-star approach allows *local* simulations with very high resolution. The price to pay for this increased resolution is the necessity to 'guess' what happens at the boundaries. But what to do in particular with the vertical boundaries? Stellar convection zones are characterized by a very strong vertical stratification due to gravitation which eventually leads to temperature and density changing by orders of magnitudes throughout the convection zone in stars such as our Sun (Sect. 2). This motivates the following strategies.

Despite they are 'artificial' closed vertical boundary conditions are often used in stellar convection simulations. This is mainly due to their mathematical simplicity. How are they applied in a stellar context? Closed vertical boundaries are implemented as solid, *slip* conditions (also called free-slip conditions), where $v_1 = 0$ and $\partial v_2/\partial x_1 = \partial v_3/\partial x_1 = 0$. Hence, no fluid can flow through the boundary ($v_1 = 0$), but the flow deflected sidewards does not create shear stresses acting on the vertical boundaries (zero vertical gradients $\partial/\partial x_1$ of horizontal components v_2 resp. v_3). Thus, they are also termed *stress-free boundaries*. This removes some of the unwanted "wall effects", but nevertheless has an impact on the flow inside the simulation box, which we discuss below. The vertical boundary conditions for (1), (2) and (3) resp. (10)

are completed in simulations with closed vertical boundaries by specifying an input heat flux at the bottom and a fixed temperature or radiative flux at the top. Since no mass is advected beyond the vertical boundaries in agreement with $v_1 = 0$, the total mass is conserved (see [24, 99, 109] and many others).

Simulations with open (vertical) boundaries also assume periodic horizontal boundary conditions. Mass flowing vertically outwards is generally left unaltered. On average it must be compensated by an equal amount of mass flowing inwards. It is usually the inflow which is modified according to physical assumptions about the surrounding fluid, since the properties of the interior are explicitly simulated anyway. For solar surface convection with a lower boundary inside the convective zone the inflow is considered to be isentropic and a fixed value for the entropy or internal energy of the inflow replaces the constant heat flux condition [124, 135, 136]. The surface effective temperature and thus surface flux and luminosity $\mathcal{L}$ cannot be predetermined and a simulation for a specific $\mathcal{L}$ requires iterating the value of the input entropy (see [135] for a complete description). Upper open boundaries are considered to be radiative and damp outwards running waves or at least avoid their reflection into the simulation domain. Examples can be found in [124, 136]. Open and closed vertical boundaries are sometimes also used just on one side [127, 135].

Figures 5, 6, 7 and 8 compare the mean structure and some higher order moments of velocity and temperature fluctuations as found in two different numerical simulations of solar granulation. In both the cases the upper boundary is within the stably stratified photosphere, while the lower boundary is inside the quasi-adiabatically stratified convection zone (cf. Fig. 5). In the simulation with vertical boundaries a second-order scheme with a Smagorinsky type sub-grid-scale viscosity [49, 120] is used. Shocks are treated by enhancing the coefficient of the sub-grid-scale viscosity proportional to $(\mathrm{div}\,\boldsymbol{v})^2$. More details on the numerical methods are given in [38, 70, 109]. The computational box has $170 \times 58 \times 58$ grid points, which are distributed equidistantly over a physical box size of $3.0 \times 2.9 \times 2.9$ Mm^3. Hence, the first (vertical) coordinate has higher resolution. The simulation itself is described in detail in [78]. The basic simulation parameters (effective temperature of 5777 K, surface gravity g of 274 m s^{-2}, chemical composition) and the initial mass distribution were taken from a standard solar structure model [109]. The horizontal width is the bare minimum to compute meaningful ensemble averages, as shown in [109].

The simulation with open boundary conditions has been presented in [127]. It uses a second-order scheme which can follow the propagation of shock fronts as described in [106], with the flux reconstruction scheme of van Leer [133]. The basic principles of this method are also discussed in [105] in the general context of higher order shock-capturing schemes. A Smagorinsky type sub-grid-scale viscosity [49, 120] is used to avoid piling up of kinetic energy at the grid scale. The simulation code was first described in [52] and in more detail in [136]. The box used for this simulation has $165 \times 400 \times 400$ grid points distributed non-equidistantly vertically and equidistantly in the horizontal directions. The physical domain size is $3.0 \times 11.2 \times 11.2$ Mm^3. The input entropy

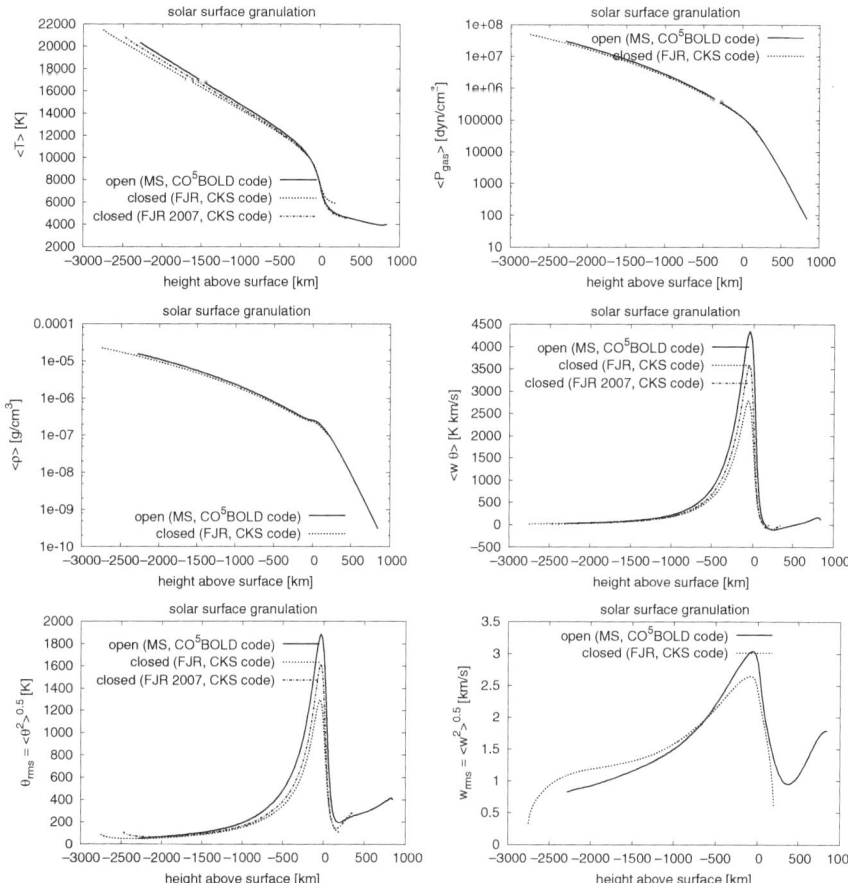

Fig. 6. Average temperature T and gas pressure P in solar simulations with open and closed vertical boundary conditions shown in the *top row*, density ρ and cross-correlation $\overline{w\theta}$ in the *middle row*, and root mean square temperature $(\overline{\theta^2})^{1/2}$ and vertical velocity $(\overline{w^2})^{1/2}$ in the *bottom row* (data by M. Steffen and F.J. Robinson, based on the CO^{5}BOLD code and the Chan–Kim–Sofia (CKS) code, see also text)

has been adjusted such that solar parameters are used as in the simulation with closed boundary conditions. Ensemble averages have been obtained from snap shots over a total of 52 min. of solar time. By comparison, the ensemble averages for the simulations with closed boundary conditions have been obtained from averages over 150 min. or $\sim 20\,\tau_c$. Note that since the box with open boundaries is located ≈ 500 km further upwards, its τ_c is somewhat shorter (cf. $\overline{w^2}$ in Fig. 6 and the definition in (27)).

Figure 6 shows that temperature, gas pressure and density as a function of depth are similar for the simulations. In particular, the steep gradient in T near the surface (at ~ 0 km) is almost identical (this leads to a similar

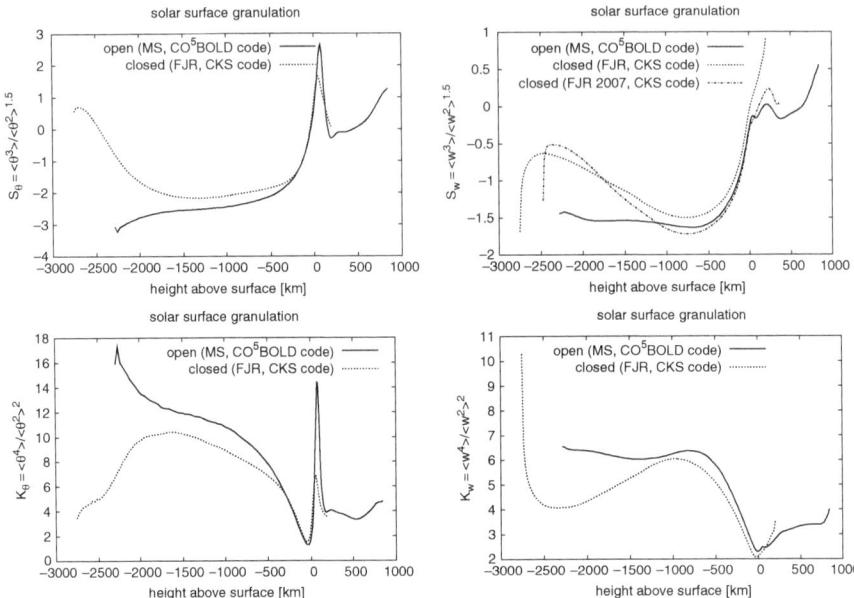

Fig. 7. Skewness and kurtosis (*upper* and *lower* panels) of temperature (S_θ, K_θ) and vertical velocity (S_w, K_w) (*left* and *right* panels) in solar granulation simulations with open and closed vertical boundary conditions (data courtesy M. Steffen and F.J. Robinson, based on the CO^{5}BOLD code and the Chan–Kim–Sofia (CKS) code)

peak in $\nabla - \nabla_{\mathrm{ad}}$, for which the results from the (FJR) simulation with closed vertical boundaries are shown in Fig. 5). The root mean square temperature deviations from the horizontal mean T and its counterpart for vertical velocity are also shown in Fig. 6. From a peak of over 1000 K, $(\overline{\theta^2})^{1/2}$ drops to less than 50 K within the simulated domain (in solar structure models it continues to drop to ~ 1 K near the bottom of the solar convection zone). This corresponds to a decrease of relative fluctuations from 10–20% to less than 0.25% (and $\sim 10^{-6}$ near the bottom of the solar convection zone). This strong decrease is a result of stratification and a constant energy flux inside the convection zone. The effects of a closed boundary are more obvious for $(\overline{w^2})^{1/2}$, as it causes a more gentle decrease of velocities towards the bottom followed by a sharp drop right near the boundary, while there is just a small increase in $(\overline{\theta^2})^{1/2}$ next to the closed lower boundary. Considering $\overline{\theta^2}$ and $\overline{w^2}$ the shape of the cross-correlation $\overline{w\theta}$ shown in Fig. 6 is not surprising. But what causes the difference in the maximum of $(\overline{\theta^2})^{1/2}$? Possible reasons could be frequency independent radiative transfer (cooling the surface more efficiently), or differences due to horizontal grid spacing, opacities, and the equation of state. But what about the vertical boundaries? To demonstrate their role averages for T, $\overline{w\theta}$ and $(\overline{\theta^2})^{1/2}$ from another simulation with closed vertical boundaries are shown in Fig. 6 (FJR 2007), which has a higher vertical

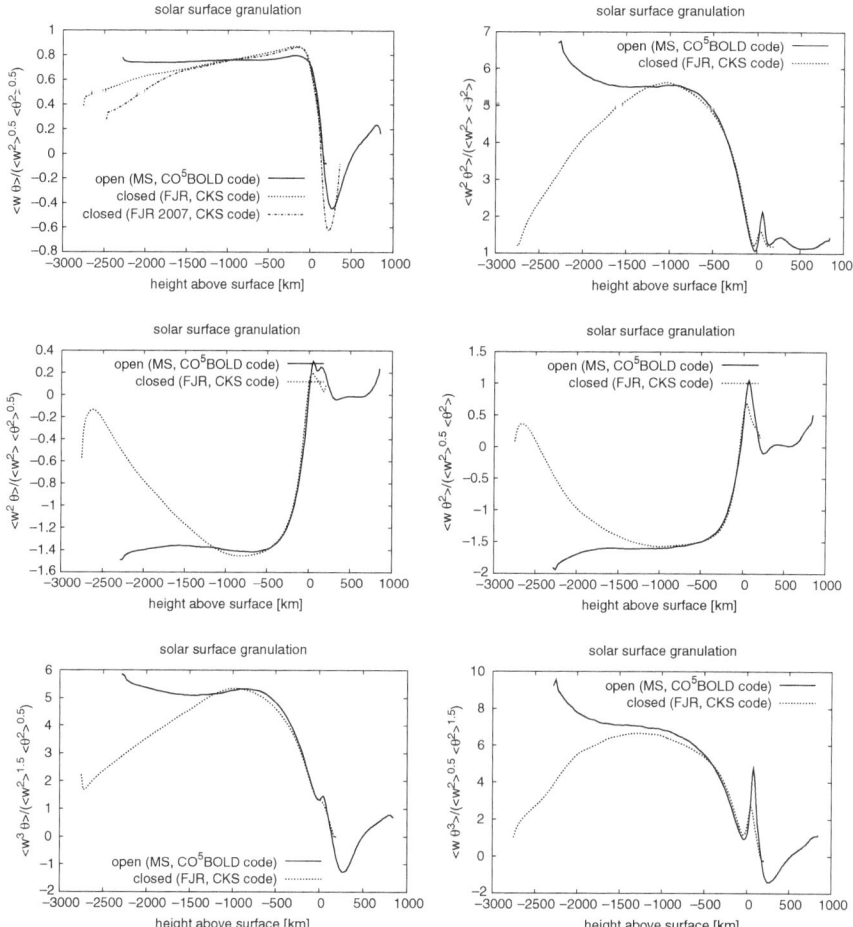

Fig. 8. Cross-correlations of vertical velocity and temperature in solar granulation simulations with open and closed vertical boundary conditions (data courtesy M. Steffen and F.J. Robinson, based on the CO^5BOLD code and the Chan–Kim–Sofia (CKS) code). Note that all correlations ($\overline{w\theta}$, $\overline{w^2\theta^2}$, $\overline{w^2\theta}$, $\overline{w\theta^2}$, $\overline{w^3\theta}$, $\overline{w\theta^3}$) have been normalized with respect to $\overline{w^2}$ and $\overline{\theta^2}$

extent above the optical surface.[8] Indeed, the distance to the top boundary is the main reason for changes in $\max(\overline{\theta^2})^{1/2}$. Nevertheless, mass and energy distributions are not very sensitive to whether vertical boundaries are open

---

[8] This is traded for a slightly smaller extent near the bottom. Moreover, the chemical composition is similar to the simulation with open boundaries and a higher horizontal resolution of 35 km was used (vertical one: 15 km) as well as a more extended box of $H_{hor} \sim 4100$ km (data from a private communication with F.J. Robinson).

or closed provided the boundaries are away from the region of interest. This relies on the fact that the simulation with closed boundaries exceeds a minimum depth, as was shown by an intercomparison of simulations with closed boundaries and different vertical extent [109]. It was concluded that simulations with closed boundaries should extend to $\gtrsim 2.8$ Mm below the observable photosphere to limit the influence of the boundaries on mean structure, the temperature gradient, and second-order moments of velocity. But what about higher order moments? In [78] it was estimated from the shape of different higher order moments as a function of depth that boundaries influence the flow and the ensemble averages up to a distance of 1.5–2 pressure scale heights (H_p, or difference in $\ln P$). Looking at their Table 1 this means that layers above $+50$ km and below -1650 km are strongly influenced by the vertical boundaries. For S_θ and K_θ shown in Fig. 7 this appears to be a reasonable estimate. The differences in the maxima of K_θ are found for the same layers as for $\overline{\theta^2}$ (Fig. 6). Comparisons with another simulation[9] with open boundary conditions [13] and a simulation with lower resolution corroborate these results. Looking at Fig. 7 we find that the differences between open and closed boundary conditions are largest for S_w followed by K_w (Fig. 7). Similar to T, $\overline{w\theta}$, and $(\overline{\theta^2})^{1/2}$ shown in Fig. 6, S_w is influenced by vertical boundaries too close to the optical surface. The influence of the lower boundary appears to reach some 500 km or 1 H_p further up for both S_w and K_w than for S_θ and K_θ. Do the cross-correlations between velocity and temperature show a different behaviour? Figure 8 compares them up to the fourth-order moments. As with $\overline{w^3}, \overline{w^4}, \overline{\theta^3}$ and $\overline{\theta^4}$, it is preferable to normalize them with second-order moments derived from the same simulation. The reason can be seen by the example of $\overline{w\theta}$ shown in both Figs. 6 and 8. The stratification leads to very pronounced maxima and minima at the solar surface ($+0$ km) with all functions rapidly dropping towards the interior, particularly for higher order moments, which makes a comparison of unnormalized quantities on a linear scale difficult. Note that while differences in $\overline{w\theta}$ in Fig. 6 are found mostly around a depth of $+0$ km, its normalized counterpart (Fig. 8) shows a linear trend for the case with closed boundaries inside the convection zone as opposed to the flat profile found for the case with open boundaries. The normalized third- and fourth-order cross-correlations all show an excellent agreement between the simulations sufficiently away from the vertical boundaries, except for the differences at the solar surface due to the enhanced temperature fluctuations in the simulation with open boundaries. The safety margins for closed boundary conditions suggested in [78] are hence acceptable for the cross-correlations and a box depth of 3 Mm.

Let us reconsider second-, third- and fourth-order moments of horizontal velocity fluctuations as shown in Fig. 5 in Sect. 4.3. Here, the deviations for the root mean square velocity are more similar in extent and overall trends to the case of S_w shown in Fig. 7. The deviations are even larger for skewness and

---

[9] From a private communication with K. Belkacem and R. Samadi.

kurtosis of horizontal velocities. However, there is also a substantial difference among the two velocity components within each simulation, particularly for the case of skewness. This indicates slow convergence of ensemble averages, probably aggravated by the limited horizontal box size of $H_{\mathrm{hor}} \sim 2.5\,L$ of the simulations with closed boundary conditions as opposed to $H_{\mathrm{hor}} \sim 9\,L$ for the simulations with open boundary conditions. In the latter, there is a deviation of horizontal velocities from a quasi-normal distribution $K_u = K_v = 3$ in the interior which is not obvious for the case with closed boundaries [78] nor for a deeper simulation with open boundaries at lower resolution.[10]

At the bottom line, mean structure and ensemble-averaged second-to fourth-order moments of velocity and temperature fields agree fairly well for simulations with either open or closed boundary conditions, provided the latter are sufficiently deep [109] and layers close to the boundary are considered to be just part of an 'extended boundary condition'. For some of the physical quantities the safety margin from a closed boundary within a convective zone appears to be slightly larger than suggested [78], about 2–2.5 pressure scale heights. If the boundary is within stably stratified layers, the safety margin in turn can be smaller. Note that also open boundary conditions have their limitations. Despite the large aspect ratio of the simulations used in this comparison, there are some trends close to the lower boundary, for instance for K_θ shown in Fig. 7, which have no apparent physical reason and are thus likely an artefact of the boundary conditions similar to the much stronger trends observed for closed boundaries. The necessary safety margin, however, appears to be substantially smaller, some 0.5–1.5 pressure scale heights depending on the quantity of interest. Away from the lower open boundary the normalized higher order moments saturate within the deep quasi-adiabatic convection zones. This agrees with the idea that not much happens there other than an increase of scales due to stratification (cf. also [126]).

It is important to remember that waves are more severely influenced particularly by upper closed boundary conditions, since within a thin medium they are easily reflected downwards. Depending on the physical question of interest this may require open boundary conditions also within stably stratified layers. Closed boundary conditions within a convective zone may have an undesirable influence on the magnetic field structure. Thus, in MHD simulations of solar surface convection open lower boundaries are clearly favoured (cf. [135]), though evidence has been reported in the literature that for this type of simulations open upper boundaries should be used as well [50].

6 Scaling Relations and Mixing in Overshooting Layers

Velocity fields in stably stratified layers are quite different from those found in convectively unstable ones. Wave motions dominate the velocity field further away from a convective zone (cf. the discussion in Sect. 4 of [89]). It is

---

[10] From a private communication with M. Steffen.

important to note that the mixing properties of waves and (advective) entrainment are very different, since a coherent oscillation mostly moves fluid just "back and forth" without much of a change of its position relative to its local environment (for a thorough study on mass transport and mixing by waves see [73, 130]). Entrainment by definition means that fluid originally belonging to a particular layer is engulfed by fluid belonging to an adjacent layer. This rapidly mixes both layers within the region of entrainment. The physical picture astrophysicists usually have in mind when they discuss convective overshooting in stars is a variant of this process where up- or downflows entrain the radiative layers as *plumes*, a term borrowed from geophysics. The mixing is particularly efficient, if the flow is turbulent all the way down to smallest scales, since *complete mixing* can then take place on very short time scales (this has important consequences for turbulent combustion, see the chapter "Turbulent Combustion in Thermonuclear Supernovae" and the chapter "One-Dimensional Turbulence Stochastic Simulation of Multi-Scale Dynamics"). Such behaviour is just the opposite of what is found for waves which contrary to other coherent structures of a turbulent flow such as plumes do not fulfill the criterion of efficient mixing used in the definition of turbulence on p. 49. In this sense waves are more akin to laminar flow, even if their excitation mechanism operates stochastically (cf. Sect. 2.4).

The different mixing capabilities of entrainment and *internal (including gravity) waves* have to be taken into account when analysing numerical simulations with layers characterized by overshooting. The upper layers (above ~ 0 km) in the solar granulation simulations as discussed in Sect. 5 are a good example: $\overline{w^2}$ first decreases also in the simulations with open boundary conditions (Fig. 6), but above $+400$ km it increases again. A similar change occurs for the horizontal velocities ($\overline{u^2}$ and $\overline{v^2}$ in Fig. 5). Lowpass filtering can be used to disentangle wave motions from advection [89]: contributions at frequencies $\omega > \omega_c(k)$ at a given wave number k are removed by a filter applied to the simulation data, where $\omega_c(k)$ is a linear function of k corresponding to a chosen constant (horizontal) phase speed. This type of analysis reveals the dominant contribution of waves to velocity fields in simulations of the upper photospheres of stars. In [89] the quantity $\overline{\rho w}_{\text{up}}/\overline{\rho}$ was used to quantify mixing in the photosphere of the Sun and in M-type dwarf stars (the latter have much lower T_{eff} than the Sun). It measures the outflow through a given layer by taking into account just regions with upwards pointing vertical velocity. An exponential decrease of the filtered velocity field as a function of distance was found in agreement with simulations for the shallow surface convection zones of A-type stars and white dwarfs [51]. Do these results corroborate a universal exponential decay law of the vertical mixing velocity?

The simulations in [51, 89, 90], despite their differences in opacity sources and the convective efficiencies attained inside the convective layers, have about the same Re_{eff}, a similar Mach number (the coolest stars have the lowest peak Ma of ~ 0.06), and rather low Peclet numbers Pe. The Peclet number measures the importance of convective relative to conductive (or radiative) heat transport. It is commonly defined as $Pe := (UL)/\chi$, in which case $Pe = Re \cdot Pr$. Since in usual stellar surface (granulation) simulations $L/h \sim 25 \ldots 50$ (cf. Sects. 2.1 and 4.2), shear instabilities inside the convection zone cannot be resolved ($Re_{\text{eff}} \lesssim 200$). With $Ma \sim 0.06 \ldots 0.6$ simulations are still a bit close to the transonic regime. But most importantly, for the solar case, $Pe \lesssim 10$ at the top of the convection zone, if we take the data from Sect. 2.1 (cf. Sect. 4.1). This value is actually the *physical Peclet number* of solar granulation. Thus, while Re_{eff} and Pr_{eff} are limited by the affordable resolution, $Pe \approx Pe_{\text{eff}}$ for simulations of stellar surface convection! In [90] is was noted that the less regular boundary of granules found for the simulations of atmospheres of M-type stars is related to the higher Peclet number at optical depth unity by comparison with the solar case. This is in agreement with the observation that convective zones in M-type stars extend much higher up into their photospheres, since molecules play a dominant role in their opacity, which in turn reduces the radiative cooling at their surfaces and thus increases Pe. Surface convection zones of A-stars as investigated in [51] represent the opposite case, as they have $Pe \lesssim 1$ and $Ma \sim O(1)$ even below the photosphere.

How close is that range of parameters to the case of stellar interiors? For the lower part of the solar convection zone representative values of Re are $\sim 10^{14}$ (Sect. 2.1) and Pr increases in the interior compared to its surface value. This results in $Pe \sim 10^6 \ldots 10^7$, whereas $Ma \sim O(10^{-4})$. Evidently, simulations of convection at the bottom of the solar convection zone such as [24] cannot reach realistic Peclet numbers: in their case $Pe \gg Pe_{\text{eff}}$. Hence, this regime is orders of magnitudes away from that one of stellar surface convection with respect to several parameters. The influence of Pe on overshooting has been studied in [23]. Although performed for idealized microphysics it provides valuable insight into the influence of Pe_{eff} (and also of Re_{eff} as well as of rotation, [22]) on overshooting. An increase of Pe_{eff} from ~ 17 to ~ 145, as evaluated for the downflows at the bottom of the convection zone, dramatically reduced overshooting (the average penetration depth was reduced by 0.6 pressure scale heights or $\gtrsim 50\%$ in terms of linear distance). The profiles of $\overline{w^2}$ and entropy changed as well in a way that is neither linear nor exponential, but clearly non-linear (Figs. 12, 13 and 14 in [23]). The complexity of vorticity was found to increase with Pe while the filling factor of the downflows became smaller and flow structures more filamentary. Due to the lower filling factor the downflowing plumes mix less efficiently in cases with larger Pe_{eff} despite the latter implies less efficient cooling and a reduced exchange of heat with the environment. This agrees qualitatively with the Reynolds stress model discussed in [29], which predicts more mixing in overshooting regions, if Pe is small. In spite of that the interpretation of both models and simulations is still under debate. Note that the maximum Re_{eff} reached in [23] is around 2200. If we take the definition used here and inspect their Figs. 1 and 11, Re_{eff} might actually be more close to ~ 1500. Nevertheless, the simulations approach the regime where shear instabilities develop.

How do the predictions for overshooting obtained from simulations compare to Reynolds stress models of stellar convection? Since a large variety of these models has been proposed, the answer is model dependent. Consider the model published and refined in [26, 27, 35, 36], since it is the most advanced one that has actually been applied to computations of stellar envelopes (it is also discussed in detail in the chapter "Turbulence in Astrophysical and Geophysical Flows"). Stationary solutions of the dynamical model equations were found for envelope models of A-stars [77], white dwarf stars [96], and for models with idealized microphysics [76, 79–81]. A linear decrease of $(\overline{w^2})^{0.5}$ was found in [77, 96] while for some of the simulations discussed in [76, 79–81] it turned out to be non-linear. No counterpart to the slow exponential decrease reported in [51, 89, 90] was found, even though we should bear in mind that the quantities used to define the region of overshooting were different. One possible reason for this difference is that the Reynolds stress models have not included a detailed representation of intermittent events, but rather describe what happens to the mean kinetic and potential energy during convective energy transport. Moreover, they do not account for waves. In this sense, they can only provide a lower limit for mixing through overshooting. It is more difficult to tell whether current simulations can provide a reliable upper limit for this process. Waves can break, create turbulence and cause mixing through a number of physical processes [130]. Thus, even for cases in which mean molecular weight gradients can be neglected, the 'definite recipe' to model overshooting in stars is still debated.

7 Conclusions

Convection is just one source of turbulence in stars. Others are related to shear stresses created by rotation or interaction with stably stratified layers (see the review in [143] and further contributions in the same volume). The theoretical modelling of convection can benefit from the advances which have been made in observational astronomy (high precision spectroscopy and photometry) and data interpretation (particularly for helio- and asteroseismology). It can also take advantage of a new generation of numerical simulations which rely not only on supercomputers but also on more refined numerical methods. The interaction among these branches of stellar astrophysical research should be complemented by exchange with other fields such as atmospheric sciences and oceanography. Stars have a number of properties which distinguish them from comparable geophysical systems: the extreme vertical stratification due to a self-sustained gravitational field and their sheer size, as well as the efficiency of radiative diffusion in comparison with molecular one represented by very low Prandtl numbers, and the absence of solid boundaries in normal stars are the most striking differences. There are also special physical conditions restricted to particular regions in stars such as the transition from high opaqueness to transparency to radiation in stellar photospheres and the dynamical properties of stellar convection introduced through its interaction with rotation and magnetic fields. In spite of those differences there is also a large body of common properties such as the coherent network of up- and downflow structures both observed for the Sun and found in numerical simulations which very closely resemble their geophysical counterparts [78]. The simulation techniques used in astrophysics and the geophysical sciences are hence very similar. This is not surprising, since the dynamics and the development of turbulence depend crucially on the approximation of non-linear advection in the hydrodynamical equations and on the internal degrees of freedom the numerical solutions have for a given, finite amount of computational resources. With stellar convection featuring both high and low Mach number flows, the numerical methods used are thus similar, as are their strengths and limitations. The scale dependence of experiments probing the treatment of unresolved scales is indeed common to laboratory, geophysical and astrophysical flows. It has to be considered with care before passing judgement on a particular simulation method. The same holds for the correct choice of scales ℓ. While the $\mathrm{Re_{eff}}$ of 2300 suggested in Sect. 4.2 should only be considered as a guideline (the actual number of grid points required per direction depends on the numerical method used!), it expresses the necessity to distinguish between low-resolution simulations, which only represent the flow structure at large scales, and high-resolution simulations, which explicitly represent different regimes of turbulence on scales larger than the finitely sized grids or volumes used in the calculations. For the same reason the dynamical range achieved in 2D and 3D simulations can be quite different at the same resolution no matter whether we deal with astrophysical or geophysical flows. Timescales and boundary conditions are more tied to the specific properties of the physical systems. Comparing them can nevertheless be instructive, since the long-term stability and dependence on boundary conditions are questions more easily analysed with a larger variety of cases. Progress in the modelling of turbulent convection can thus not only benefit from an exchange between both the astrophysical and the geophysical sciences on particular theoretical models, but also by an exchange on numerical simulation techniques.

Acknowledgements

I would like to thank Drs. F.J. Robinson and M. Steffen for their permission to use their simulation data to prepare some of the figures shown in this review. I am also grateful to Profs. H.J. Muthsam and O. von der Lühe for providing me with some of the originals of their figures.

References

1. Aerts, C.: Asteroseismology of close binary stars. In: Hartkopf W.I., Guinan E.F. Harmanec P. (eds.) Binary Stars as Critical Tools and Tests in Contemporary Astrophysics, Proc. IAU Symp., Vol. 240, p. 432. Cambridge University Press, Cambridge (2007)
2. Andersen J., Nordström B.: Critical tests in open clusters: requirements and pitfalls. In: Giménez A., Guinan E.F., Montesinos B. (eds) Theory and Tests of Convection in Stellar Structure, ASP Conf. Ser., Vol. 173, p. 31. Astron. Soc. Pacific, San Francisco (1999)
3. Antia H.M., Basu S.: Astrophys. J. **426**, 801 (1994)
4. Asplund M., Nordlund Å., Trampedach R., Allende Prieto C., Stein R.F.: Astron. Astrophys. **359**, 729 (2000)
5. Balmforth N.J.: Mon. Not. Roy. Astron. Soc. **255**, 603 (1992a)
6. Balmforth N.J.: Mon. Not. Roy. Astron. Soc. **255**, 639 (1992b)
7. Basu S., Antia H.M., Narasimha D.: Mon. Not. Roy. Astron. Soc. **267**, 209 (1994)
8. Basu S., Antia H.M.: Mon. Not. Roy. Astron. Soc. **269**, 1137 (1994)
9. Basu S., Antia H.M.: Mon. Not. Roy. Astron. Soc. **276**, 1402 (1995)
10. Basu S., Antia H.M.: Mon. Not. Roy. Astron. Soc. **287**, 189 (1997)
11. Batchelor G.K.: An Introduction to Fluid Dynamics. Cambridge University Press, Cambridge (1967)
12. Baturin V.A., Mironova I.V.: Aston. Rep. **39**, 105 (1995)
13. Belkacem K., Samadi R., Goupil M.J., Kupka F.: Astron. Astrophys. **460**, 173 (2006)
14. Belkacem K., Samadi R., Goupil M.J., Kupka F., Baudin F.: Astron Astrophys. **460**, 183 (2006)
15. Biermann L.: Z. Astrophys. **5**, 117 (1932)
16. Biskamp D.: Nonlinear Magnetohydrodynamics. Cambridge University Press, Cambridge (1993)
17. Böhm-Vitense E.: Z. Astrophys. **46**, 108 (1958)
18. Boris J.P., Grinstein F.F., Oran E.S., Kolbe R.L.: Fluid Dynam. Res. **10**, 199 (1992)
19. Borue V., Orszag S.A.: Europhys. Lett. **29**, 687 (1995)
20. Braginskii S.I.: Transport processes in a plasma. In: Leontovich M.A. (ed.) Reviews of Plasma Physics, Vol. 1, p. 205. Consultants Bureau, New York (1965)
21. Bretherton F.P.: J. Fluid Mech. **39**, 785 (1969)
22. Brummell N.H.: Turbulent compressible convection with rotation. In: Kupka F., Roxburgh I.W., Chan K.L. (eds.) Convection in Astrophysics, Proc. IAU Symp., Vol. 239, p. 417. Cambridge University Press, Cambridge (2007)
23. Brummell N.H., Clune T.L., Toomre J.: Astrophys. J. **570**, 825 (2002)
24. Brun A.S., Toomre J.: Astrophys. J. **570**, 865 (2002)
25. Canuto C., Hussaini M.Y., Quarteroni A., Zang T.A.: Spectral Methods in Fluid Dynamics. 3rd printing. Springer-Verlag, New York (1991)
26. Canuto V.M.: Astrophys. J. **392**, 218 (1992)
27. Canuto V.M.: Astrophys. J. **416**, 331 (1993)
28. Canuto V.M.: Astrophys. J. **478**, 322 (1997)
29. Canuto V.M.: Astrophys. J. **508**, 767 (1998)

30. Canuto V.M.: Astrophys. J. **524**, 311 (1999)
31. Canuto V.M.: Mon. Not. Roy. Astron. Soc. **317**, 985 (2000)
32. Canuto V.M., Mazzitelli I.: Astrophys. J. **370**, 295 (1991)
33. Canuto V.M., Mazzitelli I.: Astrophys. J. **389**, 724 (1992)
34. Canuto V.M., Goldman I., Mazzitelli I.: Astrophys. J. **473**, 550 (1996)
35. Canuto V.M., Dubovikov M.S.: Astrophys. J. **493**, 834 (1998)
36. Canuto V.M., Cheng Y., Howard A.: J. Atmos. Sci. **58**, 1169 (2001)
37. Cattaneo F., Lenz D., Weiss N.: Astrophys. J. **563**, L91 (2001)
38. Chan K.L., Sofia S.: Astrophys. J. **466**, 372 (1996)
39. Chandrasekhar S.: Hydrodynamic and Hydromagnetic Stability. Clarendon Press, Oxford (1961)
40. Chapman S.: Astrophys. J. **120**, 151 (1954)
41. Cheng Y., Canuto V.M., Howard A.M.: J. Atmos. Sci. **62**, 2189 (2005)
42. Chiosi C.: Convection and mixing in stars: theory versus observations. In: Kupka F., Roxburgh I.W., Chan K.L. (eds) Convection in Astrophysics, Proc. IAU Symp., Vol. 239, p. 235. Cambridge University Press, Cambridge (2007)
43. Christensen-Dalsgaard J.: Probing convection with helio- and asteroseismology. In: Giménez A., Guinan E.F., Montesinos B. (eds.) Theory and Tests of Convection in Stellar Structure, ASP Conf. Ser., Vol. 173, p. 51. Astron. Soc. Pacific, San Francisco (1999)
44. Christensen-Dalsgaard J.: Rev. Mod. Phys. **74**, 1073 (2002)
45. Christensen-Dalsgaard J.: Solar Phys. **220**, 137 (2004)
46. Colella P., Woodward P.R.: J. Comp. Phys. **54**, 174 (1984)
47. Cook A.W., Cabot W.H.: J. Comp. Phys. **203**, 379 (2005)
48. Cowley C.R.: Astrophys. J. **348**, 328 (1990)
49. Deardorff J.W.: J. Comp. Phys. **7**, 120 (1971)
50. Dorch S.B.F., Nordlund Å.: Astron. Astrophys. **365**, 562 (2001)
51. Freytag B., Ludwig H.-G., Steffen M.: Astron. Astrophys. **313**, 397 (1996)
52. Freytag B., Steffen M., Dorch B.: Astron. Nachr. **323**, 213 (2002)
53. Freytag B., Steffen M.: Numerical simulations of convection in A-stars. In: Zverko J., Weiss W.W., Žižňovský J. Adelman S.J. (eds.) The A-Star Puzzle, Proc. IAU Symp., Vol. 224, p. 139. Cambridge University Press, Cambridge (2004)
54. Garnier E., Mossi M., Sagaut P., Comte P., Deville M.: J. Comp. Phys. **153**, 273 (1999)
55. Gautschy A., Saio H.: Ann. Rev. Astron. Astrophys. **33**, 75 (1995)
56. Gautschy A., Saio H.: Ann. Rev. Astron. Astrophys. **34**, 551 (1996)
57. Georgobiani D., Kosovichev A.G., Nigam R., Nordlund Å., Stein R.F.: Astrophys. J. **530**, L139 (2000)
58. Germano M., Piomelli U., Moin P., Cabot W.H.: Phys. Fluids **3**, 1760 (1991)
59. Giménez A., Claret A., Ribas I., Jordi C.: Eclipsing Binaries with Accurate dimensions as a test of convective core overshooting. In: Giménez A., Guinan E.F., MonAesinos B. (eds.) Theory and Tests of Convection in Stellar Structure, ASP Conf. Ser., Vol. 173, p. 41. Astron. Soc. Pacific, San Francisco (1999)
60. Godunov S.K.: Mat. Sb. **47**, 271 (1959)
61. Goldreich P., Keeley D.A.: Astrophys. J. **212**, 243 (1977)
62. Gough D.O., Weiss N.O.: Mon. Not. Roy. Astron. Soc. **176**, 589 (1976)
63. Grossman S.A., Narayan R.: Astrophys. J. Suppl. **89**, 361 (1993)
64. Grossman S.A.: Mon. Not. Roy. Astron. Soc. **279**, 305 (1996)

65. Gryanik V.M., Hartmann J.: J. Atmos. Sci. **59**, 2729 (2002)
66. Gryanik V.M., Hartmann J., Raasch S., Schröter M.: J. Atmos. Sci. **62**, 2632 (2005)
67. Hansteen V.H., De Pontieu B., Rouppe van der Voort L., van Noort M., Carlsson M.: Astrophys. J. **647**, L73 (2006)
68. Houdek G., Balmforth N.J., Christensen-Dalsgaard J., Gough D.O.: Astron. Astrophys. **351**, 582 (1999)
69. Houdek G.: Stochastic excitation and damping of solar-type oscillations. In: Fletcher K. (ed.) Beyond the spherical Sun. SOHO 18/GONG 2006/HelAs I. ESA SP-624, p. 28.1. European Space Agency, Noordwijk (2006)
70. Kim Y.-C., Chan K.L.: Astrophys. J. **496**, L121 (1998)
71. Kippenhahn R., Weigert A.: Stellar Structure and Evolution. 3rd printing, Springer-Verlag, New York (1994)
72. Kjeldsen H., Bedding T.R.: What can we learn about convection from asteroseismology? In: Kupka F., Roxburgh I.W., Chan K.L. (eds.) Convection in Astrophysics, Proc. IAU Symp., Vol. 239, p. 113. Cambridge University Press, Cambridge (2007)
73. Knobloch E.: Mass transport and mixing by waves. In: Goupil M.-J., Zahn J.-P. (eds.) Rotation and Mixing in Stellar Interiors, Lect. Notes Phys. **366**, 109. Springer-Verlag, Berlin (1990)
74. Kosovichev A.G.: Helioseismic inferences on subsurface solar convection. In: Kupka F., Roxburgh I.W., Chan K.L. (eds.) Convection in Astrophysics, Proc. IAU Symp., Vol. 239, p. 113. Cambridge University Press, Cambridge (2007)
75. Kuhfuß R.: Astron. Astrophys. **160**, 116 (1986)
76. Kupka F.: Astrophys. J. **526**, L45 (1999)
77. Kupka F., Montgomery M.H.: Mon. Not. Roy. Astron. Soc. **330**, L6 (2002)
78. Kupka F., Robinson F.J.: Mon. Not. Roy. Astron. Soc. **374**, 305 (2007)
79. Kupka F., Muthsam H.J.: Probing Reynolds stress models of convection with numerical simulations: I. Overall properties: fluxes, mean profiles. In: Kupka F., Roxburgh I.W., Chan K.L. (eds.) Convection in Astrophysics, Proc. IAU Symp., Vol. 239, p. 80. Cambridge University Press, Cambridge (2007)
80. Kupka F., Muthsam H.J.: Probing Reynolds stress models of convection with numerical simulations: II. Non-locality and third order moments. In: Kupka F., Roxburgh I.W., Chan K.L. (eds.) Convection in Astrophysics, Proc. IAU Symp., Vol. 239, p. 83. Cambridge University Press, Cambridge (2007)
81. Kupka F., Muthsam H.J.: Probing Reynolds stress models of convection with numerical simulations: III. Compressibility modelling and dissipation. In: Kupka F., Roxburgh I.W., Chan K.L. (eds.) Convection in Astrophysics, Proc. IAU Symp., Vol. 239, p. 86. Cambridge University Press, Cambridge (2007)
82. Landau L.D., Lifshitz E.M.: The Classical Theory of Fields. Vol. 2 of Course of Theoretical Physics. Pergamon Press, London (1961)
83. Landau L.D., Lifshitz E.M.: Fluid Mechanics. Vol. 6 of Course of Theoretical Physics. Pergamon Press, Reading, MA (1963)
84. Landstreet J.D.: Observing convection in stellar atmospheres. In: Kupka F., Roxburgh I.W., Chan K.L. (eds.) Convection in Astrophysics, Proc. IAU Symp., Vol. 239, p. 103. Cambridge University Press, Cambridge (2007)
85. Lesieur M.: Turbulence in Fluids, 3rd edn. Kluwer Academic Publishers, Dordrecht (1997)
86. Lin C.-H., Antia H.M., Basu S.: Astrophys. J. **668**, 603 (2007)

87. Liu X., Osher S., Chan T.: J. Comp. Phys. **115**, 200 (1994)
88. Liu X., Osher S.: J. Comp. Phys. **142**, 304 (1998)
89. Ludwig H.-G.: Energy transport, overshoot, and mixing in the atmospheres of very cool stars. In: Piskunov N.E., Weiss W.W., Gray D.F. (eds.) Modelling of Stellar Atmospheres, Proc. IAU Symp., Vol. 210, p. 113. Astron. Soc. Pacific, San Francisco (2003)
90. Ludwig H.-G., Allard F., Hauschildt P.H.: Astron. Astrophys. **395**, 99 (2002)
91. Mihalas D., Mihalas B.W.: Foundations of radiation hydrodynamics. Oxford University Press, New York (1984)
92. Moeng C.-H., Rotunno R.: J. Atmos. Sci. **47**, 1149 (1990)
93. Montalbán J., D'Antona F., Kupka F., Heiter U.: Astron. Astrophys. **416**, 1081 (2004)
94. Monteiro M.J.P.F.G., Christensen-Dalsgaard J., Thompson M.J.: Astron. Astrophys. **283**, 247 (1994)
95. Monteiro M.J.P.F.G., Christensen-Dalsgaard J., Thompson M.J.: Astron. Astrophys. **307**, 624 (1996)
96. Montgomery M.H., Kupka F.: Mon. Not. Roy. Astron. Soc. **350**, 267 (2004)
97. Moffat H.K.: Magnetic Field Generation in Electrically Conducting Fluids. Cambridge University Press, Cambridge (1978)
98. Müller E.: Simulations of astrophysical fluid flow. In: Steiner O., Gautschy A. (eds.) Computational Methods for Astrophysical Fluid Flow, Saas-Fee Advanced Course 27, Les Diablerets 1997, Swiss Society for Astrophysics and Astronomy, Lect. Notes 1997, 343. Springer, Berlin (1998)
99. Muthsam H.J., Göb W., Kupka F., Liebich W., Zöchling J.: Astron. Astrophys. **293**, 127 (1995)
100. Muthsam, H.J. Löw-Baselli B., Obertscheider Chr., Langer M., Lenz P., Kupka F.: Mon. Not. Roy. Astron. Soc. **380**, 1335 (2007)
101. Nordlund Å.: Convective overshooting in the solar photosphere; a model granular velocity field. In: Spiegel E.A., Zahn J.-P. (eds.) Problems of Stellar Convection, Proc. IAU Coll. 38, Lect. Notes Phys. **71**, 237. Springer Verlag, Berlin (1977)
102. Nordlund Å., Spruit H.C., Ludwig H.-G., Trampedach R.: Astron. Astrophys. **328**, 229 (1997)
103. Plewa T., Linde T.J., Weirs V.G. (eds.): Adaptive Mesh Refinement – Theory and Applications: Proceedings of the Chicago Workshop on Adaptive Mesh Refinement Methods. Lect. Notes Comput. Sci. Engg., **41**. Springer-Verlag, Berlin (1994)
104. Porter D.H., Pouquet A., Woodward P.R.: Phys. Fluids **6**, 2133 (1994)
105. Quarteroni A., Valli A.: Numerical Approximation of Partial Differential Equations. Springer Ser. in Comput. Math., Vol. 23. Springer-Verlag, Berlin (1994)
106. Roe P.L.: Ann. Rev. Fluid Mech. **18**, 337 (1986)
107. Richtmyer R.D., Morton K.W.: Difference Methods for Initial Value Problems, 2nd edn. John Wiley & Sons, Inc., New York (1967)
108. Rincon F.: J. Fluid Mech. **563**, 43 (2006)
109. Robinson F.J., Demarque P., Li L.H., Sofia S., Kim Y.-C., Chan K.L., Guenther D.B.: Mon. Not. Roy. Astron. Soc. **340**, 923 (2003)
110. Rosenthal C.S., Christensen-Dalsgaard J., Nordlund Å., Stein R.F., Trampedach R.: Astron. Astrophys. **351**, 689 (1999)
111. Roxburgh I.W., Vorontsov S.V.: Mon. Not. Roy. Astron. Soc. **268**, 880 (1994)

112. Samadi R., Nordlund Å., Stein R.F., Goupil M.J., Roxburgh I.W.: Astron. Astrophys. **404**, 1129 (2003)
113. Samadi R., Kupka F., Goupil M.J., Lebreton Y., vant Veer-Menneret C.: Astron. Astrophys. **445**, 233 (2006)
114. Schatzmann E., Zahn J.-P., Morel P.: Astron. Astrophys. **364**, 876 (2000)
115. Schwarzschild K.: Göttinger Nachr. **1906**, 41 (1905)
116. Schwarzschild M.: Astrophys. J. **130**, 345 (1959)
117. Shapiro S.L., Teukolsky S.A.: Black Holes, White Dwarfs and Neutron Stars. Wiley & Sons, New York (1983)
118. Shu C.-W., Osher S.: J. Comp. Phys. **77**, 439 (1988)
119. Shu C.-W., Osher S.: J. Comp. Phys. **83**, 32 (1989)
120. Smagorinsky J.: Mon. Weather Rev. **91**, 99 (1963)
121. Spiegel E.A.: Astrophys. J. **126**, 202 (1957)
122. Spruit H.C., Nordlund Å., Title A.M.: Solar convection. Annu. Rev. Astron. Astrophys. **28**, 263 (1990)
123. Stein R.F., Nordlund Å.: Astrophys. J. **342**, L95 (1989)
124. Stein R.F., Nordlund Å.: Astrophys. J. **499**, 914 (1998)
125. Stein R.F., Nordlund Å.: Solar Phys. **192**, 91 (2000)
126. Stein R.F., Benson D., Georgobiani D., Nordlund Å.: Application of convection simulations to oscillation excitation and local helioseismology. In: Kupka F., Roxburgh I.W., Chan K.L. (eds.) Convection in Astrophysics, Proc. IAU Symp., Vol. 239, p. 331. Cambridge University Press, Cambridge (2007)
127. Steffen M.: Radiative hydrodynamics models of stellar convection. In: Kupka F., Roxburgh I.W., Chan K.L. (eds.) Convection in Astrophysics, Proc. IAU Symp. 239, p. 36. Cambridge University Press, Cambridge (2007),
128. Stix M.: The Sun. Springer-Verlag, Berlin (1989)
129. Strikwerda J.C.: Finite Difference Schemes and Partial Differential Equations. Wadsworth & Brooks/Cole, Pacific Grove, reprinted by Chapman & Hall, New York (1989)
130. Talon S.: Wave Transport in stratified media. In: Rieutord M., Dubrulle B. (eds.) Stellar Fluid Dynamics and Numerical Simulations: From the Sun to Neutron Stars, EAS Publications Series, Vol. 21, p. 105. EDP Sciences, Les Ulis (2006)
131. Trampedach R.: 3D-simulation of the outer convection-zone of an A-star. In: Zverko J., Weiss W.W., Žižňovský J., Adelman S.J. (eds.) The A-Star Puzzle, Proc. IAU Symp., Vol. 224, p. 155. Cambridge University Press, Cambridge (2004)
132. Unsöld A.: Z. Astrophys. **1**, 138 (1931)
133. van Leer B.: J. Comp. Phys. **14**, 361 (1974)
134. Verzicco R., Camussi R.: J. Fluid Mech. **477**, 19 (2003)
135. Vögler S., Shelyag S., Schüssler M., Cattaneo F., Emonet T., Linde T.: Astron. Astrophys. **429**, 335 (2005)
136. Wedemeyer S., Freytag B., Steffen M., Ludwig H.-G., Holweger H.: Astron. Astrophys. **414**, 1121 (2004)
137. Weiss A., Hillebrandt W., Thomas H.-C., Ritter H.: Cox & Giuli's Principles of Stellar Structure. Extended Second Edition. Cambridge Scientific Publ., Cambridge (2004)
138. Xiong D.R.: Chinese Astronomy **2**, 118 (1978)
139. Xiong D.R., Cheng Q.L., Deng L.: Astrophys. J. Suppl. **108**, 529 (1997)

140. Xiong D.R., Deng L.: Chinese Astron. Astrophys. **31**(3), 244 (2007)
141. Xiong D.R., Deng L.: Mon. Not. Roy. Astron. Soc. **378**, 1270 (2007)
142. Yang J.Y., Li Y.: Mon. Not. Roy. Astron. Soc. **375**, 403 (2007)
143. Zahn J.-P.: Processes competing with atomic diffusion. mass loss, turbulence, rotation, etc. In: Alecian G., Richard O., Vauclair S. (eds.) Element Stratification in Stars: 40 Years of Atomic Diffusion, EAS Publications Series Vol. 17, p. 157. EDP Sciences, Les Ulis (2005)
144. Zeman O., Lumley J.L.: J. Atmos. Sci. **33**, 1974 (1976)

Turbulence in Astrophysical and Geophysical Flows

V.M. Canuto

NASA, Goddard Institute for Space Studies, New York, NY 10025, USA; Dept. of Applied Math. and Physics, Columbia University, New York, NY, 10027
vcanuto@giss.nasa.gov

1 General Considerations

Turbulence is a physical process that enjoys some unique characteristics, first among which is the fact that it is among the most ubiquitous phenomena in the physical world and yet is one of the least understood physical processes.

Let us just consider that all existing stars (the number of which is around $\sim 10^{23}$, not far from the Avogadro's number) use or have used turbulence during different phases of their lifetime. The atmosphere of the earth, a habitat in which we spend all our life, is regularly turbulent and so are the oceans.

Much to the surprise of anyone who begins to study the subject, one finds that the most skillful practitioners at "taming" turbulence are aerospace engineers who try to minimize the unsettling effects of air turbulence on the commonly shared assumption that a smooth flight is more exciting than a bumpy one. But to advance their skills, engineers could hardly afford waiting for a theory of turbulence to be developed and tested. Wind tunnels provide much of the needed help for the effects of a turbulent flow on the wings of an aeroplane which can be examined in detail using repeated experiments. This leads to quite a satisfactory "description" of turbulence, but that is not sufficient to build a theory. For example, the huge wealth of spectroscopy data on absorption and emission properties of gases found a simple, logical explanation only when the atomic theory was invented and disparate results were explained by a set of few simple rules. There is not yet an equivalent in the field of turbulence which among other vicissitudes, is also struggling to find a proper place in science: it straddles mathematics, physics, engineering, geophysics and astrophysics. Apocryphal stories tell that W. Heisenberg, one of the founders of quantum mechanics, who pioneered with N. Kolmogorov the concept of *turbulent viscosity*, wanted to ask God what turbulence was all about, not the structure of quarks, not the true mass of neutrinos, not the unification of general relativity and quantum mechanics, not the reality of strings, but *turbulence*. More historically certain is the assertion by another great scientist, R.P. Feynman, that turbulence remains the biggest unsolved

Canuto, V.M.: *Turbulence in Astrophysical and Geophysical Flows*. Lect. Notes Phys. **756**, 107–160 (2009)
DOI 10.1007/978-3-540-78961-1_4   © Springer-Verlag Berlin Heidelberg 2009

problem of classical physics. Even among professional scientists there is a bit of conflict as to what turbulence really is. The connotation of the term used in the political–sociological lexicon gives the impression of something unwanted, unwelcome and probably to be corrected at the earliest stage, so as to buy back its opposite, most likely *laminarity*, a term that however is never heard in such conversations. And yet, it may be hard to have to realize that at the slightest provocation, flows leave laminar tranquility and "go turbulent", from the easy to observe smoke of a cigarette, to the water flowing from the tap, to the contrails in the plane's wake, that is, we see turbulence everyday, we don't need expensive laboratory setups to bring it to life. It is there for all us to see and the facility of its inception should speak volumes as to its being a most natural of physical phenomenon rather than a strange one. And yet, it is a matter of fact that all scientists (let's say of my generation) have grown up in the cozy world of Maxwell equations, Schrödinger equation, Dirac equation, the pillars of modern physics which are linear, perhaps unwittingly creating the impression that, after all, physical phenomena are well described by linearity. But it was Heisenberg himself, careful observer of linearities in physics, who tried to modify the linear Dirac equation into a non-linear one by suggesting that the "mass" say of an electron rather than being an outside input, ought to be provided by the theory itself leading him to suggest a non-linear term instead of the mass term. But the natural habitat of turbulence is not at the atomic level but at the macro level such as the earth's atmosphere, the ocean, the stellar interiors, etc., which we shall discuss.

After all, the Heisenberg–Kolmogorov (HK) model of turbulence being represented by a turbulent viscosity, relies on the very concept of "viscosity" that doesn't exist at the level of one particle but only of an ensemble of many particles. That is why the equations used to describe turbulent flows are the Navier–Stokes (NS) equations, which are the smoothed-out form of the Boltzmann equation, but the key difficulties of the latter remain intact and perhaps acquire an even clearer representation. The NS equations, describing the collective behaviour of a flow, contain non-linear terms (NLT) which are at the core of turbulence and the treatment of which has challenged all those who have dealt with them.

Difficult as it might be to account for them, the NLT have a remarkable property: they appear under the divergence operator and thus if one integrates over the volume of the system, the NLT yield zero. Turbulence is somewhat of a secretive process, an insider that wields much power but that globally acts only as the perfect transferrer-messenger that actually works for free (the zero integral just discussed). What does turbulent transfer and is this its main job? The answer to the latter is yes, as it can be seen by a simple argument. Consider the non-linear term $\nabla \cdot \mathbf{uu}$ or $\mathbf{u} \cdot \nabla \mathbf{u}$ (since we are considering incompressible flows, $\nabla \cdot \mathbf{u} = 0$). If one Fourier transforms the velocity field $\mathbf{u}(\mathbf{r}) = \Sigma_{(\mathbf{k})} \mathbf{u}(\mathbf{k}) \exp(i\mathbf{k} \cdot \mathbf{r})$, it is clear that the velocity $u_i(\mathbf{k})$ corresponding to the mode $\mathbf{k}$ will entail the non-linear term $\alpha_{ijm} \Sigma_{(\mathbf{k'})} u_j(\mathbf{k'}) u_m(\mathbf{k} - \mathbf{k'})$

where $\mathbf{k}'$ are all the modes different than $\mathbf{k}$ [104, 111]. If we visualize the mode $k \sim l^{-1}$ with a typical length scale l, it is clear that the dynamics of that length will be governed by its interactions with all the other modes or length scales. This non-linear interaction is the one responsible for the transfer process we referred to before. The extent of these "sizes" l is governed on the large size by the geometrical extent of the flow and at the opposite end by the molecular sizes which have a dynamics of their own. Contrary to a laminar flow in which the molecular sizes dictated by molecular forces cannot be changed (that is to say all molecules have the same sizes, a granularity of matter that is not under our control), the "molecules" of a turbulent flow, universally known as "eddies", span a wide range of sizes: big eddies, midsize eddies and small eddies (still larger than molecules).

While it is not physically meaningful to talk about the lifetime of a molecule (at least, not under the effect of an outside force), eddies have *dynamic sizes* and thus dynamic lifetimes, the largest eddies live longer, the smallest live shorter. The latter begin to be affected by molecular forces, the largest are still reminiscent of the way turbulence was generated, the type of geometry in which they move, etc., and therefore neither set of eddies exhibit *"universal properties"*, a feature that, if proven to exist, would help simplify the problem to treat the NLT and allow same kind of a general law to be formulated. Such a law, which goes with the HK spectrum, will be discussed shortly, but before that we need to discuss a further important feature of turbulence. Turbulence is not a physical process such as for example the gravitational attraction of two objects, something that we cannot erase, since it is written in the basic laws of nature. In principle, turbulence could be avoided entirely by avoiding its inception, a difficult and probably almost impossible process to manage, but one that has no conceptual inconsistencies. Turn your tap water very, very slowly and the water will flow laminarly, make air stand almost still and the smoke of your cigarette will be a long laminar streak, etc. This is to say, it takes the passing of some kind of a threshold to get into a turbulent regime. Stated differently, turbulence has to be nurtured into existence and kept fed with a permanent supply of energy, otherwise it will die on you. That is often cited as one of the reasons why a turbulence-based theory of galaxy generation (in spite of the more than interesting similarities between pictures of galaxies and that of frozen eddies) never got off the ground: nobody has been able to suggest a source of energy that would stir an initially laminar cosmic gas into a turbulent one long enough for galaxies to form. Unsustained turbulence will die and a laminar state will follow. While that is not difficult to prove in nature, it points to another fact, namely that in any turbulent flow the amount of energy (power to be more exact) that keeps it alive, will have to show up in all relations, one way or the other. Since, as we stated earlier, turbulence rearranges things creating a whole new picture of a flow without however using any energy (a superconductor comes to mind that transports a current without any irreversible ohmic losses), the amount of energy you put in at the largest scales is the same that gets to the smallest scales which, by contrast, via the irreversible route of friction, are fully dissipative. Turbulent eddies are energy conserving, molecules are energy dissipative. The power input-dissipation, traditionally denoted as ε, is therefore a key feature of any turbulent flow, since it represents the power that inserted at the largest scales reappears, its magnitude unchanged, as dissipation at the smallest ones. Clearly the size of the eddies that begin to be dissipation prone varies from flow to flow. In a flow characterized by a large molecular viscosity, those scales will be much larger than the ones corresponding to a low-viscosity flow. Stated differently, the more viscous the flow, the smaller will be the range in which turbulence will be able to operate (under the same stirring force). On the other hand, a low viscosity flow will allow turbulence to "operate" on many scales, thus the range of turbulence scales can be quite large. These concepts can be formulated with simple mathematical relations. First, consider the non-linear interactions we discussed earlier: when one says that they are important, one means that they are more important than other terms in the same equation. From the above arguments, it is clear that viscosity is at the top of the list of anti-turbulence agents. Thus the importance of the ratio of the non-linear interaction to the viscosity term, the so-called Reynolds number:

$$\mathrm{Re} = \frac{\overline{u}\nabla\overline{u}}{\nu\nabla^2\overline{u}} = UL\nu^{-1} \tag{1}$$

where U and L represent typical velocity and scale of the flow under consideration. Flows with $\mathrm{Re} > 10^3$ are considered to be turbulent, a rather modest value (a car in the street can reach $\mathrm{Re} \approx 10^4$ and thus the preoccupation of car manufacturers to reduce the car's friction due turbulent flow). With the two variables ε and ν (the first characterizes the rate of energy input while the second characterizes the type of fluid we deal with) we can construct (using only dimensional arguments) a length scale:

$$l_d = \left(\nu^3\varepsilon^{-1}\right)^{1/4} \tag{2}$$

which represents the typical size of an eddy that begins to feel the eroding action of viscosity. The larger the viscosity, the larger l_d so that for very large ν, l_d can be of the order of the largest possible eddy size in which case turbulence has no room to develop. Combining (1) and (2) one obtains that the ratio of the largest to the smallest eddy is given by (using relation (7) below):

$$L/l_d \sim \mathrm{Re}^{3/4} \tag{3}$$

This relation, simple as it might seem, has several consequences the most important of which is that we can't hope to solve the turbulence problem with today's computers. To do so, (3) tells that we would need to simulate a number of point (we are considering 3D turbulence):

$$N \sim (L/l_d)^3 \sim \mathrm{Re}^{9/4} \tag{4}$$

If we use values of Re corresponding to a street car Re $\approx 10^4$, $L/l_d \approx 10^3$ that is, we have a "spectrum of eddies" whose dimensions differ by a factor of 10^3 and to resolve "all the turbulent scales", we need to account for $N \sim 10^9$ degrees of freedom which is feasible with today's computers. On the other hand, in the ocean, atmosphere, stars, etc., we end up with values of N that are several orders of magnitude larger than what any modern computer can handle. Thus, one resorts to large eddy simulations (LES) in which one resolves numerically only the largest scales and models the huge number of numerically unresolved scales with a sub-grid scale model [21, 23].

2 The Heisenberg–Kolmogorov Energy Spectrum

On the assumption that a group of eddies exists that are small enough compared to the largest ones not to be affected by boundary conditions, size of the container, specific stirring mechanics, etc., but also large enough not to be yet subjected to the influence of viscosity, HK reasoned that ν cannot enter the model, neither can the specific type of stirring: only the amount of power ε that is injected into the flow to generate the turbulent process must show up. Under these conditions, and using only dimensional arguments, the turbulent kinetic energy spectrum $E(k)$ must have the form:

$$E(k) = \mathrm{Ko}\,\varepsilon^{2/3}k^{-5/3}, \qquad K = \int E(k)\mathrm{d}k \tag{5}$$

where $\mathrm{Ko} \sim 1.6$ is the Kolmogorov constant. The HK spectrum has been verified many times using different flows and its universality is now well established. Whether it can be derived from the first principle is a different matter. The kinetic energy turns out to be:

$$K = \frac{3}{2}\mathrm{Ko}\,\varepsilon^{2/3}L^{2/3}\left[1 - (l_d/L)^{2/3}\right] \sim \frac{3}{2}\mathrm{Ko}\,\varepsilon^{2/3}L^{2/3} \tag{6}$$

The inverted relation:

$$\varepsilon \sim \frac{K^{3/2}}{L} \tag{7}$$

is what we used in joining (1) and (2) to yield (3) with the proviso that $K^{1/2} \sim U$.

Next, let us consider the dissipation ε. Rather than attributing to it arbitrary values, let us consider a real case, that of the rotating but slowing down earth. The earth's rotational energy is given by:

$$E_{\mathrm{R}} = \frac{1}{2}I\Omega^2 \tag{8}$$

where I is the moment of inertia and $P = 2\pi/\Omega$ the rotational period. Due to the presence of tides, earth's rotation is slowing down and the length of the day (l.o.d) correspondingly lengthens. Astronomical data tell us that:

$$\dot{P} \cong 2 \text{ ms/cy}, \quad (1 \text{ cy} = 10^2 \text{ yr}) \tag{9}$$

which is used to estimate the rate of loss of rotational energy:

$$\dot{E}_R = I \Omega \dot{\Omega} = -2 E_R \dot{P}/P \tag{10}$$

Using the mass of the ocean 1.3×10^{24} gr and the viscosity $\nu \approx 10^{-2} \text{ cm}^2 \text{s}^{-1}$, Eq. (2) gives:

$$l_d \approx 1 \text{ cm} \tag{11}$$

Since most of the energy gets dissipated in the upper mixed layer (40–100 m) of the ocean which is stably stratified and where the eddies are not very large, we can take a representative value $L \sim 10$ m. We then have:

$$L/l_d \sim 10^3, \quad \text{Re} \sim 10^4, \quad N \sim 10^9 \tag{12}$$

Next, consider the sun's interior. Here we use the viscosity corresponding to a fully ionized gas given by Chapman [45]:

$$\nu[\text{cm}^2\text{s}^{-1}] = 1.2^* 10^{-16} T^{5/2} \rho^{-1} \tag{13}$$

If L_* and M_* denote the luminosity and the mass of a star, in the solar case we have $\varepsilon \sim L_\odot/M_\odot \sim O(1) \text{ cm}^2 \text{s}^{-3}$. Furthermore, at the bottom of the convective zone we can take the values $T \approx 10^6$ K, $\rho \approx 0.1 \text{ g cm}^{-3}$ and $\nu \approx 1 \text{ cm}^2 \text{s}^{-1}$. Thus, we obtain:

$$l_d \approx 1 \text{ cm} \tag{14}$$

which is of the same order as the one derived for a very different situation, Eq. (11).

3 Heat Fluxes: Local and Non-Local Models

Stellar structure codes or their geophysical analogues, ocean and/or atmospheric circulation models, solve the equations for the mean variables, say mean momentum, mean temperature, mean salinity, mean humidity, etc., depending on the specific problem at hand. The important point is that such codes, which by necessity have low resolution, cannot "resolve" most of the scales which must therefore be modelled and yet LES, due to their computational requirements, cannot be hooked-up to a large scale code. Thus, one must find a way to model the unresolved scales with a result that is "manageable".

Towards the end of the nineteenth century, Osborne Reynolds suggested a procedure to treat the effect of the non-linear interactions on a mean flow. He suggested that every field φ (velocity, temperature, ...) be decomposed into a mean and a fluctuating part:

$$\varphi = \overline{\varphi} + \varphi', \quad \overline{\varphi'} = 0 \tag{15}$$

Substituting the first of (15) into the equation for the full field, taking the average and using the second of (15), one obtains the dynamic equation for the fluctuating components and then finally the equation for the correlations among fluctuations. For example in the case of $\varphi = T + \theta$, the equation for the mean temperature reads:

$$\frac{\partial T}{\partial t} + \cdots = -\frac{\partial}{\partial z} \underbrace{\overline{w\theta}}_{\text{turb.heat.flux}} \tag{16}$$

where on the rhs we now have a new function, the correlation between two fluctuating variables which in this case represents the turbulent heat flux (in units of $c_p\rho$). In the mean velocity equation, one has a similar situation whereby ($\mathbf{U} = \overline{\mathbf{u}} + \mathbf{u}$):

$$\frac{\partial \overline{\mathbf{u}}}{\partial t} + \cdots = -\nabla \cdot \overline{\mathbf{uu}} \tag{17}$$

In general, correlations of the types:

$$R_{ij} = \overline{u_i u_j} \tag{18}$$

are known as *Reynolds Stresses*. In the early works, (18) and the turbulent heat fluxes were treated with phenomenological models of the type ($a_z = \partial a/\partial z$):

$$\overline{uw} = -K_m \, \overline{u}_z, \qquad \overline{w\theta} = -K_h \, T_z, \tag{19}$$

where the momentum and heat diffusivities, here denoted by $K_{m,h}$, were written on dimensional grounds as the product of a velocity times a mixing length l:

$$K_{m,h} = wl \tag{20}$$

In the stellar context, one needs to model *Turbulent Convection* and in this respect the Mixing Length Theory (MLT) constructed the heat diffusivity K_h on a phenomenological basis rather than on a Reynolds stress model [54, 69]. In that sense, MLT was less well grounded than the turbulence models used in the engineering context. However, the MLT turned out to be quite successful for reasons we now discuss. It so happens that in most stellar cases, and certainly in the sun, convection is actually governed not by heating from below but by cooling from above, much as it occurs in oceanic convection (the Labrador Sea, Gulf of Lyon and Weddell Sea where the loss of buoyancy by surface waters due to both evaporation and winds makes them heavy and thus prone to fall) and in the earth's atmosphere when the latter is cloud-capped. It turns out that in these cases, the flow is a combination of well-organized, narrow, vigorous, descending "plumes" accompanied by disordered, broad plumes which are the ones that the MLT modelled. Pioneering studies by Cattaneo et al. [42] showed that:

$$\text{Downflows:} \quad F(\text{convected}) = F_{\text{KE}} + c_p \overline{w\theta} \approx 0 \tag{21}$$

$$\text{Upflows:} \quad F(\text{convected}) = F_{\text{KE}} + c_p \overline{w\theta} \neq 0 \tag{22}$$

which tell us that the flux of the descending plumes is actually cancelled almost entirely by the flux of turbulent kinetic energy $F_{\mathrm{KE}} = \overline{wK}$ leaving only the upflows which are what the MLT describes. Later studies showed that the situation is more complicated in the sense that the convective layer should have stable layers on both sides whereas the one studied by Cattaneo et al. [42] does not (it has fixed plates as boundary conditions). In the simulation of Chan and Gigas [44], there is an extended stable layer at the bottom of the convective zone and a tiny stable layer at the top. In this case the cancellation (21) is only 30%. The conclusion seems to be that the net flux contribution from the downflows is not zero and is directed upward and since the enthalpy flux (up) is about 50% of the total flux, and the convected flux (up) is about two-thirds of the total flux, one could argue that within a factor of about 2, an MLT type convective model may be a good rough estimate for the total convective flux. To estimate the latter, one may employ the second of (19) and express K_h in terms of the main parameters of the star. One of them is the rate of radiative cooling by rising blobs characterized by a time scale t_χ and the other is the time scale t_b on which buoyancy operates:

$$t_\chi = l^2/\chi, \qquad t_b = (g\alpha\beta)^{-1/2}, \qquad S = (t_\chi/t_b)^2 = g\alpha\beta l^4\chi^{-2} \qquad (23)$$

Here, $K_{\mathrm{r}} = c_{\mathrm{p}}\rho\chi$ is the radiative conductivity (the radiative flux is $-K_{\mathrm{r}}\partial T/\partial z$), $\beta = -\partial T/\partial z + (\partial T/\partial z)_{\mathrm{ad}}$ and α is the volume expansion coefficient. When $S < 1$, convection is inefficient since the buoyancy time is long enough to give the opportunity to radiative processes to cool off the blob of gas while it rises; the case of efficient convection is when the opposite is true, the buoyancy time is so short that the blob rises before it loses its heat content due to radiative processes. In those two cases, the MLT gives the following results:

$$\text{Effective Convection:} \qquad S \gg 1, \qquad K_h/\chi \sim S^{1/2} \qquad (24)$$

which is quite natural since when radiative processes are slow and inefficient, χ should not enter the problem which implies a $S^{1/2}$ dependence. When the opposite is true and convection is inefficient, the MLT gives:

$$\text{Inefficient Convection:} \qquad S \ll 1, \qquad K_h/\chi \sim S^2 \qquad (25)$$

The full MLT [22, 54, 69] yields an expression for K_h/χ of the form:

$$K_h/\chi \quad \sim \quad S^{-1}\left[(1+S)^{1/2} - 1\right]^3 \qquad (26)$$

that embraces both limits (24, 25). We must note that (26) was derived from a turbulence model [22, 26, 28] and that improvements of (26) were also proposed [29] and tested [3, 55, 141]. The cancellation described by (21) does not mean however that in (22) we can neglect F_{KE}. To understand its physical meaning and implications, consider Fig. 1 below in which we have sketched an eddy as a spherical blob of the same size of the region in which it is formed.

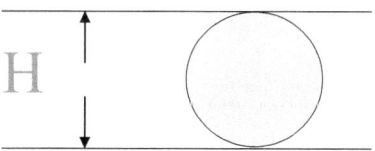

Fig. 1. An eddy as a spherical blob

Can such a large eddy exist? In an unstably stratified situation, buoyancy overpowers gravity (so to speak) and thus it is quite possible to have eddies of the size of the "container". On these grounds, one would expect that such height, called H in the sketch, should appear explicitly in the expression for the heat flux but if one looks at the second of (19) which expresses the heat flux $(J = \overline{w\theta})$ as:

$$\text{Local:} \qquad J(z) = -K_h(z)\frac{\partial T(z)}{\partial z} \qquad (27)$$

one observes that there is no H and therefore (27) cannot be complete. It is so because it represents the flux with only local variables, in fact, $J(z)$ is given by the temperature gradient and the diffusivity computed at the same z. It is a local flux that does not account for the fact that there are large eddies of the size of the "container". One way of constructing a non-local flux is to use an expression of the type:

$$\text{Non-local:} \qquad J(z) = -\int_0^z \widetilde{K}_h\,(z,z')\frac{\partial T(z')}{\partial z'}\mathrm{d}z' \qquad (28)$$

which implies that the flux at every height z is contributed by all the fluxes below it and at $z = H$, the upper limit brings in the height H of the convective region. The author is not aware of any explicit expression for the non-local "kernel" heat diffusivity $\widetilde{K}_h\,(z,z')$.

It is not difficult to find out what a non-local term looks like, a result that the RSM to be discussed later will justify. The dynamic equation for any variable (e.g. a mean temperature or a second-order correlation such as the heat flux J) has the general form:

$$\frac{\partial A}{\partial t} + \nabla \cdot \mathbf{u}A = \text{sources-sinks} \qquad (29)$$

Suppose $A = J$ and consider only the z-dimension. We have:

$$\frac{\partial J}{\partial t} + \frac{\partial \overline{w\theta w}}{\partial z} = \frac{\partial J}{\partial t} + \frac{\partial \overline{w^2\theta}}{\partial z} = \text{sources-sinks} \qquad (30)$$

Next, consider stationarity which leads us to:

$$\frac{\partial \overline{w^2\theta}}{\partial z} = \text{sources} - \text{sinks} \qquad (31)$$

The message is quite clear: the term on the lhs of (31) representing the d/dz of the *flux of the heat flux*, has a very clear physical interpretation: turbulence not only creates non-zero correlations such as J but it also transports them via the term on the lhs which represents *non-locality*: in fact, even if the "sources" on the rhs were zero, the non-local term on the lhs could act as a source that balances the sink. That is to say, in places where there may not be a local source of turbulence, there may nonetheless be mixing which, originated somewhere else in the flow, is transported there by the term on the lhs. In the case $A = K$, turbulent kinetic energy, the equivalent of (31) reads:

$$\frac{\partial F_{\mathrm{KE}}}{\partial z} = g\alpha J - \varepsilon, \qquad F_{\mathrm{KE}} \equiv \overline{wK} \tag{32}$$

where F_{KE} is the flux of kinetic energy which we already introduced in Eqs. (21) and (22). In this case we have also specified the source, buoyancy and re-introduced the rate of dissipation ε which we discussed in Sect. 2, Eq. (5). F_{KE} is a third-order moment the closure of which will be discussed in detail later. At this point, suffices it to say that the *local limit* of (32) corresponds to assuming:

$$g\alpha J = \varepsilon \tag{33}$$

which, once Eq. (7) is used, together with a "ballistic model" for K, gives rise to the MLT model, Eq. (26), the validity of which we have already discussed [22]. It is a simple exercise to reproduce the $S \gg 1$ limit Eq. (24) from (33). For the reasons given earlier in Sect. 1, the rate of dissipation must equal the rate of injection which can be written as:

$$\varepsilon \sim \int n(k)E(k)\mathrm{d}k \tag{34}$$

where $n(k)$ is the growth rate of the convective instability for which there exists a complete expression [29]. We shall limit ourselves to the strong limit in which case (see (23)):

$$n(k) \sim (g\alpha\beta)^{1/2} \tag{35}$$

Using the HK spectrum in (34), Eq. (33), once (27) is used, gives rise to (24).

A simple non-local model. Let us now consider the non-local term (31) which we express as:

$$\frac{\partial \overline{w^2\theta}}{\partial z} \sim H^{-1} w_\star J_\star \tag{36}$$

where the subscript $\star$ represents some fiducial value. What is relevant here is the presence of H which, as we discussed earlier, is the hallmark of non-locality. In the planetary boundary layer (PBL) case, Deardorff [57] suggested the expressions:

$$w_\star = [g\alpha H J_\star]^{1/3}, \qquad J_\star = J_{\mathrm{s}} \tag{37}$$

where the fiducial values are taken to be the surface values (subscript s) of the heat flux. As we shall prove later but seems physically obvious, we now have that the total heat flux is given by [76, 77]:

$$J(z) = J_{\mathrm{L}}(z) + J_{\mathrm{NL}} = -K_h \frac{\partial T}{\partial z} + \frac{c}{H} \tau w_\star J_\star \qquad (38)$$

where c is a numerical constant. Use of (38) has considerably improved the description of the PBL. Results from a coupled ocean–atmospheric models show that the non-local mixing model raises the maxima of the relative humidity and cloud cover in the tropics from the lowest atmospheric layer to about 900 mb in agreement with observations. No stellar test has yet been made of (38).

In conclusion, the LES data of Cattaneo et al. [42] and Chan and Gigas [44] suggested an interesting cancellation occurring in the downflows leaving behind the disordered upflows described by the MLT or improvements of it. That may explain the success the MLT has enjoyed in stellar structure/evolution studies over many years. The MLT is however a local theory and thus incomplete since in unstably stratified regimes nonlocality plays a major role. The next challenge is to account for nonlocality with the Reynolds Stress Model.

4 Overshooting Regions, OV

Outside the *unstably stratified* Convective Zone (CZ) there is a dynamically important *stably stratified* region referred to as overshooting region (OV) which has attracted a great deal of interest [5, 134, 135, 154]. In Fig. 2 we present a simple sketch of the stellar CZ together with the ocean (stable stratification is the norm) and the earth's atmosphere where in the daytime one has unstable stratification (heated from below) and the opposite at night.

A key question is, what is the source of turbulence in a stably stratified regime? Consider first the earth's oceans which are a large body of a stably

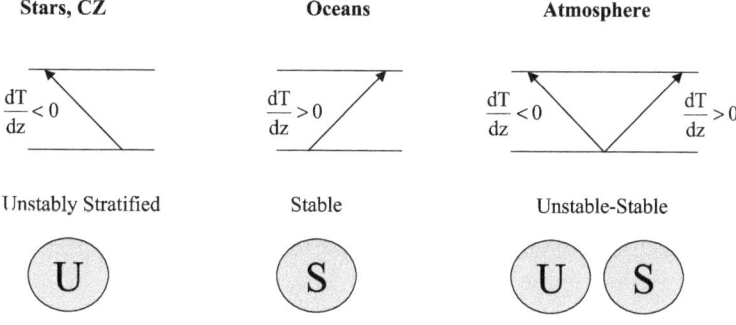

Fig. 2. Comparison of stratifications

stratified fluid since by and large cold, denser waters are at the bottom and warmer, lighter waters are on top. Without external disturbances, such a fluid would have no reason to mix and become turbulent which, on the other hand, is the true state of the ocean. The main source of the strong mixing is the shear produced by the external wind. Without mixing, there would be no upwelling of nutrients of which deep waters are rich, and a dead sea would ensue.

What about the stellar OV, say in the Sun? The two most obvious sources are: 1) differential rotation (which we call shear) and 2) non-local flow from the CZ, that is, the transport of turbulent kinetic energy represented by the term $F_{\rm KE}$ in (32) which we now generalize to:

$$\frac{\partial K}{\partial t} + \frac{\partial F_{\rm KE}}{\partial z} = -R_{ij}\overline{u}_{i,j} - gJ_\rho - \varepsilon \tag{39}$$

We have added a time dependence of K on the left, the production of K due to interaction of shear ($a_{i,j} = \partial a_i/\partial x_j$) with the Reynolds stresses and generalized the heat flux to the *mass flux:*

$$J_\rho = \rho_0^{-1}\overline{\rho w} = -\alpha_T J_h + \alpha_\mu J_\mu = -\rho_0^{-1}K_\rho\frac{\partial\overline{\rho}}{\partial z} = g^{-1}K_\rho N^2 \tag{40}$$

where $\alpha_T = -\partial\ln\rho/\partial T$, $\alpha_\mu = +\partial\ln\rho/\partial\mu$ and $N^2 = -g\rho_0^{-1}\partial\overline{\rho}/\partial z$.

Because in a stably stratified regime $N^2 > 0$, $J_\rho > 0$ in the absence of shear the rhs of (39) is negative. In this case, *the only way to satisfy (39) is by having a negative gradient of the flux of K, a non-local term.* In the presence of differential rotation, the first term on the rhs is positive (source) and, together with a negative lhs, constitutes a second source of mixing. The often discussed gravity waves (whose power $\Pi_{\rm gw}$ was computed by Kumar et al., [87]) are in effect an energy flux that originates from the CZ and thus, they may be interpreted as part of the lhs of (39). If we adopt (19) and (40), Eq. (39) becomes ($\Sigma^2 = \overline{u}_{,z}^2 + \overline{v}_{,z}^2$):

$$\frac{\partial K}{\partial t} + \frac{\partial F_{\rm KE}}{\partial z} = K_m\Sigma^2 - K_\rho N^2 - \varepsilon = K_m\Sigma^2\left(1 - \sigma_{\rm t}^{-1}{\rm Ri}\right) - \varepsilon \tag{41}$$

where the Richardson number and the turbulent Prandtl number are defined as:

$$\mathrm{Ri} = \frac{\mathrm{Sink}}{\mathrm{Source}} = \frac{N^2}{\Sigma^2}, \qquad \sigma_{\rm t} = \frac{K_m}{K_\rho} \tag{42}$$

There are physically justifiable reasons why in stably stratified flows, a local model may not be as poor an approximation as in the unstable case primarily because eddies are generally small in the stable case and thus more justifiably described by a local model. In that case, (41) becomes:

$$K_m\Sigma^2\left(1 - \sigma_{\rm t}^{-1}{\rm Ri}\right) = \varepsilon \tag{43}$$

which we can interpret by saying that the mixing caused by shear has to "work" against a naturally stable fluid and the dominance of the source over the sink (the stable stratification) is largely dictated by the flux Richardson number $\sigma_t^{-1} \mathrm{Ri}$. This has given rise to a set of confusing statements over the years about the "critical Ri" above which the source can no longer sustain the eroding action of the sink (stable stratification). Miles [114] and Howard [78], using linear stability analysis, showed that $\mathrm{Ri_L} = 1/4$ denotes the point at which *laminarity* ceases to exist, that is, when the flow becomes linearly unstable. After that, the system first enters a weakly non-linear regime and then finally a turbulent state where non-linearities dominate. Woods [180] was the first to give a physical picture of the different regimes leading to turbulence. Given a stable laminar sheet of thickness h, Kelvin–Helmholtz instabilities gradually erode and entrain fluid parcels above and below h. The process leads to an increase of h which ceases when the thickness has become four times the original value h. Woods concluded that "since the final thickness is nearly four times the original value, the final Richardson number is also four times the value prior to the instability", that is, the inception of turbulence occurs approximately at:

$$\mathrm{Ri_t} = 4\,\mathrm{Ri_L} \approx \mathrm{O}(1) \tag{44}$$

Abarbanel et al. [1] carried out a stability analysis with the inclusion of non-linearities, and concluded that the instability occurs at the value given by (44).

In oceanography, where local models are widely used [35, 37], Martin [109] showed that treating $\mathrm{Ri_{cr}}$ (defined as the value at which turbulence ceases) as an adjustable parameter, values around $1/4$ were unable to reproduce the depth of the ocean mixed layer while if one adopted values around unity, the data could be reproduced quite correctly. In other words, $\mathrm{Ri_{cr}} = 1/4$ underestimated the extent of turbulence.

In spite of this collective evidence, most authors dealing with stable stratification in stellar interiors used $\mathrm{Ri}=1/4$ as the critical value which, as just remarked, underestimates the extent of turbulent mixing [105, 106, 143, 163, 186].

Due to the physical role played by Ri, it is clear that the intensity of turbulence must decrease with Ri and many heuristic expressions were proposed over the years primarily because no good theory was available. In 2001–2002, Canuto et al. [35, 37] worked out a model for turbulence under stable stratification and shear that quite naturally reproduced (44) as the point at which turbulence has so decreased from its value as to become practically zero. The model also includes Double Diffusion (DD) processes which in the ocean can be quite important (see next section).

In conclusion, while in the ocean's mixed layer the source of mixing is well known (shear), in the stellar OV the situation is more complex since one can think of at least three possible sources and one possible sink.

Sources: Differential rotation (Shear), gravity waves, non-local K-fluxes.

Sinks: a μ-gradient $\nabla_\mu > 0$ (a positive $\nabla_\mu \sim \partial\mu/\partial P$ corresponds to a mean molecular weight that is large at the centre and low at the surface) acts to increase the local dissipation (as we shall show) thus reducing the penetration of turbulence into the OV region.

5 Semi-Convection and Salt-Fingers: Double Diffusion (DD) Processes

Double diffusion processes occur when two different fields exist which have very different kinematic diffusivities. In stars we have the fields (T, μ) where the latter is the mean molecular weight. In oceanography, one has (T, S) where the salinity field has a kinematic diffusivity that is two order of magnitude smaller than heat. Such processes are also referred to as thermohaline and/or thermosolutal convection. When both T and S increase from the ocean surface towards the bottom the result is cold, fresh water over warm, salty water. The S field is stable, the T field is unstable (heavy at the top) and one has *diffusive convection* [81, 82, 108, 145, 170–172]. Examples are lakes, water underneath an ice island and the Red Sea.

In stars, *diffusive convection* is called *semi-convection* and was studied by several authors [72, 92–94, 153, 155, 158, 159, 161, 162, 174]. Yet, there does not seem to be a generally accepted procedure to treat the phenomenon. Stothers [159] critically analysed 11 different prescriptions and concluded that only two were physically acceptable: one used by Schwarzschild and Harm [146], who adopted the Schwarzschild criterion, and the other by Sakashita and Hayashi [139] who adopted the Ledoux criterion [100]. In the absence of a turbulence model, Langer et al. [92–94], suggested a phenomenological model that we shall discuss below. Merryfield [112] found that none of his two-dimensional numerical simulations exhibited any close resemblance to the models by Stevenson [159] and/or Spruit [156] and that the closest similarity is with a Langer et al. model. Xiong [183–185] and Grossman and Taam [72] carried out non-linear studies of *semi-convection* which is characterized by the conditions ($\nabla = \partial \ln T/\partial \ln P$, $\nabla_{\mathrm{ad}} = (\partial \ln T/\partial \ln P)_{\mathrm{ad}}$, $\nabla_{\mathrm{r}} = (\partial \ln T/\partial \ln P)_{\mathrm{rad}}$, $\nabla_\mu \equiv \partial \ln \mu/\partial \ln P$):

$$\nabla - \nabla_{\mathrm{ad}} > 0, \qquad \nabla_\mu > 0, \qquad \nabla_{\mathrm{r}} > \nabla \tag{45}$$

and thus:

$$\nabla_{\mathrm{r}} > \nabla > \nabla_{\mathrm{ad}} \tag{46}$$

When both the T and S fields increase from the bottom to the top of the ocean, the result is warm, salty water over cold, fresh water. Since the T field is stable while the S field is unstable (heavy at the top), the latter causes an instability called *salt fingers*. An example is the Atlantic Ocean underneath the Mediterranean outflow of very salty water. In astrophysics, this instability occurs when a layer with a higher μ lies above a region of lower μ, for example, when the He flash does not occur at the centre of a star [166]. Salt fingers were first suggested by Stothers and Simon [160] and later studied by Ulrich [173] and Kippenhahn et al. [85]. The μ field causes the instability, while $\nabla - \nabla_{ad}$ plays the role of a stabilizing gradient. *Salt fingers* are characterized by the following conditions:

$$\nabla - \nabla_{ad} < 0, \quad \nabla_\mu < 0, \quad \nabla_r < \nabla \tag{47}$$

and thus:

$$\nabla_{ad} > \nabla > \nabla_r \tag{48}$$

For semiconvection and salt fingers, $R_\mu = \nabla_\mu (\nabla - \nabla_{ad})^{-1}$ is the stability parameter. Since:

$$N^2 = g H_p^{-1} [\nabla_\mu - (\nabla - \nabla_{ad})] \tag{49}$$

where $H_p = p/g\rho$ is the pressure scale height, we can distinguish between the following cases:

$$
\begin{array}{lllll}
\text{Ledoux stable:} & N^2 > 0, & \nabla_\mu > \nabla - \nabla_{ad}, & R_\mu > 1 & (50) \\
\text{Ledoux unstable:} & N^2 < 0, & \nabla - \nabla_{ad} > \nabla_\mu, & R_\mu < 1 &
\end{array}
$$

We further have:

$$
\begin{array}{lllll}
\text{Semi-convection:} & \nabla - \nabla_{ad} > 0, & \nabla_\mu > 0, & R_\mu > 0 & \\
\text{Ledoux stable:} & N^2 > 0, & \nabla_\mu > \nabla - \nabla_{ad}, & R_\mu > 1 & (51) \\
\text{Ledoux unstable:} & N^2 < 0, & \nabla - \nabla_{ad} > \nabla_\mu, & R_\mu < 1 &
\end{array}
$$

$$
\begin{array}{lllll}
\text{Salt fingers:} & \nabla_\mu < 0, & \nabla_{ad} - \nabla > 0, & R_\mu > 0 & \\
\text{Ledoux stable:} & N^2 > 0, & \nabla_{ad} - \nabla > |\nabla_\mu|, & R_\mu < 1 & (52) \\
\text{Ledoux unstable:} & N^2 < 0, & |\nabla_\mu| > \nabla_{ad} - \nabla, & R_\mu > 1 &
\end{array}
$$

The Ledoux vs. Schwarzschild criteria were discussed in Canuto [24] who developed and solved a RSM that includes:

1. salt-fingers
2. semi-convection
3. solid and differential rotation

Some interesting results will be discussed here. Using (40) and (49), Eq. (39) *without shear* becomes:

$$\frac{\partial K}{\partial t} + \frac{\partial F_{KE}}{\partial z} = g H_p^{-1} [K_h (\nabla - \nabla_{ad}) - K_\mu \nabla_\mu] - \varepsilon \tag{53}$$

whose local limit is then:

$$g H_p^{-1} [K_h (\nabla - \nabla_{ad}) - K_\mu \nabla_\mu] = \varepsilon \tag{54}$$

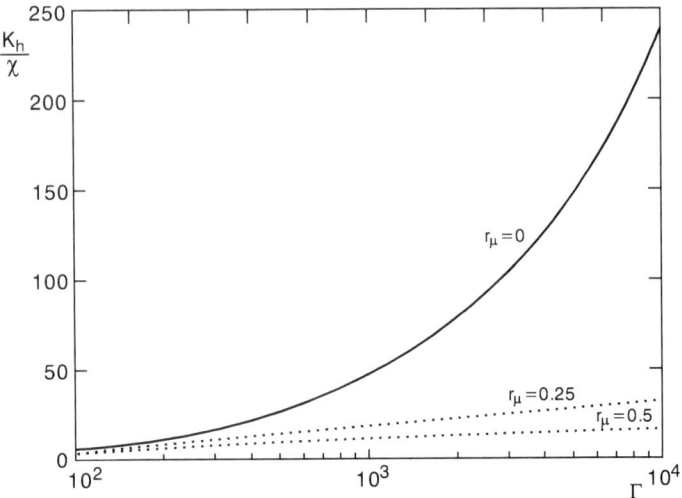

Fig. 3. Semiconvection. The ratio K_h/χ vs. Γ for different values of r_μ. The $r_\mu = 0$ case corresponds to the standard local model of convection

which is the generalization of (33) to include a μ-gradient. The algebraic expressions for the two diffusivities $K_{h,\mu}$ were derived in Canuto [24]. From (45, 47) wee see that in the presence of a μ-gradient, semi-convection acts like a sink while in the case of salt-fingers, it acts like a source. In Fig. 3 we plot the ratio K_h/χ which is now a function of the two parameters r_μ, Γ defined as follows:

$$r_\mu = \nabla_\mu (\nabla_r - \nabla_{ad})^{-1}, \; \Gamma = \frac{8\pi^2}{125} [A_p (\nabla_r - \nabla_{ad})]^{1/2}, \; A_p = g\Lambda^4 H_p^{-1} \chi^{-2} \;\; (55)$$

where Γ can be viewed as a convective efficiency (within a factor of order unity, Λ is the same as L in (7). As expected on physical grounds, in the $r_\mu = 0$ case (no semi-convection), the heat diffusivity grows quite rapidly with Γ. On the other hand, in the case of semi-convection, the growth with Γ is considerably reduced. In Fig. 4 we plot K_μ vs. r_μ, Γ. One can compare the results of Fig. 4 with the empirical relations suggested by Langer et al. [92] and Woosley et al. [181]:

$$K_\mu/\chi = \frac{1}{6}\alpha_{sc}(R_\mu - 1)^{-1} \qquad (56)$$

where the efficiency factor α_{sc} was determined to be $0.008 < \alpha_{sc} < 0.05$. Salasnich et al. [140] suggested the expression:

$$K_\mu/\chi = \alpha_2^{-1} = (50 - 100)^{-1} \qquad (57)$$

while Eggleton [61, 62] proposed the law:

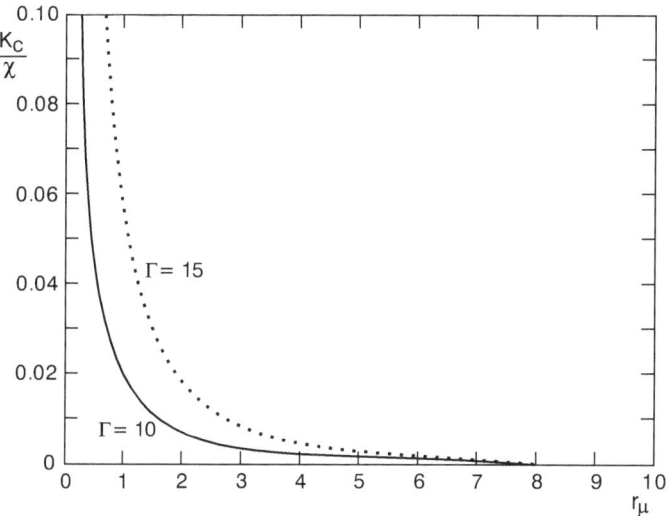

Fig. 4. Semiconvection. Turbulent concentration diffusivity ($K_c \equiv K_\mu$)

$$K_\mu \sim r_\mu^{-n}, \quad n > 1 \tag{58}$$

It is clear from Fig. 4 that the RSM can reproduce these empirical laws and gives more information. For example, to reproduce (57), we read from Fig. 4 that $r_\mu \sim 2 - 3$ which imposes a further constraint on the model.

6 OV and DD

Quantifying the effect of DD on the extent of OV is an interesting and thus far quantitatively unexplored problem that was analysed in Canuto [24]. The gist of the qualitative argument can be seen rather directly by considering (41) written as:

$$\frac{\partial K}{\partial t} + \frac{\partial F_{\mathrm{KE}}}{\partial z} = K_m \Sigma^2 + g H_\mathrm{p}^{-1} K_h (\nabla - \nabla_{\mathrm{ad}}) - \varepsilon_{\mathrm{eff}} \tag{59}$$

where:

$$\varepsilon_{\mathrm{eff}} = \varepsilon + g H_\mathrm{p}^{-1} K_\mu \nabla_\mu \equiv Q \varepsilon \tag{60}$$

Since $\varepsilon_{\mathrm{eff}} > \varepsilon$, semi-convection increases dissipation and causes a smaller OV extent.

7 Effect of Rotation on DD and the OV

Though rotation is an important factor in stellar structure and evolution, its effects on mixing is still not fully understood. In unstably stratified regions where $\nabla - \nabla_{ad} > 0$, the presence of shear in the rhs of (59) helps boost the turbulent mixing while the presence of a positive μ-gradient increases dissipation and thus reduces the mixing. In stably stratified regimes $\nabla - \nabla_{ad} < 0$ such as the OV, all the terms in the rhs of (59) except the first, are negative and act like sinks. It is therefore important to have both shear and the kinetic energy flux.

8 The Extent of the OV Region

One question that has remained theoretically unanswered is the extent of the OV which, among other things, depends on how it is defined. Though several suggestions were made, none is particularly attractive since they do not take into account the physical fact that the OV is primarily a region of *extended mass flux*. We shall therefore suggest the following picture. Consider the mean density equation:

$$\frac{\partial \overline{\rho}}{\partial t} + \overline{\mathbf{u}}_h \nabla_h \overline{\rho} + \overline{w} \frac{\partial \overline{\rho}}{\partial z} = -\rho_0 \frac{\partial J_\rho}{\partial z} \tag{61}$$

where $\overline{\mathbf{u}}_h, \overline{w}$ are the mean flow velocities. In the stationary case and considering only the vertical velocity, we have:

$$\overline{w} = -\frac{\rho_0}{\overline{\rho}_z} \frac{\partial J_\rho}{\partial z} = \frac{\partial K_\rho}{\partial z} + K_\rho \frac{\overline{\rho}_{zz}}{\overline{\rho}_z} \tag{62}$$

where we have used (40). We suggest to view the OV as a region of additional mass transport which ceases when the vertical mass flux velocity $\overline{w}$ vanishes, which occurs at a $z_\star$ where:

$$\frac{\partial}{\partial z}(\ln K_\rho + \ln \overline{\rho}_z) = 0 \tag{63}$$

a proposition that has not yet been tested in a stellar code.

9 Transport of Angular Momentum: The ^{7}Li Problem

Thus far, this important problem has been treated in a way that is not fully satisfactory for the following reasons. Consider the angular momentum equation [63]:

$$\frac{\partial}{\partial t}(\rho L) + \nabla \cdot (\rho \mathbf{F}_L) = 0 \tag{64}$$

where $\mathbf{F}_L$ is the vector flux of the angular momentum L while Ω_0 is the solid-body rotation ($\Gamma \equiv \sin \theta$):

$$F_L^r = L\overline{u}_r + r\Gamma R_{r\varphi}, \quad F_L^\theta = L\overline{u}_\theta + r\Gamma R_{\theta\varphi}, \quad L = \Gamma r^2 \Omega_0 + \Gamma r \overline{u}_\varphi \tag{65}$$

As one can see, one needs two Reynolds stresses $R_{r\varphi} = \overline{u_r u_\varphi}$, $R_{\theta\varphi} = \overline{u_\theta u_\varphi}$. After integration (or averaging) over θ, the equation that is usually considered is the following [43, 127, 187]:

$$\frac{\partial}{\partial t}\left(r^2 \Omega\right) = r^{-2}\frac{\partial}{\partial r}\left(r^4 K_m \frac{\partial \Omega}{\partial r}\right) + \cdots \tag{66}$$

where the momentum diffusivity was introduced in the first of (19). Though (66) is generally referred to as a "*diffusion equation*", it is not since the latter has the form:

$$r^{-2}\frac{\partial}{\partial r}\left[r^2 K_m \frac{\partial}{\partial r}\left(r^2\Omega\right)\right] \tag{67}$$

Equations (66) and (67) give similar results only for an Ω that varies with r like a power law. From helio-seismological data we have however learned that Ω is differential in the CZ but below it, it becomes Ω=constant [32, 167]. In that region, also known as the *tachocline* [154], Eq. (66) yields a zero rhs while (67) does not. To understand the origin of such different equations, consider the more general equation in terms of the Reynolds stresses:

$$\frac{\partial}{\partial t}\left(r^2\Omega\right) = -r^{-2}\frac{\partial}{\partial r}\left(r^3 R_{r\varphi}\right) + \cdots \tag{68}$$

If one employs the first of (19) which in this case becomes:

$$R_{r\varphi} = -K_m \Gamma\, r\, \frac{\partial \Omega}{\partial r} \tag{69}$$

and substitutes it into (68), one recovers (66). Is there anything wrong with (19) or (69)? Consider the following. Since the mean velocity $\overline{u}$ is a vector, one can construct with it two independent tensors that represent *shear and vorticity:*

$$\Sigma_{ij} = \frac{1}{2}\left(\overline{u}_{i,j} + \overline{u}_{j,i}\right) = \text{Shear}, \qquad V_{ij} = \frac{1}{2}\left(\overline{u}_{i,j} - \overline{u}_{j,i}\right) = \text{Vorticity} \tag{70}$$

and therefore, the Reynolds stresses must be of the form:

$$R_{ij} = f\left(\Sigma_{ij}, V_{ij}\right) \tag{71}$$

Equation (69) corresponds to having used shear but not of vorticity, while the inclusion of both gives rise to a new angular momentum equation:

$$\frac{\partial}{\partial t}\left(r^2\Omega\right) = A r^{-2}\frac{\partial}{\partial r}\left(r^4 K_m \frac{\partial \Omega}{\partial r}\right) + B r^{-2}\frac{\partial}{\partial r}\left(r^2 K_m \frac{\partial}{\partial r}\left(r^2\Omega\right)\right) + \cdots \tag{72}$$

where the coefficients A and B can only be provided by a complete model of the Reynolds stresses. The last term in Eq. (72) is just (67) and represents the first modification Eq. (66). There are other modifications as well and thus we extend (71) to the more general form:

$$R_{ij} = f\left(\Sigma_{ij}, V_{ij}; \text{buoyancy, gravity waves}; \underbrace{\bar{u}_r, \bar{u}_\theta}_{\text{mer.curr.}}\right) \tag{73}$$

where the buoyancy flux (or mass flux) is defined as:

$$B_i = -g\overline{\rho u_i} \quad \rightarrow \quad (\nabla T, \nabla \mu) \tag{74}$$

and is characterized by the two gradients of temperature and mean molecular weight. Finally, the presence of gravity waves can be accounted for by adding the flux Π_{gw} [87] to the source on the rhs (39).

The complete expression (73) was derived and expressed in algebraic form by Canuto and Minotti [36]. For example, the new form of the Reynolds stress $R_{r\varphi} = \overline{u_r u_\varphi}$ reads:

$$R_{r\varphi} = A_1 \left[\Omega_0 + \Omega(r, \theta)\right] + A_2 \Gamma \frac{\partial \Omega}{\partial \theta} + A_3 \Gamma r \frac{\partial \Omega}{\partial r} + A_4 B_{r\varphi} + E_{\text{mer}} \tag{75}$$

This expression contrasts quite significantly with (69) which has only the A_3-term. It is important to stress that the second Reynolds stress $R_{\theta\varphi} = \overline{u_\theta u_\varphi}$ exhibits the same structure (75). Salient new features in (75) are the presence of rigid rotation (first term), of the meridional currents (last term) and the buoyancy flux (last but one term) which depends on both the T and μ fluxes thus including the transport by DD processes. Some of the new terms in (75) have been accounted for in heuristic models [137].

These considerations may be relevant to the important problem of ^{7}Li [46, 86]. The basic facts are well known: big bang nucleosynthesis predicts a ^{7}Li abundance that is too high compared to what is observed in the oldest stars in the galaxy. Recent measurements by Korn et al. [86] suggest a solution: these old stars have destroyed part of their pristine ^{7}Li [46] via diffusive processes that have brought ^{7}Li to regions hot enough to have caused its burning. In other words, turbulent mixing is now deemed responsible for the discrepancy between big bang predictions and stellar observations. Ad hoc mixing models have been suggested that can explain the data but the problem remains since the physical processes underlying the mixing (one or many) are still unclear. The inclusion of several physical processes, as formally written in (73), may be a good starting point to sort out which of these processes or a combination of them, is capable of explaining the new data.

10 Reynolds Stress Model: Buoyancy Only

It was the Russian mathematician A.A. Friedmann (the same of the Friedmann universe) who at a mathematical conference towards the end of the 1920s suggested that if the NSE yield dynamical equations for the mean components, they also yield the equations governing the fluctuation's correlations such as the Reynolds stresses, the turbulent heat flux, etc. The Reynolds Stress Model (RSM) could have been born then, but that was not to be. One had to wait until 1940 when the Chinese physicist P.Y. Chou [52] published the first dynamical equations for the Momentum Reynolds Stresses. He treated mostly shear flows and the engineering community has since then used the RSM as a working tool. The successes of the RSM in that field are well documented and there is no need to dwell on them. Suffices to say that closure problems still exist especially concerning the pressure correlations, but the work of many groups has considerably narrowed the uncertainties.

In Canuto [20], a detailed derivation of the RSM equations was presented. For the *buoyancy only case*, the second-order moments of interest to stellar structure studies are the turbulent kinetic energy K, the heat flux $J = \overline{w\theta}$, the temperature variance $\overline{\theta^2}$ and the kinetic energy in the z-direction $1/2\,\overline{w^2}$ whose dynamic equations are [20, 33, 34, 37]:

$$\frac{\partial K}{\partial t} + \underbrace{\frac{\partial F_{\mathrm{KE}}}{\partial z}}_{\text{non-locality}} = g\alpha J - \varepsilon \tag{76}$$

$$\frac{\partial}{\partial t}\overline{w^2} + \underbrace{\frac{\partial \overline{w^3}}{\partial z}}_{\text{non-locality}} = \frac{2}{3}(1 + 2\beta_5)g\alpha J - \frac{2}{3}\varepsilon - 5\tau^{-1}\left(\overline{w^2} - \frac{2K}{3}\right) \tag{77}$$

$$\frac{\partial J}{\partial t} + \underbrace{\frac{\partial \overline{w^2\theta}}{\partial z}}_{\text{non-locality}} = -\overline{w^2}T_z + (1 - \gamma_1)g\alpha\overline{\theta^2} - \tau^{-1}\pi_4^{-1}J \tag{78}$$

$$\frac{\partial \overline{\theta^2}}{\partial t} + \underbrace{\frac{\partial \overline{w\theta^2}}{\partial z}}_{\text{non-locality}} = -2JT_z - 2\overline{\theta^2}\pi_5^{-1}\tau^{-1} \tag{79}$$

where the dynamical time scale is defined as $\tau = 2K/\varepsilon$ and the equation for dissipation is given by ($c_1 = 2.88$, $c_2 = 3.8$):

$$\frac{\partial \varepsilon}{\partial t} + \underbrace{\frac{\partial \overline{w\varepsilon}}{\partial z}}_{\text{non-locality}} = c_1 g\alpha J\tau^{-1} - c_2\varepsilon\tau^{-1} \tag{80}$$

with $\overline{w\varepsilon} = 3/2\tau^{-1}F_{\mathrm{KE}}$ with $F_{\mathrm{KE}} = \overline{wK}$. The suggested values of the constants are:

$$\beta_5 = 1/2, \ \gamma_1 = 1/3, \ \pi_4 = 0.084, \ \pi_5 = 0.72 \tag{81}$$

The first consideration to be made is that these equations are all linked together. To solve the equation for the heat flux J, one needs to know the temperature variance $\overline{\theta^2}$ and $\overline{w^2}$ which are given by two other equations. In the stationary and local limit, Eqs. (76), (77), (78) and (79) become algebraic and the solution has the MLT form [33]. To carry out the next step, consider

(78) and neglect the temperature variance $\overline{\theta^2}$ since in an unstably stratified situation, the potential energy, which is proportional to $\overline{\theta^2}$, transforms into kinetic energy. We have:

$$J = -\pi_4 \tau \overline{w^2} \frac{\partial T}{\partial z} - \pi_4 \tau \frac{\partial}{\partial z} \overline{w^2 \theta} = J_{\mathrm{L}} + J_{\mathrm{NL}} \tag{82}$$

We can observe that (38) is just a simplified form of Eq. (82). For the use of these equations in stars, see, Kupka [89], Kupka and Montgomery [90] and Montgomery and Kupka [117].

11 Non-locality: Third-Order Moments

Clearly, each of (76), (77), (78), (79) and (80) entails a third-order moment (TOM) and (36) can only be a rough approximation. To obtain a more physical expression for the TOMs, one begins with the TOMs dynamical equations ([20], Eq. 55). For example, in the case of buoyancy forces only, the equation for $\overline{w^3}$ reads:

$$\frac{\partial}{\partial t} \overline{w^3} = -\frac{\partial}{\partial z} \underbrace{\overline{w^4}}_{\mathrm{FOM}} + 3\overline{w^2} \frac{\partial \overline{w^2}}{\partial z} + 3g\alpha \overline{w^2 \theta} - 2c_8 \tau^{-1} \overline{w^3} \tag{83}$$

which shows that, to proceed, we need to model the fourth-order moment (FOM), $\overline{w^4}$.

11.1 FOMs: Previous Models

Most previous FOM models [6, 7, 13, 20, 30, 50, 116, 123, 124, 164, 188] employed the quasi-normal approximation, QNA, whereby $\overline{abcd} = \overline{ab}\,\overline{cd} + \overline{ac}\,\overline{bd} + \overline{ad}\,\overline{bc}$. For example, we have:

$$\overline{w^4}|_{\mathrm{QN}} = 3\overline{w^2}^2, \quad \overline{w^3\theta}|_{\mathrm{QN}} = 3\overline{w^2}\,\overline{w\theta}, \quad \overline{w^2\theta^2}|_{\mathrm{QN}} = \overline{w^2}\,\overline{\theta^2} + 2\overline{w\theta}^2 \tag{84}$$

In the *convectively unstable* case, the QNA is known to suffer from realizability problems, that is, the resulting TOMs contain denominators that become zero at some critical $\tau^2 N^2 \sim -20$, $N^2 = g\alpha\partial T/\partial z$, which easily attains in a convective PBL. To prevent this from happening, Canuto et al. [34] proposed an ad hoc procedure to limit the value of $\tau^2 N^2$ in the unstable case; as a result, the eddy sizes are chopped down and the transport is weakened. In the *stable case*, Moeng and Randall [116] pointed out that (83) under QNA leads to a "wave equation":

$$\frac{\partial^2}{\partial t^2} \overline{w^3} = 3g\alpha|\partial T/\partial z|\overline{w^3} + \text{other terms} \tag{85}$$

with an oscillation frequency of

$$f = (3g\alpha|\partial T/\partial z|)^{1/2} \tag{86}$$

which occurs in the upper part of the convective PBL. Similar "wave equations" resulted from other TOM equations. The oscillations generated by these "wave equations" are not observed in nature and are therefore spurious.

11.2 FOMs: New Model

Since the QNA (with zero-cumulants) causes singular behaviours of the TOMs, a more physical FOM model with nonzero cumulants was proposed and tested by Cheng et al. [51]. In principle, to formulate a new FOM model, one could try to solve the dynamic equations of the FOMs, but this would bring about a new set of parameterizations for the pressure and dissipation terms, and most of all, the need to model the fifth-order moments. A new model was therefore proposed [51] which we briefly sketch here. First, from the TOMs dynamic equations one subtracts the QNA part leaving behind the dynamic equations for the cumulants. For example, one has:

$$\frac{\partial}{\partial z}\left(\overline{w^4} - \overline{w^4}|_{QN}\right) = -2c_8\tau^{-1}\overline{w^3} + 3g\alpha\overline{w^2\theta} - 3\overline{w^2}\frac{\partial}{\partial z}\overline{w^2} \tag{87}$$

Next, it was assumed that the FOMs can be modelled by linear combinations of the TOMs, an assumption that assures that in the Gaussian limit, the TOMs vanish and the FOMs acquire the QNA form. For example, it was assumed that:

$$\frac{\partial}{\partial z}\left(\overline{w^4} - \overline{w^4}|_{QN}\right) = p_1\tau\overline{w^3} \tag{88}$$

The constants that appear in expressions like (88) were chosen so that (87) and (88) best match the full expressions (83), using as input the LES simulation data for the TOMs and SOMs of Mironov et al. [115]. Most importantly, use was made of new aircraft data on the FOMs by Hartmann et al. [75], to further determine these constants. The "best" values are listed in Table 1 of Cheng et al. [51]. The choice of such constants helps provide adequate damping that was lacking in previous models and effectively cancels the $\beta \sim \partial T/\partial z$ terms in the TOM equations, a choice supported by the TOM equations and the DNS data [89]. In addition, the cancellation of the β terms not only greatly simplifies the TOM equations, but also avoids the singularities in the unstable case and eliminates the source of the spurious oscillations in the stable case.

To assess their validity, the new FOMs were compared with measured data by plotting the modelled FOMs with the SOMs and TOMs from the LES data [115] as input, vs. z/h (h is the PBL height). In Fig. 5, the thick solid lines represent the new model results, the filled circles represent the aircraft data of Hartmann et al. [75], the dashed and dotted lines represent the model results

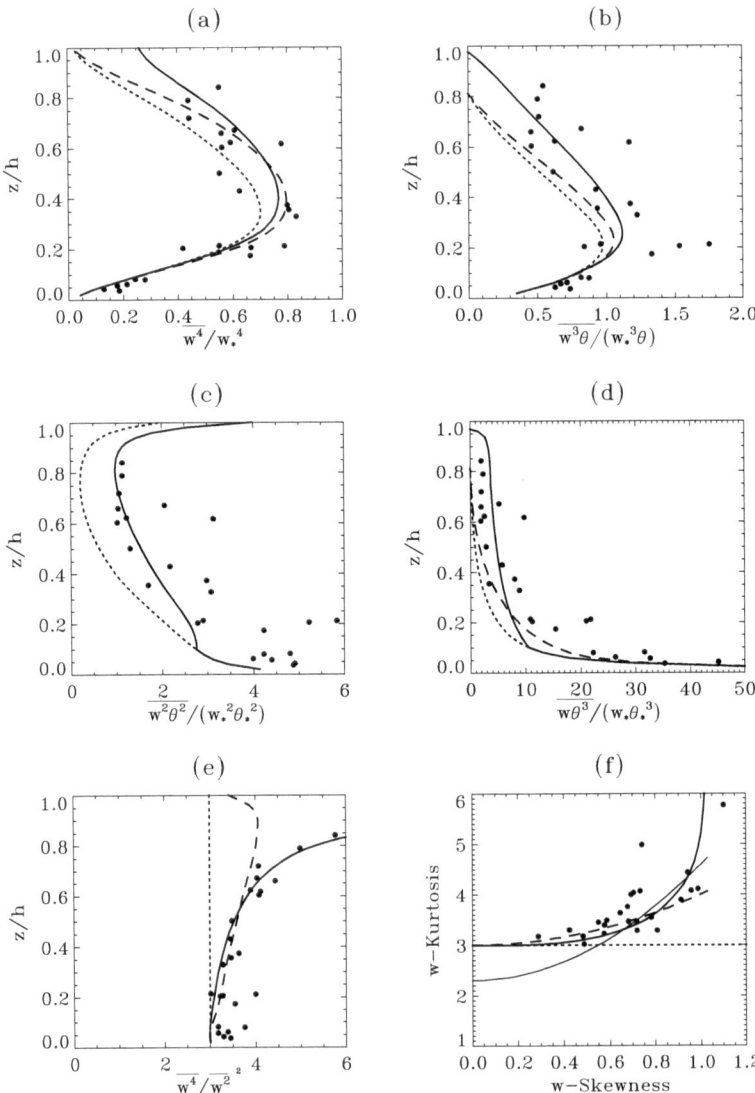

Fig. 5. In (**a**–**d**), the normalized FOMs are plotted vs. z/h in a convective PBL, using the LES data of Mironov et al. [115] for lower order moments as input. The solid lines represent results from the present FOM model, the dashed lines represent results from the recent model of Gryanik and Hartmann [73] and the dotted lines represent the QNA. The filled circles are the aircraft data of Hartmann et al. [75]. In (**e**) the kurtosis of w is plotted vs. z/h. The thick solid line represents the result from the new model, the *dashed line* represents the result from the model of Gryanik and Hartmann [73] and the *dotted line* represents the QNA, for comparison the aircraft data are the filled circles. In (**f**) the w-kurtosis K_w is plotted vs. w-skewness S_w, using the new model (thick *solid line*), the model of Gryanik and Hartmann ([73], *dashed line*) and the QNA (*dotted line*), for comparison the aircraft data are the filled circles and the empirical formula $K_w = 2.3\,(S_w^2 + 1)$ is the thin solid line

of Gryanik and Hartmann [73] and QNA respectively. The kurtosis of w from the models and from the aircraft data is plotted in Fig. 5e. To help assess the improvement shown in Fig. 5e, we refer the reader to the measurements of w kurtosis by Lenschow et al. [102, 103] who stated that "The kurtosis increases with height from around 3 to about 5 near 0.9 z/z_i. Above it, the kurtosis increases sharply". In Fig. 5f we plot the w-kurtosis K_w vs. the skewness S_w from the new model (thick solid line) and from Gryanik and Hartmann [73] (dashed line) to be compared with the aircraft data (filled circles) and with the empirical formula ([2], thin solid line)

$$K_w = 2.3 \left(S_w^2 + 1 \right). \tag{89}$$

Judging from the comparisons with these data, the new model exhibits significant improvements over the QNA and the Gryanik and Hartmann [73] model.

11.3 New TOM Model with New FOMs

Next, one employs the new FOMs into the TOM equations. The resulting equations are simpler than in previous models and more importantly, *they are singularity-free*. They are given by Eqs. (9a–f) of Cheng et al. [51] which we don't reproduce here. Suffices it to say that in the stationary limit, the new model for the TOMs reads as follows:

$$\overline{w^3} = -A_1 \frac{\partial}{\partial z} \overline{w^2} - A_2 \frac{\partial}{\partial z} \overline{w\theta} - A_3 \frac{\partial}{\partial z} \overline{\theta^2} \tag{90}$$

$$\overline{w^2\theta} = -A_4 \frac{\partial}{\partial z} \overline{w^2} - A_5 \frac{\partial}{\partial z} \overline{w\theta} - A_6 \frac{\partial}{\partial z} \overline{\theta^2} \tag{91}$$

$$\overline{w\theta^2} = -A_7 \frac{\partial}{\partial z} \overline{w\theta} - A_8 \frac{\partial}{\partial z} \overline{\theta^2}, \qquad \overline{\theta^3} = -A_9 \frac{\partial}{\partial z} \overline{\theta^2} \tag{92}$$

Equations (90), (91) and (92) exhibit the same structure of a linear combination of the z-derivatives of the SOMs first discussed in Canuto et al. [30, 34]. In (90), (91) and (92), the "diffusivities" A_i (with $\lambda = g\alpha$) are given by:

$$A_1 = \left(a_1 \overline{w^2} + a_2 \lambda \tau \overline{w\theta} \right) \tau, \quad A_2 = \left(a_3 \overline{w^2} + a_4 \lambda \tau \overline{w\theta} \right) \lambda \tau^2,$$

$$A_3 = \left(a_5 \overline{w^2} + a_6 \lambda \tau \overline{w\theta} \right) \lambda^2 \tau^3, \quad A_4 = a_7 \tau \overline{w\theta},$$

$$A_5 = \left(a_8 \overline{w^2} + a_9 \lambda \tau \overline{w\theta} \right) \tau, \quad A_6 = \left(a_{10} \overline{w^2} + a_{11} \lambda \tau \overline{w\theta} \right) \lambda \tau^2,$$

$$A_7 = a_{12} \tau \overline{w\theta}, \quad A_8 = \left(a_{13} \overline{w^2} + a_{14} \lambda \tau \overline{w\theta} \right) \tau, \quad A_9 = a_{15} \tau \overline{w\theta} \tag{93}$$

The coefficients a_i in (93) are given in Appendix B of Cheng et al. [51]. In Figs. 6 and 7 we exhibit the new TOMs and FOMs compared with LES data and aircraft data.

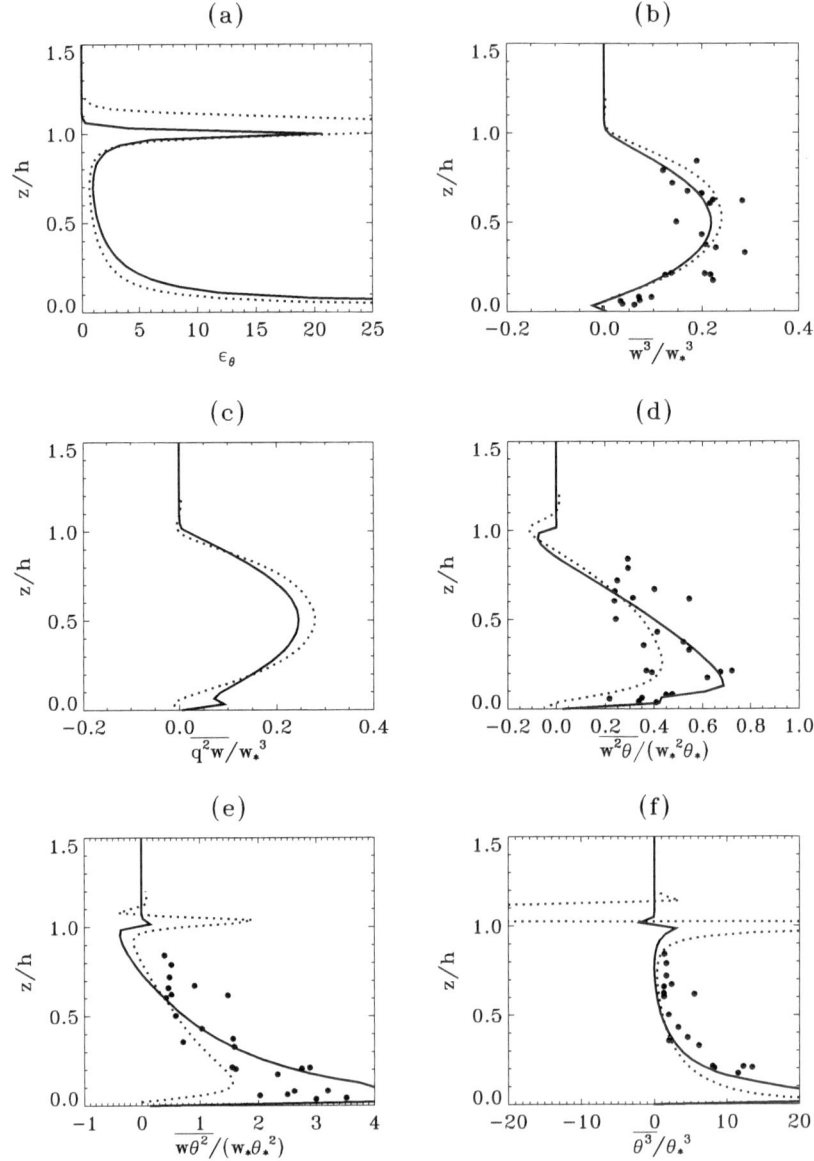

Fig. 6. Temperature variance dissipation rate ε_θ and normalized TOM vs. z/h resulting from the numerical model of a convective PBL. The *solid lines* represent the new model, the *dot-dashed lines* represent the LES data of Mironov et al. [115] and the filled circles represent the aircraft data of Hartmann et al. [75]

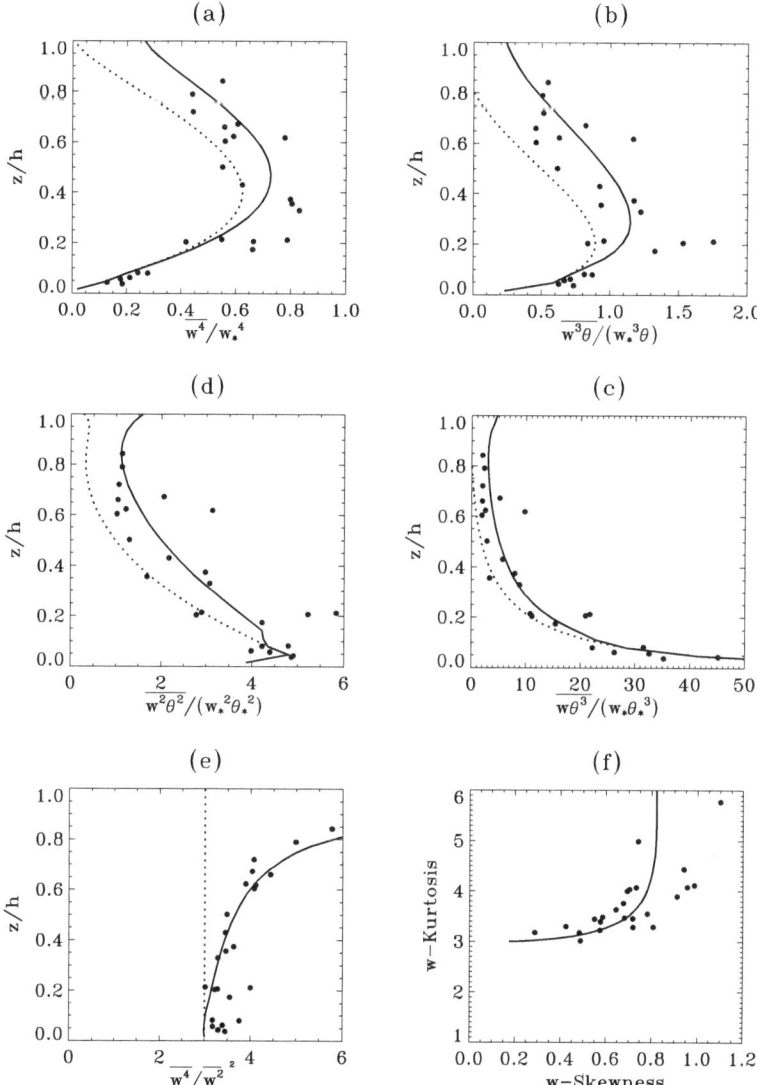

Fig. 7. Normalized FOMs vs. z/h according to the FOM model, as *solid lines*, using T, SOMs and TOMs resulting from the numerical simulation of a convective PBL as input, QNA FOMs as *dotted lines*. The *filled circles* represent the aircraft data of Hartmann et al. [75]

Even though Eqs. (90), (91) and (92) are relatively simple and have been successfully tested against LES data [51], more recently we have succeeded in reducing them even further without deteriorating the comparison with LES data. In fact, we have found the following simplified version:

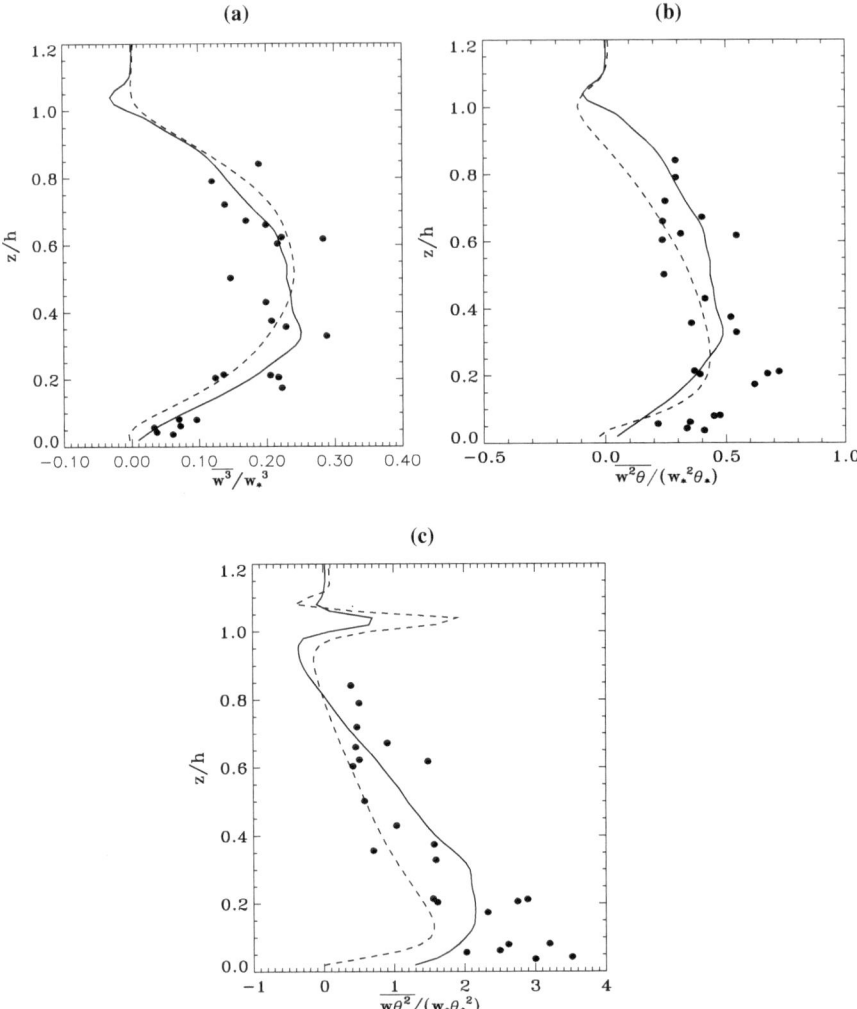

Fig. 8. (a) The third moment $\overline{w^3}$ normalized by w_*^3 vs. height normalized by the PBL depth h. The filled circles represent the aircraft data of Hartmann et al. [75]. The *dashed line* shows the LES data of Mironov et al. [115]. The solid line represents the result of the new, simple model, using the lower order moments from LES data as input. (b) Same as in (a) but for $\overline{w^2\theta}$ normalized by $w_*^2\theta_*$. (c) Same as in (a) but for $\overline{w\theta^2}$ normalized by $w_*\theta_*^2$

$$\overline{w^3} = -0.06g\alpha\tau^2\overline{w^2}\frac{\partial\overline{w\theta}}{\partial z}, \quad \overline{w^2\theta} = -0.3\tau\overline{w^2}\frac{\partial\overline{w\theta}}{\partial z}, \quad \overline{w\theta^2} = -\tau\overline{w\theta}\frac{\partial\overline{w\theta}}{\partial z} \quad (94)$$

which are compared in Figs. 8a–c to the LES data of Mironov et al. [115] and to the aircraft data of Hartmann et al. [75]. The data are reproduced quite well. The first of (94) correctly yields a negative skewness below the cooling ocean surface (or equivalently below the cloud top in the PBL case, see Stevens et al. [157]) where $\partial B/\partial z > 0$ (B is the buoyancy), while it yields a positive skewness near a surface heated from below where $\partial B/\partial z < 0$. By contrast, a down-gradient approximation which corresponds to retaining only the first term in (90):

$$\overline{w^3} \approx -\tau \overline{w^2} \frac{\partial \overline{w^2}}{\partial z} \tag{95}$$

yields the wrong sign of the skewness in both the above cases. To further highlight the physical content of (94), we can re-write them as follows:

$$g\alpha \overline{w^2 \theta} = 5\tau^{-1} \left(\overline{w^2}\right)^{3/2} S_w, \qquad g\alpha \overline{w\theta^2} = \frac{50}{3}\tau^{-1} J \left(\overline{w^2}\right)^{1/2} S_w, \tag{96}$$

so as to exhibit the skewness $S_w = \overline{w^3}/(\overline{w^3})^{3/2}$, as emphasized by previous authors [74, 182]. The development of this new model has also benefitted from the test of nonlocal models in stars [89, 90, 117].

12 Plumes and Turbulence

Consider Eq. (16) which in the more general form reads:

$$\frac{\partial T}{\partial z} + \cdots = \frac{\partial}{\partial x_j}\left(K_{ij}\frac{\partial T}{\partial x_i}\right) \tag{97}$$

Here, K_{ij} represents the heat diffusivity tensor which we split into its symmetric and anti-symmetric parts:

$$K_{ij} = 1/2\left(K_{ij} + K_{ji}\right) + 1/2\left(K_{ij} - K_{ji}\right) \equiv K_{ij}^{\text{s}} + K_{ij}^{\text{a}} \tag{98}$$

and let us define the divergence-free velocity field:

$$u_i^\star = -\frac{\partial K_{ij}^{\text{a}}}{\partial x_j}, \qquad \partial_i u_i^\star = 0 \tag{99}$$

Equation (97) then becomes:

$$\frac{\partial T}{\partial z} + u_i^\star \frac{\partial T}{\partial x_i} = \frac{\partial}{\partial x_j}\left(K_{ij}^{\text{s}}\frac{\partial T}{\partial x_i}\right) + \cdots \tag{100}$$

This means that the symmetric part of K_{ij} gives rise to the standard *Diffusion* while the anti-symmetric part is instead an *Advection*. As Lappen and Randall [95] have pointed out, in diffusive transport, information flows both upward and downward; by contrast, in advective transport, the information is either up or down depending on the time evolution, as we shall discuss below.

If one employs a local turbulence model, one only accounts for the rhs of (100) and the model is *purely diffusive*. On the other hand, the plume model

(PM) that has been widely used in the literature ([119], cited as MTT) is *purely advective*, since in fact, the mean T equation reads:

$$\frac{\partial T}{\partial t} + w^\star \frac{\partial T}{\partial z} = 0 \tag{101}$$

Clearly, neither a purely diffusive nor a purely advective model is satisfactory since both advection and diffusion must be accounted for since they represent different stages of the dynamical evolution of the system. Since the formal derivation just presented underlines the fact that advection and diffusion are not separate processes for they are both described by the same general diffusivity tensor, one must conclude that provided one includes *nonlocality*, *turbulence models have all the ingredients to account for both diffusion and advection*. The further task is to "plumenize" that is, to reformulate a nonlocal RSM so as to exhibit the up–down drafts [40].

Plume models are attractive for they provide an intuitive visualization of narrow descending plumes and wide ascending plumes exhibited by LES studies of convection cooled from above. There are however difficulties, the first of which is that the MTT is purely advective and its extension to include diffusion is far from obvious.

The second problem is that MTT contains two equations representing conservation of momentum and buoyancy. However, since there are three unknowns, the third being the fraction of space occupied by the plumes that varies with z (or the plume's radius), Taylor suggested a phenomenological "*entrainment equation*" which contains an entrainment parameter α that MTT is unable to determine. The parameter α has thus far been treated as an adjustable coefficient but in reality it is a function of the large scale features of the flow. Ellison and Turner [64] used laboratory data to determine $\alpha = \alpha(\mathrm{Ri})$, where Ri is the Richardson number, but this function provides a poor fit to the Mediterranean outflow data [131, 176]. A more complete formulation of α that includes nonlocal transport and which leads to a better representation of the newest data has recently been proposed [39]. The third problem is that MTT assumes that σ, the fractional area occupied by a plume, is much smaller than unity:

$$\sigma \ll 1 \qquad \sigma = \Sigma^p \left(\Sigma^p + \Sigma^e \right)^{-1} \tag{102}$$

where $\Sigma^p(\Sigma^e)$ is the total cross-section of the plumes (environment) at a given depth. However, since during their evolution, plumes entrain fluid from the environment, σ is bound to increase with depth to the point where (102) becomes invalid. Specifically, *entrainment* causes the plume's mass flux $w_p \Sigma^p \propto \sigma w_p$ to grow while stable stratification decreases w_p, the net result being an increase of σ to the point where (102) breaks down. In addition, a small σ model cannot satisfy the *zero mass flux relation* ($w_{u,d}$ are the velocities of the up–down drafts and z is considered upward):

$$\sigma w_d + (1 - \sigma) w_u = 0 \tag{103}$$

which, in the small σ limit, implies that:

$$|w_d| >> w_u \tag{104}$$

On the other hand, for the argument given above, when $\sigma=1/2$, Eq. (103) implies that:

$$|w_u| = |w_d| \tag{105}$$

which is not allowed under (102). Finally, the mass conservation (103) is invariant under the transformation:

$$w_u \to w_d, \qquad \sigma \to 1 - \sigma \tag{106}$$

and so should be any PM. The MTT model is not invariant under (106) since it is valid only in the plumes' early development stages when the fraction of space occupied by the plumes is still small (see Table 1 below).

Table 1. Development of plumes in Ocean Convection

Time	S_w	σ	Up/Down	Diff	Adv
Init phase	< 0	small	$|w_d| \gg w_u$	No	Yes
Final phase	≈ 0	1/2	$w_u \approx |w_d|$	Yes	No

In the fourth column we indicate the up/down interplay. In the early stages, downdrafts dominate over updrafts, while in the final stages, updrafts and downdrafts are equally important. The last two columns show that the initial stages are governed by advection while the final stages are governed by diffusion. Paluszkiewicz et al. [125], Alves [4] and Paluszkiewicz and Romea [126], used the MTT to study Ocean Convection. See also The Labrador Sea Deep Convection Experiment [165].

In summary, the MTT model has the advantage of simplicity but at present: 1) it is restricted by (102), 2) it depends on the undetermined rate of entrainment α and 3) it is only advective and leaves out diffusion.

13 New Plume Model

To correct the limitations of the MTT model, we proceed as follows.

1. We employ the RSM in which nonlocality is represented by the TOMs for which we employ the new model discussed above.
2. We write the nonlocal TOMs in the "plume approximation" which assumes a top hat profile that consists of two delta functions for the pdf of each state variable, corresponding to ascending and descending plumes. This implies [95] that such a profile has 100% probability of having one of

just two possible values, the two allowed states being up-drafts and down-drafts. This introduces a considerable simplification to the problem since it reduces substantially the number of higher order moments that are required, it assures the realizability condition of the higher order moments and requires fewer prognostic equations.

3. The new turbulence-based PM is such that all relations are invariant under (106) and thus the model is valid throughout the entire plume's development

4. In the small σ limit, the new model reproduces the MTT model.

To "plumenize" the TOMs using the up–down draft notation, we begin with the following relations [33]:

$$\overline{w^2} = \sigma(1-\sigma)(w_u - w_d)^2 = \beta_\sigma w^2, \quad \beta_\sigma = \sigma(1-\sigma)^{-1} \tag{107}$$

$$J = \sigma(1-\sigma)(w_u - w_d)(\theta_u - \theta_d) \tag{108}$$

$$\overline{\theta^2} = \sigma(1-\sigma)(\theta_u - \theta_d)^2 = \beta_\sigma^{-1} w^{-2} J^2 \tag{109}$$

where $w \equiv w_d$. Analogous relations hold for the salinity field. These relations are invariant under (106). Then, the plumenized TOMs read:

$$\overline{w^3} = -\sigma(1-\sigma)(1-2\sigma)(w_u - w_d)^3 = \overline{w^2}^{3/2} S_w$$

$$\overline{w^2\theta} = -\sigma(1-\sigma)(1-2\sigma)(\theta_u - \theta_d)(w_u - w_d)^2 = \overline{w^2}^{1/2} S_w J$$

$$\overline{w\theta^2} = -\sigma(1-\sigma)(1-2\sigma)(\theta_u - \theta_d)^2(w_u - w_d) = \overline{\theta^2}^{1/2} S_\theta J \tag{110}$$

where the skewness is taken to be:

$$S_{\theta,w} \equiv (2\sigma - 1)[\sigma(1-\sigma)]^{-1/2} \tag{111}$$

With the additional relation:

$$\overline{w^3} = -0.06 \, g\alpha\tau^2 \, \overline{w^2} \frac{\partial J}{\partial z} \tag{112}$$

the dynamic Eqs. (76), (77), (78), (79) and (80), together with (110), (111) and (112), constitute a new PM.

14 The Morton–Turner–Taylor Plume Model

The PM just derived is valid for an arbitrary σ and must reduce to the MTT model in the small σ limit. In Turner ([170], with $b^2 = \sigma l^2$ and for rising plumes with z pointing upward, $w > 0$), the MTT model contains three equations representing the plume's kinetic energy $1/2 \, w^2$, the fractional area

σ occupied by the plume and the buoyancy $B(\text{cm}^2\text{s}^{-3})$. The MTT dynamic equations are:

$$\frac{\partial w^2}{\partial z} = \frac{2B}{\sigma w} - \frac{4\alpha}{l}\frac{w^2}{\sigma^{1/2}}$$

$$\frac{\partial \sigma}{\partial z} = -\frac{B}{w^3} + \frac{4\alpha\sigma^{1/2}}{l} \qquad (113)$$

$$\frac{\partial B}{\partial z} = -\sigma w N^2$$

where α is the entrainment coefficient discussed earlier. Eq. (113) are not invariant under (106) since they are only applicable in the regime (102). If one substitutes the buoyancy equation into the mean temperature equation, one obtains:

$$\frac{\partial T}{\partial t} + (\overline{w} + w_{\text{adv}})\frac{\partial T}{\partial z} = 0, \qquad w_{\text{adv}} = -\sigma w \qquad (114)$$

As one can see, there is no diffusion which shows what we stated earlier that the MTT is purely advective with an advection velocity w_{adv} that is σ times the plume's velocity. An interesting variable is the plume's "mass flux" defined as:

$$M = \sigma w \qquad (115)$$

Using (113), one obtains the equation:

$$M^{-1}\frac{\partial M}{\partial z} = E - D = \frac{2\alpha}{l}\sigma^{-1/2} > 0 \qquad (116)$$

where E and D stand for the rates of entrainment and detrainment respectively. Since the rhs of (116) is positive, MTT *accounts only for entrainment but not detrainment* which is understandable since detrainment requires a dynamical environment which is excluded in the MTT model which assumes the environment to be quiescent.

In the ocean case ($w < 0$) and $\sigma << 1$, we have from Eq. (107) and (111) that:

$$\beta_\sigma = \sigma, \quad S_w = \sigma^{-1/2}, \quad \overline{w^2} = \sigma w^2 \qquad (117)$$

and thus from Eq. (110) it follows that:

$$\overline{w^3} = \sigma w^3, \qquad \overline{w^2\theta} = wJ, \qquad \overline{w\theta^2} = \sigma^{-1/2}J\left(\overline{\theta^2}\right)^{1/2} \qquad (118)$$

Using (117) and (118), the first of Eq. (94) becomes ($C = 0.06$):

$$\frac{\partial J_h}{\partial z} = -\frac{w}{Cg\alpha\tau^2} \qquad (119)$$

Next, using (111) and (112), together with (118), we obtain:

$$w^3\frac{\partial\sigma}{\partial z} + \frac{3\sigma w}{2}\frac{\partial w^2}{\partial z} = \frac{2}{3}(1 + 2\beta_5)B - \frac{4\sigma w^2}{\tau} \tag{120}$$

$$w\frac{\partial J}{\partial z} + \frac{J}{2w}\frac{\partial w^2}{\partial z} = -\sigma w^2 T_z + (1 - \gamma_1)g\alpha\sigma^{-1}w^{-2}J^2 - \tau^{-1}\pi_4^{-1}J \tag{121}$$

Solving Eqs. (119), (120) and (121), and using (the coefficient C_0 will be discussed later):

$$\tau = -C_0\sigma^{1/2}w^{-1}l \tag{122}$$

we obtain:

$$\frac{\partial w^2}{\partial z} = 2(1 - \gamma_1)w^{-1}\sigma^{-1}B + \Gamma_1$$
$$\Gamma_1 = 2\,C_0^{-1}\pi_4^{-1}w^2\sigma^{-1/2}l^{-1} - 2\,B^{-1}w^3\sigma\left(N_h^2 - C^{-1}C_0^{-2}\sigma^{-2}l^{-2}w^2\right) \tag{123}$$

$$\frac{\partial\sigma}{\partial z} = (3\,\gamma_1 + 4\,\beta_5/3 - 7/3)Bw^{-3} + \Gamma_2$$
$$\Gamma_2 = (C_0\pi_4)^{-1}(4\,\pi_4 - 3)\,\sigma^{1/2}l^{-1} + 3\,wB^{-1}\left(\sigma^2 N_h^2 - C^{-1}C_0^{-2}w^2l^{-2}\right) \tag{124}$$

Using (81), we further have:

$$\frac{\partial w^2}{\partial z} = 1.3\,w^{-1}\sigma^{-1}B + \Gamma_1,$$
$$\Gamma_1 = 24\,C_0^{-1}w^2\sigma^{-1/2}l^{-1} - 2\,B^{-1}w^3\sigma(N_h^2 - 17\,C_0^{-2}\sigma^{-2}l^{-2}w^2) \tag{125}$$

$$\frac{\partial\sigma}{\partial z} = -0.7\,Bw^{-3} + \Gamma_2$$
$$\Gamma_2 = -32\,C_0^{-1}\sigma^{1/2}l^{-1} + 3\,wB^{-1}(\sigma^2 N_h^2 - 17\,C_0^{-2}w^2l^{-2}) \tag{126}$$

Since in our system z is positive upward, and the descending plume is small near the surface but becomes progressively larger at depth, $d\sigma/dz$, the second term in the rhs of (126) that represents entrainment, must be negative. By the same token, the second term in (125) is positive since $dw^2/dz > 0$. If $C_0 = 6$, Eqs. (125) and (126) compare well with Eq. (113) of the MTT model (their Eq. 6.1.4) and Eqs. (10) and (11) of Paluszkiewicz and Romea [126].

15 Compressibility and Magnetic Fields

In the earth's atmosphere, the height of the PBL is about 1 km while the pressure scale height is about 8 km, yielding a ratio less than unity that ensures the validity of an incompressible treatment. Quite different is the situation in stars where the convective zone may be several pressure scale heights, just the opposite of the PBL situation. This implies that compressibility effects are important [14]. A RSM for compressible flows has been developed [23] and the compressible counterparts of equations (76), (77), (78), (79) and (80) are available. It would be quite instructive to consider the stationary and local limits of these new equations so as to sort out the compressible equivalent of the standard MLT.

As for the effect of magnetic fields on the heat transport and the possible combination of magnetic fields and rotation, the study by Canuto and Hartke [27] leads to analytic results. Depending on the angle between the vector $\mathbf{H}$ and the z-axis, as well as on the magnetic Rayleigh number and the ratio of magnetic energy (density) to kinetic energy (density), the heat flux exhibits different dependence on the convective efficiency S defined in (23). In other words, the heat flux can be either enhanced or reduced depending on those parameters. In Figs. 4, 5, 6, 7, 8, 9, 10, 11, 12, 13, 14, 15, 16 and 17 of the reference just cited one can find a set of heat flux vs. S results for different cases.

16 Helioseismology

The advent of helioseismology and the wealth of information that it has brought to the fore can and has been used to assess the validity of mixing models. For an assessment, the reader can consult the review article by Canuto and Christensen-Dalsgaard [32].

17 Reynolds Stress Model: Buoyancy and Shear

In the case of stable stratification (which is of interest to the astrophysical OV regions, the ocean and the nocturnal PBL), one must consider three fields, velocity, temperature and salinity (the latter becomes the mean molecular weight field in the stellar case and the moisture field in the PBL case). The RSM prescribes the rules to derive the dynamic equations for the second-order moments as discussed in detail in several papers [20, 35–37]. Here, we shall only quote the final results:

Reynolds stresses, $R_{ij} = \overline{u_i u_j}, \ \ b_{ij} = R_{ij} - \frac{2}{3}\delta_{ij}K, \ \ \mathrm{D/Dt} = \partial/\partial t + \overline{u}_i \partial_i:$

$$\frac{\mathrm{D}b_{ij}}{\mathrm{D}t} = -\frac{8K}{15}\Sigma_{ij} - (1 - p_1)\Omega_{ij} + (1 - p_2)Z_{ij} + \frac{1}{2}g(\alpha_T L_{ij} - \alpha_s M_{ij}) - 5\tau^{-1}b_{ij}$$

(127)

Since the lhs has zero trace, all the terms in the rhs have the same feature. The new terms are defined as follows:

$$\Omega_{ij} = b_{ik}\Sigma_{jk} + b_{jk}\Sigma_{ik} - 2/3\,\delta_{ij}b_{km}\Sigma_{km}, \quad Z_{ij} = b_{ik}V_{jk} + b_{jk}V_{ik} \quad (128)$$

where the mean shear and vorticity were defined in Eq. (70). Furthermore,

$$L_{ij} = \lambda_i J_j^h + \lambda_j J_i^h - 2/3\,\delta_{ij}\lambda_k J_k^h, \quad M_{ij} = \lambda_i J_j^s + \lambda_j J_i^s - 2/3\,\delta_{ij}\lambda_k J_k^s \quad (129)$$

Here, $\lambda_i = -(g\bar\rho)^{-1}\partial_i p$, $\alpha_{T,s}$ are the thermal expansion and haline contraction coefficients which require an equation of state to be computed. The coefficients $p_{1,2}$ are 0.832 and 0.545 respectively.

Heat flux, $J_i^h = \overline{u_i\theta}$:

$$\frac{\mathrm{D}J_i^h}{\mathrm{D}t} = -R_{ij}T_j - J_j^h \bar u_{i,j} - \left(2\alpha_T\Psi - \alpha_s\overline{\sigma\theta}\right)\partial_i\bar p - \pi_4^{-1}\tau^{-1}J_i^h \quad (130)$$

Salinity flux, $J_i^s = \overline{u_i\sigma}$:

$$\frac{\mathrm{D}J_i^s}{\mathrm{D}t} = -R_{ij}S_j - J_j^s \bar u_{i,j} - \left(\alpha_T\overline{\sigma\theta} - 2\alpha_s\Phi\right)\partial_i\bar p - \pi_1^{-1}\tau^{-1}J_i^s \quad (131)$$

Temperature variance, $\Psi = \frac{1}{2}\overline{\theta^2}$, Salinity variance, $\Phi = \frac{1}{2}\overline{\sigma^2}$:

$$\frac{\mathrm{D}\Psi}{\mathrm{D}t} = -J_i^h T_i - 2\pi_5^{-1}\tau^{-1}\Psi, \quad \frac{\mathrm{D}\Phi}{\mathrm{D}t} = -J_i^s S_i - 2\pi_3^{-1}\tau^{-1}\Phi \quad (132)$$

T-S correlation, $\overline{\theta\sigma}$:

$$\overline{\theta\sigma} = -\pi_2\tau\left(J_i^h S_i + J_i^s T_i\right) \quad (133)$$

where $T_i = \partial_i T$, $S_i = \partial_i S$, T and S being the mean temperature and salinity fields. To these equations we must add the equation for K given by Eqs. (39) and (40) which were written for the mean molecular field but which have the same formal structure for the salinity field. The dissipation time scales here were written in terms of the dynamical time scale $\tau = 2K/\varepsilon$ and the proportionality coefficients were denoted by π_k. Without the help of an outside model, the RSM per se is incapable of determining such coefficients and that may be one of the reasons why in the past the RSM was not extended to include the salinity field. Without the knowledge of such constants, the above equations would be quite useless. In Canuto et al. [35–37] it was shown that the outside model is provided by the RNG, the renormalization group, and we refer the reader to the discussion in the original papers. The numerical values are presented in Eq. (22d) of Canuto et al. [37]:

$$\pi_1 = \pi_4 = \left(27\,\mathrm{Ko}^3/5\right)^{-1/2}\left(1 + \sigma_t^{-1}\right)^{-1}, \quad \pi_3 = \pi_5 = \sigma_t, \quad \pi_2 = 1/3 \quad (134)$$

with a suggested valued of 0.72 for the turbulent Prandtl number σ_t.

Of course, it is quite cumbersome to hook up the above turbulence equations to a large scale code for stars, PBL and/or the ocean. Thus, we present the solutions of Eqs. (127), (128), (129), (130), (131), (132) and (133) in the stationary case. In that limit, Eqs. (127) (128), (129), (130), (131), (132) and (133) become algebraic and can be solved analytically though admittedly with the help of a symbolic algebra code. Quite interestingly, the results are simple as the following expressions show:

$$\overline{w\theta} = -K_h\frac{\partial T}{\partial z}, \quad \overline{w\sigma} = -K_s\frac{\partial S}{\partial z}, \quad \overline{uw} = -K_m\frac{\partial \overline{u}}{\partial z}, \quad \overline{vw} = -K_m\frac{\partial \overline{v}}{\partial z} \quad (135)$$

where all the diffusivities have the same general form:

$$K_\alpha = \frac{2K^2}{\varepsilon}S_\alpha, \quad S_\alpha = S_\alpha(\text{Ri}, R_\rho) \quad (136)$$

where Ri was defined in (42) and the density ratio is given by:

$$R_\rho = \frac{\alpha_s S_z}{\alpha_T T_z} \quad (137)$$

which in the stellar case was introduced before (49). The dimensionless "structure functions" S's are algebraic expressions given in Canuto et al. ([37], see also Figs. 3, 4 and 5). Relation (136) points to a clear division of labour, the K–ε equations must be solved to determine these two variables but, as Eqs. (39) and (40) show, such equations require (135), (136) and (137). As an example of the role played by the dissipation time scale represented by the π_k, we consider the ratio heat/salinity diffusivities which turns out to have a rather simple form:

$$\frac{K_h}{K_s} = \frac{1 - \pi_1\pi_3 x R_\rho + \pi_1\pi_2 x(1 + R_\rho)}{1 + \pi_1\pi_3 x - \pi_1\pi_2 x(1 + R_\rho)} \quad (138)$$

where $x = (\tau N)^2(1 - R_\rho)^{-1}$. This expression exhibits the correct symmetry: when $R_\rho = -1$, heat and salt diffusivity coincide, as they indeed must. Furthermore, in the $\tau \gg N^{-1}$ limit turbulence is not strong and Eq. (138) becomes independent of x:

$$\frac{K_h}{K_s} = \frac{(\pi_2 - \pi_3)R_\rho + \pi_2}{\pi_3 - \pi_2(1 + R_\rho)} \quad (139)$$

In the opposite limit of $\tau \ll N^{-1}$, we obtain instead:

$$K_h = K_s \quad (140)$$

as expected since this corresponds to strong turbulence. For future use, we shall rewrite Eq. (136) in the form:

$$K_\alpha = \Gamma_\alpha\frac{\varepsilon}{N^2}, \quad \Gamma_\alpha = \frac{1}{2}(\tau N)^2 S_\alpha \quad (141)$$

where the Γ_α are called *mixing efficiencies*. Lacking a predictive mixing model, in the past the practice has been to assume $\Gamma_\rho = 0.2$. The mixing model presented here shows that the Γ_α are not universal constants and in Canuto et al. [37, 39] it is shown that they increase strongly near $R_\rho = 0.6$, a prediction consistent with the recent observations of much larger mixing at Barbados ($R_\rho = 0.6$) than at the NATRE location [98] where $R_\rho = 0.56$. In addition, the model predicts $\Gamma_\alpha = O(1)$, in agreement with LES data [177].

18 Mixing in the Ocean: Mixed layer, Gravity Waves and Tides

18.1 Mixed Layer

Broadly speaking, the vertical structure of the ocean can be characterized by three distinct regimes: a mixed layer which is stirred by external forces, primarily the external wind and in which mixing is very robust leading to a very well-mixed regime. In this mixed layer (ML) heat and salt diffusivity are the same but they differ from the momentum diffusivity. Since this regime is stably stratified, $N^2 > 0$, mixing can occur only as long as Ri < Ri(cr), as discussed in Sect. 4. Below the ML, stable stratification is "stronger" than the stirring due to the wind whose effect largely subsides.

Clearly, the depth of the ML varies with both latitude and seasons. In tropical regions, where surface waters are warm, the temperature gradient with respect to deep waters is large leading to a large Ri, small mixing and shallow ML. By contrast, high latitude regions in winter have much cooler surface waters whose temperature gradient with respect to deeper waters is not so large, leading to a small Ri, strong mixing and a deeper ML. In general, ML depths are hardly deeper than say a few hundred meters which is a minute fraction of the ocean's depth (3 km on average) but, due to its interaction with the atmosphere, the ML plays a critical role in climate studies.

18.2 Internal Gravity Waves

Below the ML there is an extensive region where diffusivities are of the order of 0.1 cm^2 s^{-1} which is thousand times smaller than the one characterizing the ML. This weakly mixed regime can exhibit DD processes.

It is generally believed that the mixing is due to the presence of internal gravity waves that are known to permeate the whole ocean. Although a fundamental theory for non-linear wave interactions is still lacking, the available information is considerable [70, 71, 88, 120, 128, 129, 169]. Kunze and Sanford [88], using Sargasso Sea data, suggested an expression called the Gregg–Heyney–Polzin parameterization whereby one has (cgs units):

$$\varepsilon_{\mathrm{igw}} = 2.88AN^2 \tag{142}$$

which can therefore be used in (141). The dimensionless factor A (related to the ratio of the shear variance to that computed from the Garrett and Munk model [67]) varies at most by a factor of 2 over the entire depth of the ocean. Recently, direct measurements of such diffusivities were made in the open ocean [97, 98] and one has therefore a set of data to compare with the predictions of a mixing model, like the one presented in the previous section.

In Figs. 9a–d we present the results of a 3D Global Ocean Model in which the mixing model described above was used with $\varepsilon \equiv \varepsilon_{\mathrm{ML}} + \varepsilon_{\mathrm{igw}}$. The North

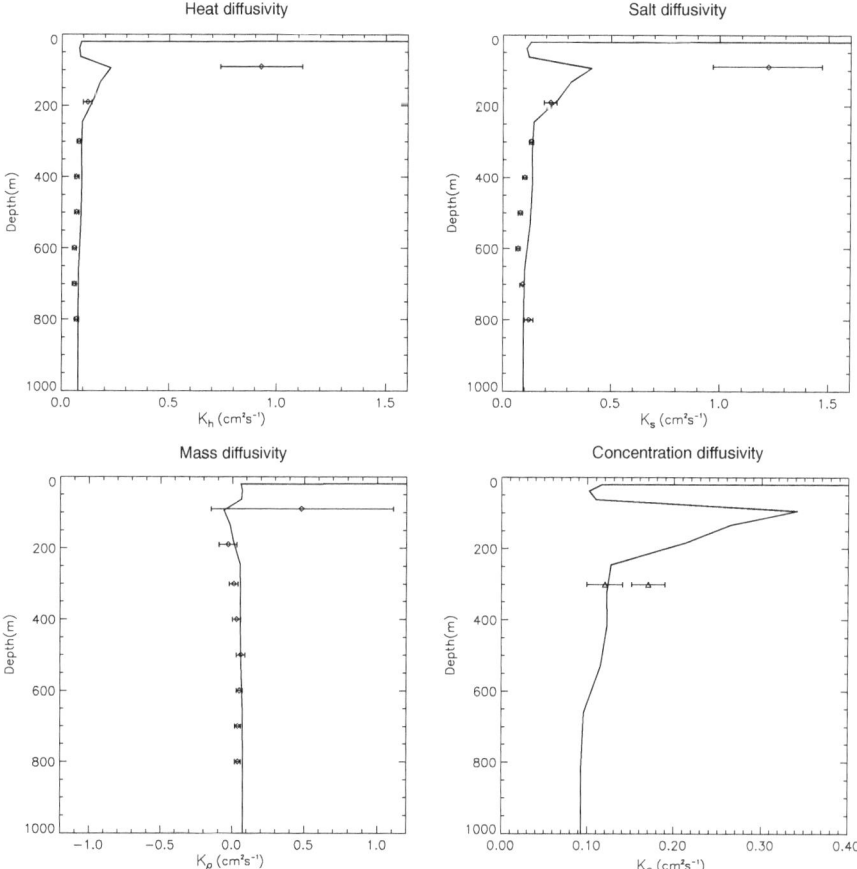

Fig. 9. The diffusivities predicted by the mixing model vs. the NATRE data (see text in Sect. 18)

Atlantic Tracer Release Experiment (NATRE) data are represented as diamonds with errors bars [149] while for the concentration diffusivity, Fig. 9d, the triangles are from Ledwell et al. [97, 98].

18.3 Tides

There is a well-documented evidence from the Topex/Poseidon altimetry data [60, 79] that tidal energy in the amount of approximately 3.5 terawatts (TW) is dissipated in the ocean. While most of it gets dissipated in shallow waters, a fraction amounting to 25–30%, that is, about 1 TW, is left for deep conversion to internal tides. Thus far, the only tidal effect on global ocean properties that has been investigated is the *enhanced bottom mixing* due to the break-up of tides against a rough bottom topography resulting in a fraction q of the tidal energy to act as a source of additional mixing. This picture has been confirmed by many studies [65, 99, 118, 122, 130, 168]. Since modelling studies have shown long ago [19] that oceanic heat transport is quite sensitive to the value of the vertical diffusivity which also affects quite significantly the uptake and storage of the ocean's heat [152], this additional mixing must be quantified and its consequences on the oceanic global properties assessed. Two recent pioneering studies [138, 148] have done so but they reach somewhat different conclusions. In this case, (141) must be employed with:

$$\varepsilon \equiv \varepsilon_{\mathrm{ML}} + \varepsilon_{\mathrm{igw}} + \varepsilon_{\mathrm{tides}} \tag{143}$$

To derive the dissipation due to tides, we begin by first considering the *internal tidal energy flux* $E(x, y)$ (in W m^{-2}) which is derived in a similar manner as in Jayne and St. Laurent [79] from a parameterization of the conversion of barotropic tidal energy into internal waves. A number of analytical models have been used to predict this energy conversion rate. The main theoretical approach used ideas first developed by Bell ([10, 11]; see also [83, 101, 150]). In the parameterization of Jayne and St. Laurent [79], the conversion of barotropic tidal kinetic energy is given by:

$$E(x, y) = \frac{1}{2}\rho_0 N k h^2 \overline{u_t^2} \tag{144}$$

where ρ_0 is a reference density of seawater, (k, h) are the wave number and amplitude that characterize the bathymetry and $\mathbf{u}$ is the barotropic tidal velocity solution of the Laplace tidal equation. It should be emphasized that (144) is a scale relation, and not a precise specification of the internal tide energy flux. In the barotropic tidal model, the value of k was tuned to give the best fit to the observed tides. Following Jayne and St. Laurent [79], the tidal dissipation is then given by:

$$\varepsilon_{\mathrm{tides}} = qE(x, y)\rho^{-1}F(z) \tag{145}$$

where the vertical structure function $F(z)$ was assumed to have the form:

$$F(z) = \zeta^{-1}(1 - \exp - H/\zeta)\exp - [(H + z)/\zeta], \quad \int_{-H}^{0} F(z)\,\mathrm{d}z = 1 \tag{146}$$

The vertical scale ζ was taken to be 500 m. The quantity q represents the amount of tidal energy that is used to produce mixing and clearly $0 \leq q \leq 1$. Thus far only the value $q = 0.3$ was employed but a wider range of values was recently investigated [41].

18.4 Energy Considerations

As we said earlier, the tidal energy amounts to about 3.5 TW [121]. About 75% of it, about 2.6 TW is dissipated in shallow waters while the remaining

Table 2. Partition of tidal energy (in TW) in the Ocean

	Drag	Internal tides	Total
Shallow waters (< 1 km)	1.49	0.98	2.47
Deep ocean	0.01	1.01	1.02
Total	1.5	1.98	**3.49**

25%, that is, about 1 TW, goes into the deep ocean. In the work of [41], the partition of energy turned out as shown in Table 2.

The global dissipation and conversion of tidal energy in the barotropic tide model is 3.49 TW, of which about 1.50 TW is dissipated by drag, largely in shallow seas, and 1.98 TW is scattered by conversion of the barotropic tide into internal waves. Of the 1.98 TW of internal wave conversion energy, about 1.00 TW occurs in the ocean deeper than 1000 m, while 0.98 TW occurs in areas $< 1,000$ m. As one can see, the overall distribution of tidal energy used is in good accord with the measured data.

18.5 Mechanical Power

The problem regarding the power required for driving mixing processes is an old one which has recently been reviewed and restudied by St. Laurent and Simmons [151]. By definition, we have:

$$P_m = \int K_m \Sigma^2 \, dV = \overline{N^2} \, \overline{K_\rho (1 + \Gamma_\rho^{-1})} \tag{147}$$

where we have used production = dissipation which reads $K_m \Sigma^2 - K_\rho N^2 = \varepsilon$ and where the averages are defined as:

$$\overline{f} = \left(\int f N^2 \, dV \right) / \left(\int N^2 \, dV \right) \tag{148}$$

The results for P_m (in TW) are shown in Table 3.

In each case, the background contribution is always 0.31 TW. As discussed in Sect. 1 of St. Laurent and Simmons [151], using the water mass budgets, various authors arrived at results which can be bracketed between (0.5–2) TW. On the other hand, St. Laurent and Simmons [151] employed a

Table 3. Contribution of tides to power available for driving mixing in the Ocean

No tides	0.31 TW	(background only)
$q = 0.3$	0.59 TW	
$q = 0.7$	0.87 TW	
$q = 1.0$	1.07 TW	

(a)   (b)

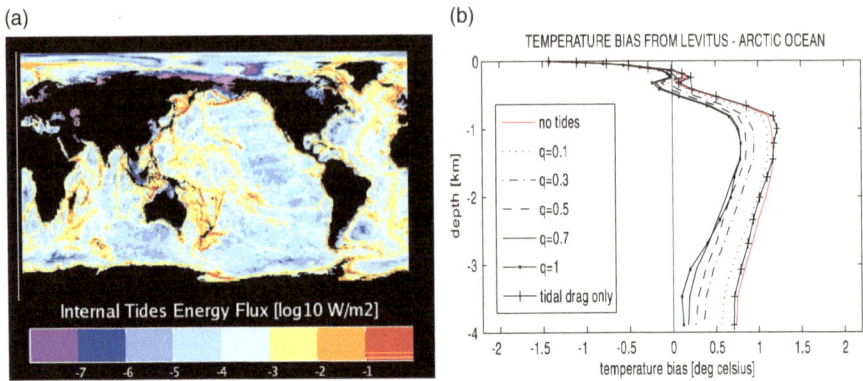

Fig. 10. (**a**) World map of the tidal internal energy in W/m^{-2}. (**b**) The temperature vertical profile in the Arctic ocean with and without the effect of tides

different approach based on diffusivities measurements and thermodynamic arguments and obtained (2.7 ± 0.7) TW. It must be stressed however that such value depends on the mixing efficiency which was taken to be $\Gamma_\rho = 0.2$ (for further details, see Sect. 4c of St. Laurent and Simmons, [151]). In the work of Canuto et al. [41], there is no such a freedom since Γ_ρ is computed from within the model.

In Fig. 10a we show a map of the internal tidal energy given by Eq. (144) while in Fig. 10b we show the $T(z)$ profile in the Arctic ocean for several values of the parameter q. As one can see, the bias (predicted temperature minus the Lavitus data) is considerably reduced by the presence of tidal effects. The same effect was also found for the salinity field.

19 Mixing in the Ocean: Deep Convection

Earth's atmosphere interacts with the ocean in two ways. As already discussed in the previous section, wind stresses cause strong mixing in the first $\sim 100\,m$ of the ocean (ML) but hardly affect water masses below the ML where lies the largest portion of the ocean characterized by stable stratification and weak mixing. To the effect of communicating with the deep ocean, e.g. in the process of absorbing atmospheric CO_2, stable stratification acts like a rigid lid that insulates the strongly mixed ML from the weakly mixed deep ocean. If such a configuration were to prevail, the deep ocean would be shielded from climatic events; deep waters would be dynamically decoupled from surface phenomena and the deep ocean currents would be considerably weaker than observed. Differential solar heating between low and high latitudes would not result in a poleward flow of warm waters and the Atlantic would look more like the Pacific where there are no deep convective regions [178]. Earth's climate would be quite different from what is observed today.

Deep Convection is the process through which surface phenomena such as buoyancy losses, brine rejection, etc., pierce the lid of strong stable stratification that characterizes the thermocline. Ultimately, this leads to the formation of deep waters [53, 107, 110, 165]. Open deep ocean convection occurs in a small number of locations, Labrador, Greenland, Weddell and Western Mediterranean Seas, and yet it is one of the ocean's major features since it represents the initial stage of the global-scale ventilation loops of the world ocean [142]. In fact, it is a dominant mechanism for the production of North Atlantic Deep Water and of the Antarctic Bottom Water [178], both of which play a major role in earth's climate.

Loss of surface buoyancy and/or brine rejection lead to a top-heavy, unstable configuration which acts as precursor of turbulent motion that ultimately leads to deep convection. The latter upwells warmer waters that can melt ice and reduce the albedo resulting in a "negative feedback" that affects climate [84]. Regrettably, however, deep convection is still poorly modelled in coarse resolution ocean general circulation models (for details, see [38]). While laboratory and numerical simulations [59, 107, 110, 142, 147] have brought to light several key features of convective processes, the translation of this information into a reliable model for coarse resolution OGCMs has not yet been achieved but, given the complexity of the phenomenon, this is hardly surprising. Before we test models for deep convection, it is important to discuss some of its key features:

1. *Deep convection is a highly turbulent process.* This is exhibited by the large vertical diffusivities K_v:

$$K_v \sim lw \sim (1-10)\ 10^5\ \mathrm{cm^2\ s^{-1}} \qquad (149)$$

where $l \sim (1-2)$ km and $w \sim (1-5)$ cm s^{-1}, as discussed by Marshall and Schott [107]. Equivalently, one may consider the large value of the Reynolds number:

$$\mathrm{Re} \sim K_v/\nu \sim 10^7 \qquad (150)$$

where $\nu \sim 10^{-2}$ cm^2 s^{-1} is the kinematic viscosity of seawater. Thus, a high-Re turbulence model is needed to describe deep convective processes.

2. *Deep convection is affected by rotation.* Consider the characteristic length scale [68]:

$$l(\mathrm{rot}) \sim (B_s f^{-3})^{1/2} \sim (0.15-0.56)\ \mathrm{km} \qquad (151)$$

where B_s is the surface buoyancy and f is the Coriolis parameter. The numerical values in (151) correspond to the Greenland and Mediterranean Seas ([107], Table 3.4.1). Contrary to the atmospheric case whose $l(\mathrm{rot})$ is much larger than the height ≈ 1 km of the planetary boundary layer, in the ocean case the reverse is often true, namely $l(\mathrm{rot})$ may be considerably smaller than the ocean depth H that, for the two cases just cited, is 1.5 and 1.8 km. This yields small Rossby numbers Ro=$l(\mathrm{rot})/H = 0.1 - 0.3$, an indication of the importance of rotation.

A turbulence model must be able to incorporate rotational effects and more specifically, it must be able to reproduce key features like the Golystin's scale (151). Rotation enters the turbulence equations not only through the Coriolis term in the dynamic equations for the velocity but, more important, it affects the very structure of the non-linear interactions that are at the heart of turbulence. In the presence of rotation, velocity components with different vectors are rotated by the Coriolis force around different axes that coincide with the directions of the corresponding wave-vectors. Thus, the energy cascade from large to small eddies is inhibited. An inertial range is still present but only for wavenumbers larger than $k(\text{rot})$ where the latter is the inverse of Eq. (151). For wavenumbers $k < k(\text{rot})$, the spectrum is no longer of the Kolmogorov type, that is, one has two regimes:

$$l > l(\text{rot}) : E(k) \sim (\varepsilon \Omega)^{1/2} k^{-2}, \qquad l < l(\text{rot}) : E(k) \sim \varepsilon^{3/2} k^{-5/3} \qquad (152)$$

Integrating the two spectra, one derives that the corresponding velocities with and without rotation are given by:

$$w(\text{rot}) \sim [l/l(\text{rot})]^{1/2} (B_s f^{-1})^{1/2}, \qquad w_0 \sim (B_s H)^{1/3} \qquad (153)$$

where we have taken the dissipation rate ε equal to the surface buoyancy flux B_s. Values of $w(\text{rot})$ and w_0 for the Mediterranean, Labrador and Greenland

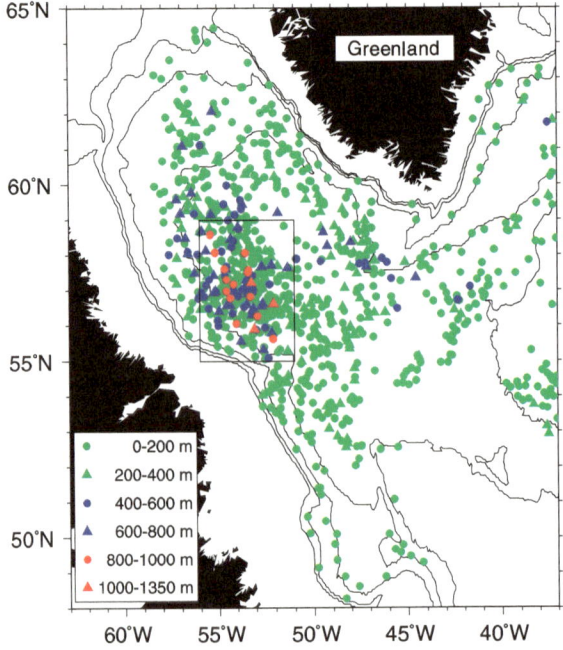

Fig. 11. Measured ML depths from Lavender et al. [96]

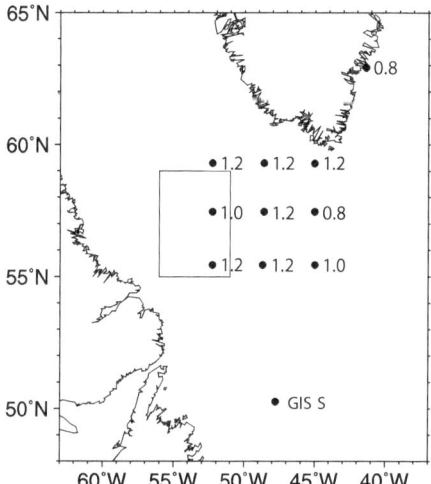

Fig. 12. The Labrador ML depths resulting from the local vertical mixing model discussed in Sects. 17 and 18 and in Canuto et al. [37]

Seas can be found in Table 3.4.1 of Marshall and Schott [107]. A turbulence model capable of predicting the two regimes of the energy spectrum (152) has recently been constructed and its implications tested against Direct Numerical Simulations [31].

Although for the reasons already discussed in Sects. 12 and 13, one should employ the new PM to study ocean's Deep Convection, this has not yet been done. The results that are available are nonetheless interesting since they highlight the limitations of a local model without rotational effects. The model results are taken from the work of Canuto et al. [38] and they are displayed in Figs. 11 and 12. In the first figure we reproduce the measured ML depths while in Fig. 12 we present the model results. Measurements show that the deepest convection with depths deeper than 800 m, is confined to a rather limited region inside the box. The model results, while still lacking key ingredients, seem to capture at least some of the main features of the Deep Convection processes. The limitations of the model are clearly visible in the regions outside the box where ML depths up to 1.2 km are predicted which are however not observed. It must be stressed however that other mixing models fare worse than the one used in Fig. 12, as discussed in Canuto et al. [38]. The message seems clear: to obtain better model results, one needs to employ a PM in which non-locality is accounted for.

20 Dissipation: An Open Problem

Consider Eqs. (76) and (80) in the homogeneous and stationary case. The first gives $P = \varepsilon$ while the second gives $c_1 = c_2$, which contradicts the values of the two coefficients given before Eq. (80), namely $c_1 = 2.88$ and $c_2 = 3.8$. Eqs. (76) and (80) are therefore inconsistent, at least in that limit. The reason is that while Eq. (76) is derived exactly from the RSM method and contains no adjustable parameters, the equation for the dissipation is highly parameterized since an a priori derivation of (80) is still not available [136, 144]. This may come as surprise since after all the definition of dissipation given by:

$$\varepsilon_{ij} = 2\nu\overline{\partial_i u_k \partial_j u_k} = \frac{2}{3}\delta_{ij}\varepsilon \qquad (154)$$

entails only the velocity field and in principle one should be able to derive a dynamic equation for ε. Davidov [56] was the first to derive such an equation but the result was unmanageable. Over the years people tried to come up with an equation that would mimic the equation for K and indeed (80) has a strong similarity with (76). A thorough review of such an equation, the variants that have been suggested over the years and an extensive list of references has recently been given by Kantha [80]. The two coefficients $c_{1,2}$ have been the subject of much speculation [9, 80, 175]. For example: are they really constant or are they functions? We don't have an answer yet and we can only highlight how c_2 is generally arrived at. Consider the case of freely decaying turbulence in which case Eqs. (76) and (80) become:

$$\frac{\partial K}{\partial t} = -\varepsilon, \qquad \frac{\partial \varepsilon}{\partial t} = -c_2 \, \varepsilon \, \tau^{-1} \qquad (155)$$

Using simple power laws of the type $K \sim t^{-n}$, $\varepsilon \sim t^{-m}$, one obtains $m = n + 1$ and:

$$c_2 = 2\left(1 + n^{-1}\right) \qquad (156)$$

Since numerical simulations [47–49] indicate that $n = 6/5$–$10/7$, one obtains the value of c_2 cited above. Once that value is accepted, the coefficient c_1 is determined using other data which include a nonzero production term. The procedure is hardly without a fault since there is no reason why the c_2 determined in the freely decaying case should also hold in the case when production is present. For example, one could suggest a generalization of (156) of the type:

$$c_2 = 2(1 + n^{-1})(1 + aP/\varepsilon)^{-1} \qquad (157)$$

that satisfies (156) when there is no production $P = 0$ while in the $P = \varepsilon$ case considered before, a proper choice of the parameter "a" could yield $c_1 = c_2$ and make the inconsistency disappear. This patch-up exercise is clearly open to other suggestions, for example, the exponential -1 in the last term in (157) could be changed to an additional adjustable parameter and so on. None of these procedures is very satisfactory and thus Eq. (80) must be viewed as the weakest point in any turbulence model. So much so that often (80) is not even used and instead one employs a more heuristic approach on the grounds that to fix the problems just mentioned one has to resort to empirical models anyway and thus (80) does not seem to offer great advantages. One very common approach is to use a Kolmogorov's type of formula (see Eq. (7)):

$$\varepsilon = K^{3/2}\Lambda^{-1}, \quad \Lambda - c\,l(z), \quad l(z) = \kappa\,z\,l_0\,(l_0 + \kappa\,z)^{-1} \tag{158}$$

with $c = 5.87$. These expressions were suggested by Blackadar [12] and Deardorff [58] and are still widely used [7, 66]. Here z is the distance from the nearest "wall" and κ is the von Karman constant whose value is around 0.4. One can observe that for small z's, $l(z) = \kappa\,z$ which is the so-called "law of the wall", while for larger z's, one obtains $l(z) = l_0$ which, on the basis of LES data, is taken to be $0.17H$, where H is the extent of the mixing zone. Furthermore, since in the case of stable stratification, turbulent kinetic energy transforms into eddy potential energy, one can define [58] a length scale $l_s = c_s(KN^{-2})^{1/2}$ with $c_s = 0.2$ and rewrite the last relation (158) as:

$$\frac{1}{l(z)} = \frac{1}{\kappa\,z} + \frac{1}{l_0} + \frac{1}{l_s} \tag{159}$$

The universality of these relations is of course unknown. To this, we may add the fact that the non-local term in (80) is a further source of uncertainty. The closure suggested by Canuto ([20], Eq. (49)):

$$\overline{w\varepsilon} = 3/2\,\tau^{-1}F_{\mathrm{KE}}, \quad F_{\mathrm{KE}} = \overline{wK} \tag{160}$$

was recently assessed quite successfully by Kupka et al. [91] using DNS data for buoyant convection, Fig. 13. Other authors [17, 18, 175] have instead used a closure of the down-gradient type:

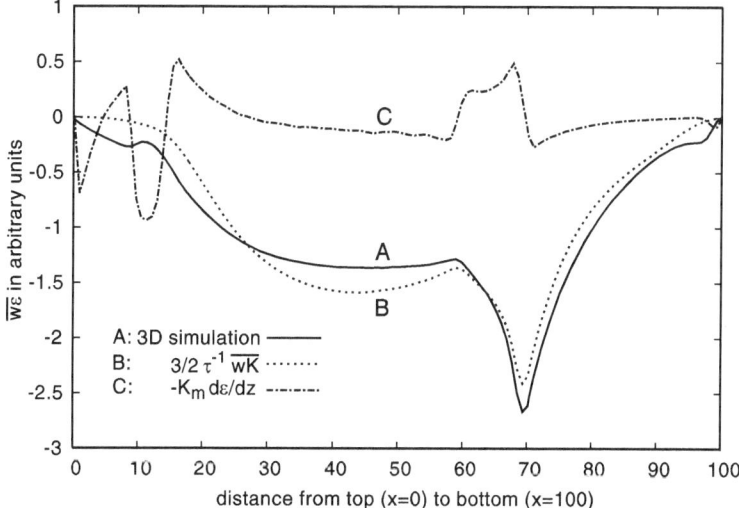

Fig. 13. The function $\overline{w\varepsilon}$ (in arbitrary units) from the 3D simulation of Kupka and Muthsam [91]. As one can observe, the down-gradient closure (161) fails to reproduce the data while (160) yields better results

$$\overline{w\varepsilon} = -K_m \frac{\partial \varepsilon}{\partial z} \tag{161}$$

but, at least in the unstable buoyancy case, the results of Kupka et al. [91] shown in Fig. 13 cast severe doubts about (161). What is the reason underlying all these difficulties? A basic one is the following: in k-space (154) gives:

$$\varepsilon = 2\nu \int k^2 E(k) \, \mathrm{d}k \tag{162}$$

The first consideration is that the largest contribution to the integral comes from the high wave number region, which corresponds to the smallest eddies which are very difficult to model since they contain little energy but a large vorticity and a short lifetime. As an example, it is easy to verify the inapplicability of the HK spectrum: if (162) were to be integrated over all wave numbers, Eq. (162) would diverge and the kinematic viscosity ν would not disappear while it is known that ε does not depend on ν. This is true because the non-linear interactions enter the dynamic equation under a divergence and yield zero when one integrates over the whole volume leading to the physical relation: energy input = energy output which alternatively means that the non-linear interactions do not use any energy, they simply transfer it from large to small eddies. Thus, what "gets" to the small eddies is the same energy that was put into the system which clearly has nothing to do with how viscous the system is, being an arbitrary external input. So, how should one read Eq. (162)? From left to right: given a fixed amount of energy input, which is identical to ε, the right hand side tells us at which k_d the dissipation occurs: the smaller the viscosity, the larger k_d has to be, that is, the smaller are the eddies that operate the dissipation process. Thus, one can say that Eq. (162) fixes the upper limit of integration. For example, if one uses the HK spectrum *only* up to k_d, integration of (162) gives:

$$k_d = \left(\frac{\varepsilon}{\nu^3}\right)^{1/4} \tag{163}$$

which is just Eq. (2). Seen from this viewpoint, it is perhaps not surprising that a satisfactory dynamical equation for ε has thus far eluded us. To compute (162), say from right to left, one needs a two-point closure that yields an energy spectrum $E(k)$ reliable in the high wavenumber regime which is hard to obtain. This is not surprising since in the dissipation region the Reynolds number is of order unity while all turbulence models have been constructed for large Re. To justify Re $= O(1)$, consider that in the dissipation region the typical time scale is $(\nu/\varepsilon)^{1/2}$ and the velocity is $v = (\varepsilon\nu)^{1/4}$. Since Re $= v/(\nu k_d)$, use of (162) leads to Re $= O(1)$. We can only conclude that for the time being we must live with these uncertainties about the dissipation ε.

21 Conclusions

It is a matter of record that especially in astrophysical settings one witnesses a surprising dichotomy: on the one hand, LES data are rich in content, information, details etc., while in stellar structure-evolution codes the models used to parameterize turbulent processes are still quite rudimentary, as the MLT and/or Eqs. (56), (57) and (58) clearly demonstrate. In accretion disks studies, which we have not discussed in this review, the situation is rather similar for most studies still employ the Shakura–Sunyaev α-model even though an effort was made to derive such a phenomenological parameter from more basic principles [25]. Due to the large computational requirements that any LES entails, there is little hope that astrophysical LES codes (e.g. [8, 15, 16, 113, 132, 133, 156, 179]) will be hooked-up to a stellar structure and evolution code any time soon. *The only route out of the perennially unpredictive nature of phenomenological models is the RSM.* It solves the basic NSE and scalar equations, it can accommodate diverse physical processes without having to change the rules of the game every time a new process is included, it gives very frequently algebraic results, it can be local and nonlocal, it can be hooked-up to a general code, it has an extensive and successful pedigree for it has been used for several decades to describe turbulent phenomena that exhibit regimes quite similar to those occurring in stars, as we have discussed in the case of cooling- from- above flows. While it is surprising that in stellar structure and evolution studies the RSM are just beginning to be used, on the other hand one can benefit from these advances to treat similar processes in other settings.

Though fully aware of having asked more questions than the answers we have offered, our goal was to show that much advantage can be had once one switches from the plethora of ad hoc phenomenological models to the unified treatment represented by the RSM which is the only available methodology capable of including the great many physical processes that are present in turbulent flows in stars.

Acknowledgements

I would like to thank Miss Angela Cheng and Dr. F. Kupka for their careful reading of the manuscript and for several constructive suggestions and Mrs. M. Depner for help in typesetting the LaTeX version of this manuscript.

References

1. Abarbanel, H.D., Holm, D.D., Marsden, J.E., Ratiu, T.: Phys. Rev. Lett. **52**, 2352 (1984)
2. Alberighi, S., Maurizi, A., Tampieri, F.: J. App. Meteor. **41**, 885 (2002)

3. Althaus, L.G., Benvenuto, O.G.: MNRAS **278**, 981 (1996)
4. Alves, J.O.S.: Ph.D. Thesis, University of Reading, UK (1995)
5. Anderson, J., Nordstrom, B., Clausen, J.V.: ApJ **363**, L33 (1990)
6. André, J.C., De Moor, G., Lacarrère, P., du Vachat, R.: J. Atmos. Sci. **33**, 476 (1976)
7. André, J.C., DeMoor, G., Lacarrèere, P., Therry, G., du Vachat, R.: J. Atmos. Sci. **35**, 1861 (1978)
8. Asplund, M., Ludwig, H.-G., Nordlund, Å., Stein, R.F.: A&A **359**, 669 (2000)
9. Baumert, H. Peters, H.: J.Geophys. Res. **105**, 6453 (2000)
10. Bell, T.H.: J. Fluid Mech. **67**, 705 (1975a)
11. Bell, T.H.: J. Geophys. Res. **80**, 320 (1975b)
12. Blackadar, A.K.: J.Geophys. Res. **67**, 3095 (1962)
13. Bougeault, P.: J. Atmos. Sci. **38**, 2429 (1981)
14. Brummel, N.H., Hurlburt, N.E., Toomre, J.: ApJ **473**, 494 (1996)
15. Brun, A.S., Toomre, J.: ApJ **570**, 865 (2002)
16. Brun, A.S., Miesch, M.S., Toomre, J.: ApJ **614**, 1073 (2004)
17. Burchard, H., Bolding, K.: J. Phys. Oceanogr. **31**, 1943 (2001a)
18. Burchard, H., Deleersnijder, E.: Ocean Modelling **3**, 33 (2001b)
19. Bryan, K.: Tellus **34**, 104 (1991)
20. Canuto, V.M.: ApJ **392**, 218 (1992)
21. Canuto, V.M.: ApJ **428**, 279 (1994)
22. Canuto, V.M.: ApJ **467**, 385 (1996)
23. Canuto, V.M.: ApJ **478**, 322 (1997)
24. Canuto, V.M.: ApJ **524**, 311 (1999)
25. Canuto, V.M., Goldman, I., Hubicky, O.: ApJ Lett. **280**, L55 (1984)
26. Canuto, V.M., Goldman, I.: Phys. Rev. Lett. **54**, 430 (1985)
27. Canuto, V.M., Hartke, G.J.: A&A **168**, 89 (1986)
28. Canuto, V.M., Goldman, I., Chasnov, J.: Phys. Fluids **30**, 339 (1987)
29. Canuto, V.M., Mazzitelli, I.: ApJ **370**, 295 (1991)
30. Canuto, V.M., Minotti, F., Ronchi, C., Ypma, R.M., Zeman, O.: J. Atmos. Sci. **51**, 1605 (1994)
31. Canuto, V.M., Dubovikov, M.S.: Phys. Fluids **9**, 2132 (1997)
32. Canuto, V.M., Christensen-Dalsgaard, J.: Ann Rev. Fluid Mech. **30**, 167 (1998)
33. Canuto, V.M., Dubovikov, M.S.: ApJ **493**, 834 (1998)
34. Canuto, V.M., Cheng, Y., Howard, A.: J. Atmos. Sci. **58**, 1169 (2001)
35. Canuto, V.M., Howard, A., Cheng, Y., Dubovikov, M.S.: J. Phys. Oceanogr. **31** 1413 (2001)
36. Canuto, V.M., Minotti, F.: MNRAS **328**, 829 (2001)
37. Canuto, V.M., Howard, A., Cheng, Y., Dubovikov, M.S.: J. Phys. Oceanogr. **32**, 240 (2002)
38. Canuto, V.M., Howard, A., Hogan, P., Cheng, Y., Dubovikov, M.S., Montenegro, L.M.: Ocean Model. **7**, 75 (2004)
39. Canuto, V.M., Dubovikov, M.S., Cheng, Y.: Geophys. Res. Lett. **32**, L22604 (2005)
40. Canuto, V.M., Cheng, Y., Howard, A.M.: Ocean Model. 16, 28 (2007)
41. Canuto, V.M., Howard, A.M., Muller, C.J., Cheng, Y., Jayne, Ocean Mod. (2008)
42. Cattaneo, F., Brummell, N.H., Toomre, J., Malagoli, A. Hurlburt, N.E.: ApJ **370**, 282 (1991)

43. Chaboyer, B., Zahn, J.-P.: A&A **253**, 173 (1992)
44. Chan, K.L., Gigas, D.: ApJ **389**, L87 (1992)
45. Chapman, S.: ApJ **120**, 151 (1954)
46. Charbonnel, C.: Nature **442**, 636 (2006)
47. Chasnov, J.R.: The Phys. Fluids **7**, 600 (1995)
48. Chasnov, J.R.: J. Fluid Mech. **342**, 335 (1997a)
49. Chasnov, J.R.: The Phys. Fluids **9**, 171 (1997b)
50. Chen, C., Cotton, W.R.: Bound.-Layer Met. **25**, 375 (1983)
51. Cheng, Y., Canuto, V.M., Howard, A.M.: J. Atmos. Sci. **62**, 2189 (2005)
52. Chou, P.Y.: Quart. J. Applied Math., **3**, 38 (1945) **52**
53. Chu, P.C., Gascard, J.C.: Deep Convection and Deep Water Formation in the Oceans, pp. 397. Elsevier, New York (1991)
54. Cox, J.P. Giuli, R.T.: Principles of Stellar Structure, Gordon and Breach, New York (1968)
55. D'Antona, F., Mazzitelli, I.: ApJ **470**, 1093 (1996)
56. Davidov, B.I.: Dokl. Akad. Nauk SSSR **136**, 472 (1961)
57. Deardorff, J.W.: J. Geophys. Res. **77**, 5900 (1972)
58. Deardorff, J.W.: Bound.-Layer Met. **18**, 495 (1980)
59. Denbo, D.W., Skyllingstad, E.D.: J. Geophys. Res. **101**, 1095 (1996)
60. Egbert, G.D., Ray, R.D.: Nature, **405**, 775 (2000)
61. Eggleton, P.P.: MNRAS **151**, 351(1971)
62. Eggleton, P.P.: MNRAS, **156**, 361 (1972)
63. Elliott, J.R., Miesch, M.S., Toomre, J.: ApJ **533**, 546 (2000)
64. Ellison, T.H., Turner, J.S.: J. Fluid Mech. **6**, 423 (1959)
65. Finnigan, T.D., Luther, D.S., Lukas, R.: J. Phys. Oceanogr. **32**, 2899 (2002)
66. Galperin, B., Kantha, L.H., Hassid, S., Rosati, A.: J. Atmos. Sci. **45**, 55 (1988)
67. Garret, C.J.R., Munk, W.H.: J. Geophys. Res. **80**, 291 (1975)
68. G.S. Golystin: Dokl. Akad. Nauk SSSR 251, 1356 (1980)
69. Gough, D.O., Weiss, N.O.: MNRAS **176**, 589 (1976)
70. Gregg, M.C.: J. Geophys. Res. **94**, 9689 (1989)
71. Gregg, M.C., Winkel, D.P., Sanford, T.S., Peters, H.: Dyn. Atm. Oceans **24**, 1 (1996)
72. Grossman, S.A., Taam, R.E.: MNRAS **283**, 1165 (1996)
73. Gryanik, V.M., Hartmann, J.: J. Atmos. Sci. **59**, 2729 (2002)
74. Hamba, F.: J. Atmos. Sci. **52**, 1084 (1995)
75. Hartmann, J., et al.: Reports on Polar Research. Alfred Wegener Institute for Polar and Marine Research, Bremerhaven, **305**, 81 (1999)
76. Holtslag, A.A.M., Boville, B.A.: J. Climate **6**, 1825 (1993)
77. Holtslag, A.A.M., Moeng, C-H.: J. Atmos. Sci. **48**, 1690 (1991)
78. Howard, L.N.: J. Fluid Mech. **10**, 509 (1961)
79. Jayne, S.R., St. Laurent, L.C.: Geophys. Res. Lett. **28**, 811 (2001)
80. Kantha, L.H.: Nonlinear Processes in Geophys. **11**, 83 (2004)
81. Kelley, D.: J. Geophys. Res. **89**, 10484 (1984)
82. Kelley, D.: J. Geophys. Res. **95**, 3365 (1990)
83. Khatiwala, S.: Deep Sea Res. **50**, 3 (2003)
84. Killworth, P.D.: Rev. Geophys. Space Phys. **21**, 1 (1983)
85. Kippenhahn, R., Ruschenplatt, G., Thomas, H.C.: A&A **91**, 175 (1980)
86. Korn, A.J., Grundahl, F., Richard, O., Barklem, P.S., Mashonkina, L., Collet, R., Piskunov, N., Gustafsson, B.: Nature **442**, 657 (2006)

87. Kumar, P., Talon, S., Zahn, J.-P.: ApJ **520**, 859 (1999)
88. Kunze, E., Sanford, T.S.: J. Phys. Ocean **26**, 2286 (1996)
89. Kupka, F.: ApJL **526**, L45 (1999)
90. Kupka, F., Montgomery, M.H.: MNRAS **330**, L6 (2002)
91. Kupka, F., Muthsam, H.J.: In: IAU Symosium 239, Kupka, F., Roxburgh, I.W., Chan, K.L., (eds.), p. 86 Cambridge University Press, Cambridge (2007)
92. Langer, N., Sugimoto, D., Fricke, K.: A&A **126**, 207 (1983)
93. Langer, N., El Eid, M.F., Fricke, K.: A&A **145**, 179 (1985)
94. Langer, N., El Eid, M.F., Baraffe, I.: A&A **224**, L17 (1989)
95. Lappen, C.L., Randall, D.: J. Atmos. Sci. **58**, 2021 (2001)
96. Lavender, K.L., Davis, R.E., Owens, W.B.: J. Phys. Ocean. **32**, 511 (2002)
97. Ledwell, J.R., Watson, A.J., Law, C.S.: Nature **364**, 701 (1993)
98. Ledwell, J.R., Watson, A.J., Law, C.S.: J. Geophy. Res. **103**, 21499 (1998)
99. Ledwell, J.R., Montgomery, E.T., Polzin, K.L., St. Laurent, L.C., Schmitt, R.W., Toole, J.M.: Nature **403**, 179 (2000)
100. Ledoux, P.: ApJ **105**, 305 (1947)
101. Llewellyn Smith, S.G., Young, W.R.: J. Phys. Ocean. **32**, 1554 (2002)
102. Lenschow, D.H., Mann, J., Kristensen, L.: J. Atmos. Oc. Technol. **11**, 661 (1994)
103. Lenschow, D.H., Wulfmeyer, V., Senff, C.: J. Atmos. Oc. Technol. **17**, 1330 (2000)
104. Leslie, D.C.: Development in the Theory of Turbulence, Clarendon Press, Oxford (1973)
105. Maeder, A.: A&A **299**, 84 (1995)
106. Maeder, A., Meynet, G.: A&A **313**, 140 (1996)
107. Marshall, J., Schott, F.: Rev. Geophys. **37**, 1 (1999)
108. Marmorino, G.O., Caldwell, D.R.: Deep Sea Res. **23**, 59 (1976)
109. Martin, P.J.: J. Geophys. Res. **90**, 903 (1985)
110. Maxworthy, T.: Ann. Rev. Fluid Dyn. **29**, 327 (1997)
111. McComb, W.D.: The Physics of Fluid Turbulence, Clarendon Press, Oxford (1990)
112. Merryfield, W.J.: ApJ **444**, 318 (1995)
113. Miesch, M.S., Elliott, J.R., Toomre, J., Clune, T.L., Glatzmeier, G.A., Gilman, P.A.: ApJ **532**, 593 (2000)
114. Miles, J.W.: J. Fluid Mech. **10**, 496 (1961)
115. Mironov, D.V., Gryanik, V.M., Moeng, C.-H., Olbers, D.J., Warncke, T.H.: Quart. J. Roy. Meteor. Soc. **126**, 477 (2000)
116. Moeng, C.-H., Randall, D.A.: J. Atmos. Sci. **41**, 1588 (1984)
117. Montgomery, M.H., Kupka, F.: MNRAS **350**, 267 (2004)
118. Morris, M.Y., Hall, M.M., St. Laurent, L.C. Hogg, N.G.: J. Phys. Oceanogr. **31**, 3331 (2001)
119. Morton, B.R., Taylor, G.I., Turner, J.S.: Proc. Roy. Soc. London A **234**, 1 (1956)
120. Moum, J.N., Osborne, T.R.: J. Phys. Ocean., **16**, 125 (1986)
121. Munk, W., Wunsch, C.: Deep Sea Res. I. **45**, 1977 (1998)
122. Naveira Garabato, A., Polzin, K.L., King, B.A., Haywood, K.J., Visbeck, M.: Science **303**, 210 (2004)
123. O'Brien, E.E., Francis, G.C.: J. Fluid Mech. **13**, 369 (1962)
124. Ogura, Y.: J. Geophys. Res. **67**, 3143 (1962)

125. Paluszkiewicz, T., Garwood, R.W., Denbo, D.W.: Oceanography **7**, 37 (1994)
126. Paluszkiewicz, T., Romea, R.D.: Dyn. Atmos. Oceans **26**, 95 (1997)
127. Pinsonneault, M.H., Kawaler, S.D., Sofia, S., Demarque, P.: ApJ **338**, 424 (1989)
128. Polzin, K.: J. Phys. Ocean. **26**, 1409 (1996)
129. Polzin, K., Toole, J.M., Schmitt, R.W.: J. Phys. Ocean. **25**, 306 (1995)
130. Polzin, K., Oakey, N.S., Toole, J.M., Schmitt, R.W.: J. Geophys. Res. **101**, 14111 (1996)
131. Price, J.F., Yang, J.: In: Chassignet, E.P., Verron, J. (eds.) Ocean Modeling and Parameterizations, pp. 55. Kluwer, Dordrecht (1998)
132. Robinson, F.J., Demarque, P., Li, L.H., Sofia, S., Kim, Y.-C., Chan, K.L., Guenther, D.B.: MNRAS 340, **923** (2003)
133. Rosenthal, C.S., Christensen-Dalsgaard, J., Nordlund, Å. Stein, R.F., Trampedach, R.: A&A, **351**, 689 (1999)
134. Roxburgh, I.: A&A **65**, 281 (1978)
135. Roxburgh, I.: A&A **211**, 361 (1989)
136. Rubinstein, R., Zhou, Y.: Phys. Fluids **8**(11), 3172 (1996)
137. Rudiger, G.: Differential Rotation and Stellar Convection, Gordon and Breach, New York (1989)
138. Saenko, O.A., Merryfield, W.J.: J. Phys. Ocean. **35**, 826 (2005)
139. Sakashita, S., Hayashi, C.: Prog. Theor. Phys. **22**, 830 (1959)
140. Salasnich, B., Bressan, A., Chiosi, C.: A&A **342**, 131 (1998)
141. Samadi, R., Kupka, F., Goupil, M.-J., Lebreton, Y., van't Veer-Menneret C.: A&A 445, **233** (2006)
142. Sander, J., Wolf-Gladrow, D., Olbers, D.: J. Geophy. Res. **100**, 20579 (1995)
143. Schatzman, E., Zahn, J.-P., Morel, P.: A&A **364**, 876 (2000)
144. Schiestel, R.: Phys. Fluids **30** (3), 722 (1987)
145. Schmitt, R.W.: Ann. Rev. Fluid Mech. **26**, 255 (1994)
146. Schwarzschild, M., Harm, R.: ApJ **128**, 348 (1958)
147. Skyllingstad, E.D., Paluszkiewicz, T., Denbo, D.W., Smyth, W.D.: Physica D **98**, p. ?!? (1996)
148. Simmons, H.L., Jayne, S.R., St. Laurent, L.C., Weaver, A.J.: Ocean Model. **6**, 245 (2004)
149. St. Laurent, L.C., Schmitt, R.W.: J. Phys. Oceanogr. **29**, 1404 (1999)
150. St. Laurent, L.C., Garrett, C.: J. Phys. Oceanogr. **32**, 2882 (2002)
151. St. Laurent, L.C., Simmons, H.: J. Climate **19**, 4877 (2006)
152. Sokolov, A., Stone, P.H.: Climate Dynamics **14**, 291 (1998)
153. Spiegel, E.A.: Ann. Rev. Astron. and Astroph. **10**, 261 (1972)
154. Spiegel, E.A., Zahn, J.-P.: A&A **265**, 106 (1992)
155. Spruit, H.C.: A&A **253**, 131 (1992)
156. Stein, R.F., Nordlund, Å.: ApJ **499**, 914 (1998)
157. Stevens, B., et al. (16 co-authors): Monthly Weather Rev. **133**, 1443 (2005)
158. Stevenson, D.J.: MNRAS **187**, 129 (1979)
159. Stothers, R.: MNRAS **151**, 65 (1970)
160. Stothers, R., Simon, N.R.: ApJ **157**, 673 (1969)
161. Stothers, R., Chin, C.W.: ApJ **198**, 407 (1975)
162. Stothers, R., Chin, C.W.: ApJ **204**, 472 (1976)
163. Talon, S., Zahn, J.-P.: A&A **317**, 749 (1997)
164. Tatsumi, T.: Proc. Roy. Soc. London **A239**, 16 (1957)

165. The Labrador Sea Deep Convection experiment, J. Phys. Oceanogr. (February 2002)
166. Thomas, H.-C.: Z. für Astrophysik **67**, 420 (1967)
167. Tomczyk, S., Schou, J., Thompson, M.J.: ApJL **448**, L57 (1995)
168. Toole, J.M., Schmitt, R.W., Polzin, K.L., Kunze, E.: J. Geophys. Res. **102** C1, 947 (1997)
169. Toole, J.M.: In: Chassignet, E.P., Verron, J. (eds.) Ocean Modeling and Parameterization, NATO Science Series, Vol. 516, pp. 171–190. Kluwer Academic Publisher, Dordrecht (1998)
170. Turner, J.S.: Buoyancy Effects in Fluids, Cambridge University Press, Cambridge (1973)
171. Turner, J.S.: Ann. Rev. Fluid Mech. **17**, 11 (1985)
172. Turner, J.S., Stommel, H.: Proc. Nat. Acad. Sci. **52**, 49 (1964)
173. Ulrich, R.K.: ApJ **172**, 165 (1972)
174. Umezu, M.: MNRAS **298**, 193 (1998)
175. Umlauf, L., Burchard, H.: Contin. Shelf Res. **25**, 795 (2005)
176. Wahlin, A.K., Cenedese, C.: Deep Sea Res. **53**(1–2), 157 (2006)
177. Wang, D., Large, W.G., McWillims, J.C.: J. Geophy. Res. **101**, 3649 (1996)
178. Weaver, A.J., Bitz, C.M., Fanning, A.F., Holland, M.M.: Ann. Rev. Earth Planet. Sci. **27**, 231 (1999)
179. Wedemeyer, S., Freytag, B., Steffen, M., Ludwig, H.-G., Holweger, H.: A&A **414**, 1121 (2004)
180. Woods, J.D.: Radio Sci. **4**, 1289 (1969)
181. Woosley, S.E., Heger, A., Weaver, T.A., Langer, N.: In SN 1987A: Ten Years After, ed. M.M. Phillips & N.B. (1999)
182. Wyngaard, J.C., Weil, J.C.: Phys. Fluids **A3**, 155 (1991)
183. Xiong, D.R.: Sci. Sinica **28**, 764 (1985a)
184. Xiong, D.R.: A&A **150**, 133 (1985b)
185. Xiong, D.R.: A&A **167**, 239 (1986)
186. Zahn, J.-P.: In: Ledoux, P., Noels, A., Rodgers, W. (eds.) Stellar Instability and Evolution, IAU Symp., Vol. 59, p. 185. Reidell, Dordrecht (1974)
187. Zahn, J.-P.: A&A **265**, 115 (1992)
188. Zeman, O., Lumley, J.L.: J. Atmos. Sci. **33**, 1974 (1976)

Turbulence in the Lower Troposphere: Second-Order Closure and Mass–Flux Modelling Frameworks

D.V. Mironov

Deutscher Wetterdienst, Forschung und Entwicklung, FE14, Kaiserleistr. 29/35, D-63067 Offenbach am Main, Germany
dmitrii.mironov@dwd.de

1 Introduction

Turbulence is ubiquitous in the earth's atmosphere. Its spatial scales range from many tens of kilometres to a few millimetres. A wide assortment of turbulent eddies of different size and shape lies within this range. These include breeze circulations, storms, clouds, plumes and rolls in the planetary boundary layer (PBL), and eddies in urban street canyons and in plant canopies, to mention a few. It is this large variety of turbulent motions that numerical weather prediction (NWP) models, as well as other numerical models of the atmosphere, have to deal with.

Atmospheric turbulence is a notoriously difficult and extensive subject. Even a cursory examination of its most important aspects would require a voluminous account that goes far beyond the scope of the present chapter. We restrict our consideration to modelling turbulence in the lower troposphere as it is practised nowadays in NWP and related applications, e.g. climate modelling and air pollution dispersion studies. Turbulence modelling in this context means the representation of the effect of turbulent motions, which are not explicitly computed by a numerical model of the atmosphere, on the explicitly computed fields. Other topics, such as measurements and numerical simulations (large-eddy and direct numerical simulations) of turbulence in the atmosphere, although very important, are not considered here. Readers are referred to review articles and books [60, 61, 66–68, 99, 119, 131, 159, 183, 208], where further references can be found.

Before proceeding any further, we recall a plain point that a numerical model of the atmosphere solves the evolution equations in the form

$$\frac{\partial \langle f \rangle}{\partial t} + \langle u_i \rangle \frac{\partial \langle f \rangle}{\partial x_i} = -\frac{\partial \langle u_i' f' \rangle}{\partial x_i} + F_f, \qquad (1)$$

Mironov, D.V.: *Turbulence in the Lower Troposphere: Second-Order Closure and Mass–Flux Modelling Frameworks.* Lect. Notes Phys. **756**, 161–221 (2009)
DOI 10.1007/978-3-540-78961-1_5   © Springer-Verlag Berlin Heidelberg 2009

where t is the time, x_i are the space co-ordinates, and u_i are the velocity components. The Einstein summation convention for repeated indices is adopted. A generic variable f refers to any quantity treated by an atmospheric model, and F_f symbolizes the source of f due to various processes, such as radiation and precipitation. The angle brackets denote the quantities that are explicitly computed (resolved) by a numerical model, and primes denote deviations therefrom. The incompressibility is assumed which is a fairly accurate approximation for the lower troposphere. Equation (1) is obtained by applying a spatial filter to the governing momentum and scalar equations (see the chapter "Turbulent Combustion in Thermonuclear Supernovae" of this volume). The quantity $\langle u_i' f' \rangle$ represents the flux of f due to subfilter scale motions.[1] In what follows, the subfilter scale quantities will be referred to, perhaps somewhat loosely, as the sub-grid scale (SGS) quantities, considering that the latter term has been universally accepted. As the SGS motions are not explicitly computed by a numerical model, the SGS flux must be modelled, or parameterized, in terms of resolved quantities. The terms "model" and "parameterization scheme" may be used interchangeably in this context. The term "parameterization scheme" is more often used in the NWP community.

The SGS flux divergence term on the right-hand side (r.h.s.) of Eq. (1) should in principle represent the effect of all SGS motions down to the smallest scales where turbulence kinetic energy (TKE) eventually dissipates. It is, however, customary for NWP models to split this term into contributions due to various processes. The contributions due to "turbulence", that is thought to represent quasi-random small-scale motions, and due to "convection", that is thought to represent quasi-organized motions of larger scales, are usually distinguished. That is,

$$\langle u_i' f' \rangle = \langle u_i' f' \rangle_{\text{turb}} + \langle u_i' f' \rangle_{\text{conv}} . \tag{2}$$

The SGS momentum flux may also contain a contribution due to the orographic drag, i.e. the unresolved drag caused by the effects of sub-grid scale orography, such as the absorption and reflection of orographically induced gravity waves. The orographic drag parameterization issues will not be considered in what follows. Readers are referred to [23, 36, 129, 139, 185], where further references can be found. Although the decomposition (2) is commonly accepted, it is not as innocent as it might seem. Caution is required to ensure that the sum of the above two contributions actually represents the total SGS flux. Otherwise, serious problems may be encountered, for example, double-counting of some energetically relevant modes of SGS motions, or their loss.

It is general practice in NWP and related applications to model the two contributions on the r.h.s. of Eq. (2) in different ways. Turbulence

---

[1] Strictly speaking, the Lilly [123] notation with no primes should be used to emphasize that the filter does not necessarily satisfy the Reynolds averaging assumptions. The subfilter scale flux is then given by $\langle u_i f \rangle - \langle u_i \rangle \langle f \rangle$. See [109, 183] for further discussion.

parameterization schemes are usually developed on the basis of the ensemble-mean second-order turbulence closure approach. Convection parameterization schemes are usually developed on the basis of the mass–flux approach. In the subsequent text (Sects. 2 and 3), we will consider these parameterization schemes in some detail. We attempt to show that the two approaches have much in common, although differences remain and they may be important.

Convection in the atmosphere manifests itself in many different forms. Dry convection is usually driven by the surface buoyancy flux and is confined to the PBL. Regimes of moist convection are many and varied. As nicely stated by Stevens [187] "moist convection is many, rather than one thing". This author presents a comprehensive account of many essential features of moist atmospheric convection, including phenomenology and theoretical frameworks to describe its major regimes, namely, stratocumulus, shallow non-precipitating cumulus, and deep precipitating cumulus convection. Driven by a powerful engine – the latent heat release due to water vapour condensation in rising air parcels, moist convection can penetrate all the way up to the top of the troposphere. Such deep penetrative convection is typically associated with heavy precipitation. With their horizontal grid size of about 40 km or so, the present-day global NWP models are unable to resolve deep convection. A parameterization is required to describe convective fluxes of scalar quantities and of momentum, convection source terms due to condensation/evaporation, due to release/consumption of latent heat and due to precipitation fall-out, as well as a sophisticated interplay of convective, radiative and microphysical processes. Improving deep convection parameterizations represents one of the major challenges in NWP and related applications [12]. Details of this fascinating phenomenon are discussed in [60, 61, 91, 179, 187]. The present-day limited-area NWP models have a horizontal grid size of 10 to 5 km, and there is a strong tendency to achieve an even higher resolution with a horizontal grid size of order 1 km. With such a grid size, deep convection will likely be computed explicitly. However, the PBL turbulence and shallow convection will still remain at the sub-grid scale and will require an adequate parameterization scheme. Some theoretical problems related to parameterizing SGS motions in high-resolution atmospheric models are discussed in [210]. Notice that an increased resolution will alleviate some parameterization issues but will make other issues even more complicated. One example is the atmospheric radiation. High-resolution NWP models will need to account for three-dimensional effects of radiation transfer, inevitably making the radiation parameterization schemes computationally very expensive.

Noteworthy are a number of issues that complicate the development of physical parameterizations for NWP models, of turbulence–convection parameterizations in particular. Since NWP models are used operationally, their quality is judged by the quality of the final product – the weather forecast. The implementation of any innovation into an NWP model can only be justified if a new version of the NWP model that incorporates the innovation beats an older version of the model in terms of the forecast quality. This is by no means guaranteed. The NWP models are very complex non-linear systems where a very sophisticated interplay of their various components takes place. Apart from the physics and numerics, initialization and data assimilation should also be borne in mind. Every component of an NWP model has its own deficiencies. Working together, they produce errors that may amplify (a very unwanted situation), but may partially compensate each other as well. Then, the incorporation of a new physical parameterization scheme does not necessarily lead to an improved forecast, no matter how advanced and physically sound that new scheme is as compared to an old scheme, i.e. to the scheme currently used in the operational version of a given NWP model. Putting it differently, it is not sufficient to prove that a new parameterization scheme is superior to an old one in that it is more physically sound and performs well as a stand-alone physical model, e.g. as judged by a comparison of results from idealized one-dimensional experiments with observational and numerical data. It should also be ensured that a new parameterization scheme works in harmony with the other components of a particular NWP model.

One more aspect of great significance is that in order to be useful a turbulence-convection parameterization for NWP should be computationally efficient. Since the NWP products must be delivered to end users in due time, it is simply not possible to apply parameterizations whose high computational cost may lead to a forecast delay. There are many physically sound turbulence–convection models which proved to be very useful research tools. However, there is no way to use them in operational NWP models for they are computationally prohibitively expensive. On the other hand, a useful parameterization scheme should account for much of the essential physics of atmospheric turbulence and convection. The key to the success in developing turbulence–convection parameterization schemes for NWP and related applications is, therefore, to find the best possible compromise between physical realism and computational economy.

Recent advances in observations and numerical simulations of atmospheric flows along with new theoretical ideas have led to considerable progress in representing turbulence and convection in NWP models. The progress, however, is somewhat slower than one might wish. Given the severe constraints mentioned above, it seems likely that comparatively simple second-order closure schemes and mass–flux schemes will be further used in NWP models for some years to come to parameterize turbulence and convection respectively. The question then arises whether more regime-dependent sub-models (parameterization schemes) should be developed, or some unification of various parameterization frameworks is possible (see discussions in [12, 145, 187]). A definitive answer to this question does not seem to exist at present. Some attempts have already been made to achieve a more unified description of several types of fluctuating motions (see Sect. 4). It is also the author's opinion that a more unified description is desirable. There are several ways to do so, however, and it is not a priory clear which way should be preferred.

In the next section, we outline the ensemble-mean second-order modelling framework and briefly discuss parameterization assumptions that should be made in order to arrive at a reasonably simple turbulence closure. As we shall see subsequently, only a small fraction of what is presented in Sects. 2.1 and 2.2 is actually used in applications. A rather extended discussion is necessary, however, in order to understand how simplified parameterisation schemes are obtained and what is lost on the way (Sect. 2.3). In particular, comprehending the role of the third-order moments in maintaining the second-order moment budgets is the key to understanding how non-local transport properties of convective motions can be accounted for within the second-order modelling framework. A consideration of parameterizations of the pressure redistribution terms is required, among other things, to understand how the down-gradient diffusion approximation for fluxes are derived from the second-moment budget equations. Furthermore, a systematic consideration of the second-order modelling framework demonstrates the limits of applicability of simplified turbulence closures. The mass–flux parameterization schemes are outlined in Sect. 3. In Sect. 3.3, we explore analogies between the ensemble–mean and the mass–flux approaches. This exercise helps to elucidate the essential physics behind various parameterization assumptions and suggests possible ways towards their improvement. It also shows that the two approaches have much in common and suggests how the mass–flux parameterization ideas can be translated into the language of the ensemble–mean second-order closures and vice versa. In Sect. 4, we discuss the steps that should be made towards an improved description of turbulence and shallow non-precipitating convection within a unified parameterization framework. Conclusions are presented in Sect. 5. The discussion below inevitably reflects the author's personal experience, and, to some (hopefully minor) extent, his preferences. The author apologises for omissions that are unavoidable in any effort to address such an extensive and difficult subject as atmospheric turbulence.

2 Turbulence Parameterisation Schemes

2.1 Governing Equations

The basis for the development of turbulence parameterization schemes is the set of transport equations for the second-order turbulence moments (see e.g. [152]). Those equations are derived using the Reynolds averaging and are thought to describe the ensemble–mean statistical moments of fluctuating fields. As already noted, a filter applied to the governing momentum and scalar equations does not generally coincide with the Reynolds averaging. Hence, the sub-grid scale (sub-filter scale) moments of fluctuating fields do not coincide with the ensemble–mean moments. It is, however, assumed (often tacitly) that the two sets of moments are not too different. The ensemble–mean second-order closure approach is commonly taken to parameterize the SGS fluxes and variances as if these were truly ensemble–mean quantities. Caution should be exercised since the validity of this assumption deteriorates as the resolution of atmospheric models is refined.

For the sake of clarity, we first consider the case of the unsaturated atmosphere, treating the atmospheric air as a two-component medium characterized by the two thermodynamic variables, viz., potential temperature θ and specific humidity q. The real atmosphere is of course more complicated as it also contains water in liquid and solid forms. Some issues related to modelling turbulence in the cloudy atmosphere are briefly discussed in Sect. 2.5.

The set of governing equations, hereafter referred to as the ensemble–mean equations, consists of the transport equations for the Reynolds stress $\langle u_i' u_j' \rangle$, for the scalar fluxes $\langle u_i' \theta' \rangle$ and $\langle u_i' q' \rangle$, for the scalar variances $\langle \theta'^2 \rangle$ and $\langle q'^2 \rangle$, and for the temperature–humidity covariance $\langle \theta' q' \rangle$. Using the Boussinesq approximation and assuming that the Reynolds number is sufficiently high to neglect the molecular diffusion terms in the second-moment budget equations (a good approximation for the majority of atmospheric flows), they read

$$
\left(\frac{\partial}{\partial t} + \langle u_k \rangle \frac{\partial}{\partial x_k} \right) \langle u_i' u_j' \rangle = - \left(\langle u_i' u_k' \rangle \frac{\partial \langle u_j \rangle}{\partial x_k} + \langle u_j' u_k' \rangle \frac{\partial \langle u_i \rangle}{\partial x_k} \right)
$$
$$
- \left(\beta_i \langle u_j' \theta_v' \rangle + \beta_j \langle u_i' \theta_v' \rangle \right)
$$
$$
- 2 \left(\epsilon_{ilk} \Omega_l \langle u_k' u_j' \rangle + \epsilon_{jlk} \Omega_l \langle u_k' u_i' \rangle \right)
$$
$$
- \left(\left\langle u_i' \frac{\partial p'}{\partial x_j} \right\rangle + \left\langle u_j' \frac{\partial p'}{\partial x_i} \right\rangle - \frac{2}{3} \delta_{ij} \frac{\partial}{\partial x_k} \langle u_k' p' \rangle \right)
$$
$$
- \frac{\partial}{\partial x_k} \left(\langle u_k' u_i' u_j' \rangle + \frac{2}{3} \delta_{ij} \langle u_k' p' \rangle \right) - \epsilon_{ij}, \qquad (3)
$$

$$
\left(\frac{\partial}{\partial t} + \langle u_k \rangle \frac{\partial}{\partial x_k} \right) \langle u_i' \theta' \rangle = - \langle u_k' \theta' \rangle \frac{\partial \langle u_i \rangle}{\partial x_k} - \langle u_i' u_k' \rangle \frac{\partial \langle \theta \rangle}{\partial x_k}
$$
$$
- \beta_i \langle \theta' \theta_v' \rangle - 2 \epsilon_{ijk} \Omega_j \langle u_k' \theta' \rangle - \left\langle \theta' \frac{\partial p'}{\partial x_i} \right\rangle
$$
$$
- \frac{\partial}{\partial x_k} \langle u_k' u_i' \theta' \rangle - \epsilon_{i\theta}, \qquad (4)
$$

$$
\left(\frac{\partial}{\partial t} + \langle u_k \rangle \frac{\partial}{\partial x_k} \right) \langle u_i' q' \rangle = - \langle u_k' q' \rangle \frac{\partial \langle u_i \rangle}{\partial x_k} - \langle u_i' u_k' \rangle \frac{\partial \langle q \rangle}{\partial x_k}
$$
$$
- \beta_i \langle q' \theta_v' \rangle - 2 \epsilon_{ijk} \Omega_j \langle u_k' q' \rangle - \left\langle q' \frac{\partial p'}{\partial x_i} \right\rangle
$$
$$
- \frac{\partial}{\partial x_k} \langle u_k' u_i' q' \rangle - \epsilon_{iq}, \qquad (5)
$$

$$
\frac{1}{2} \left(\frac{\partial}{\partial t} + \langle u_k \rangle \frac{\partial}{\partial x_k} \right) \langle \theta'^2 \rangle = - \langle u_k' \theta' \rangle \frac{\partial \langle \theta \rangle}{\partial x_k} - \frac{1}{2} \frac{\partial}{\partial x_k} \langle u_k' \theta'^2 \rangle - \epsilon_\theta, \qquad (6)
$$

$$\frac{1}{2}\left(\frac{\partial}{\partial t} + \langle u_k \rangle \frac{\partial}{\partial x_k}\right)\langle q'^2 \rangle = -\langle u_k' q' \rangle \frac{\partial \langle q \rangle}{\partial x_k} - \frac{1}{2}\frac{\partial}{\partial x_k}\langle u_k' q'^2 \rangle - \epsilon_q, \quad (7)$$

$$\left(\frac{\partial}{\partial t} + \langle u_k \rangle \frac{\partial}{\partial x_k}\right)\langle \theta' q' \rangle = -\langle u_k' \theta' \rangle \frac{\partial \langle q \rangle}{\partial x_k} - \langle u_k' q' \rangle \frac{\partial \langle \theta \rangle}{\partial x_k}$$

$$- \frac{\partial}{\partial x_k}\langle u_k' \theta' q' \rangle - \epsilon_{\theta q}. \quad (8)$$

Here, $\theta_v = \theta[1 + (R_v/R_d - 1)q] \approx \theta(1 + 0.608q)$ is the virtual potential temperature, R_v and R_d are the gas constants for water vapour and for dry air, respectively, $\beta_i = g_i/\theta_r$ is the buoyancy parameter, g_i is the acceleration due to gravity, θ_r is the reference value of temperature, Ω_i is the angular velocity of the earth's rotation, and p is the kinematic pressure (deviation of pressure from the hydrostatically balanced pressure divided by the reference density ρ_r). The dissipation rates of various quantities are denoted by ϵ_{ij}, $\epsilon_{i\theta}$, ϵ_{iq}, ϵ_θ, ϵ_q, and $\epsilon_{\theta q}$.

Taking the trace of Eq. (3) yields the budget equation for the TKE $e = \frac{1}{2}\langle u_i'^2 \rangle$,

$$\frac{1}{2}\left(\frac{\partial}{\partial t} + \langle u_k \rangle \frac{\partial}{\partial x_k}\right)\langle u_i'^2 \rangle = -\langle u_i' u_k' \rangle \frac{\partial \langle u_i \rangle}{\partial x_k} - \beta_i \langle u_i' \theta_v' \rangle$$

$$- \frac{\partial}{\partial x_k}\left(\frac{1}{2}\langle u_k' u_i'^2 \rangle + \langle u_k' p' \rangle\right) - \epsilon. \quad (9)$$

where $\epsilon = \frac{1}{2}\epsilon_{ii}$ is the TKE dissipation rate.

It should be noted that the decomposition of the pressure gradient–velocity correlation $(\langle u_j' \partial p'/\partial x_i \rangle + \langle u_i' \partial p'/\partial x_j \rangle)$ that appears in the Reynolds stress equation (3) is not unique. Along with a more traditional decomposition into pressure–strain and pressure–diffusion, the decomposition into deviatoric and isotropic parts has also been advocated (e.g. [184]). Keeping in mind that the issue is not resolved, the former decomposition is adopted here.

2.2 Closure Assumptions

The second-order equations (3, 4, 5, 6, 7, 8 and 9) are not closed as they contain a number of unknown quantities. There are three groups of unknowns, namely, the pressure–velocity and pressure–scalar covariances, the third-order velocity–velocity and velocity–scalar covariances, and the dissipation rates of the second-order moments. Parameterizations (closure assumptions) are required for these quantities to express them in terms of the first-order and the second-order moments involved and thereby close the system of governing equations. A large number of parameterizations have been developed to date that vary greatly in terms of their complexity and field of application. These parameterizations are the subject of an extremely voluminous

literature. General scope reviews are given in e.g. [85, 120, 130, 160, 194].
Reviews of second-order closures for geophysical applications are given in
[16, 37, 61, 99, 142, 154, 169, 185, 202], among others. In this section, we
consider parameterizations typically utilized in the second-order modelling of
geophysical flows. We briefly discuss their advantages and limitations, empha-
sizing their utility for modelling turbulence in the lower troposphere.

One remark is in order. Plausible parameterizations used in the second-
order equations should satisfy a number of physical and mathematical requirements. Apart from the requirements of proper physical dimensions, tensor
invariance and symmetry, the so-called realizability requirements should also
be met. The concept of realizability [57, 130, 170] states that the Schwarz' inequalities for all turbulence moments must always be satisfied. For the second-
order moments, this means that the velocity variances and the scalar variances
must always be non-negative and that the magnitude of the correlation coef-
ficient $\langle a'b' \rangle / (\langle a'^2 \rangle \langle b'^2 \rangle)^{1/2}$ between any two fluctuating quantities a and b
must not exceed 1. Intrinsically realisable models do not generate, by virtue of
their construction, physically impossible results. Notice that the realizability
constraints not only help to develop turbulence models that possess desired
physical and mathematical properties, but they are also useful in that they
provide additional relations between the model coefficients, thus reducing the
number of undetermined coefficients to be evaluated (tuned) on the basis of
empirical and/or numerical data.

Pressure Terms

Rotta [168] proposed a return-to-isotropy parameterization for the pressure–
velocity gradient covariance in turbulent shear flows. That parameterization
states that the rate of return of turbulence to isotropy is proportional to the
degree of anisotropy and inversely proportional to a certain time scale called
"return-to-isotropy" time scale. Then, the pressure redistribution term, the
fourth term on the r.h.s. of Eq. (3) that we denote by Π_{ij}, is given by

$$\left(\left\langle u_i' \frac{\partial p'}{\partial x_j} \right\rangle + \left\langle u_j' \frac{\partial p'}{\partial x_i} \right\rangle - \frac{2}{3} \delta_{ij} \frac{\partial}{\partial x_k} \langle u_k' p' \rangle \right) \equiv$$

$$\Pi_{ij} = \frac{\langle u_i' u_j' \rangle - \frac{1}{3} \delta_{ij} \langle u_k' u_k' \rangle}{\tau_{u*}}, \quad (10)$$

where τ_{u*} is the return-to-isotropy time scale. The return-to-isotropy formulation was extended to the pressure gradient–scalar covariance, assuming that
the rate of destruction of turbulent scalar flux is related to the flux in question through a certain relaxation time scale. By way of example, consider the
formulation for the potential temperature. Other scalars can be treated in
much the same way. Denoting the pressure gradient–temperature covariance,
the fifth term on the r.h.s. of Eq. (4), by $\Pi_{\theta i}$, we get

$$\left\langle \theta' \frac{\partial p'}{\partial x_i} \right\rangle \equiv \Pi_{\theta i} = \frac{\langle u_i' \theta' \rangle}{\tau_{\theta *}}, \tag{11}$$

where $\tau_{\theta *}$ is the relaxation "return-to-isotropy" time scale for potential temperature.

Applying return-to-isotropy parameterizations to the entire pressure terms, one lumps all the uncertainties on the relaxation time scales. A more common approach nowadays is to decompose Π_{ij} and $\Pi_{\theta i}$ into the contributions due to the non-linear turbulence interactions, mean shear, buoyancy, and the Coriolis effects, and to model these contributions separately. Then, the return-to-isotropy parameterisation is applied to the non-linear turbulence contributions only. The approximations for the pressure terms Π_{ij} and $\Pi_{\theta i}$ can be written in the form (see e.g. [144, 202, 213]):

$$\Pi_{ij} = C_t^u \frac{a_{ij}}{\tau_u} e$$

$$+ \left[C_{s1}^u S_{ij} + C_{s2}^u \left(a_{ik} S_{kj} + a_{jk} S_{ki} - \frac{2}{3} \delta_{ij} a_{kl} S_{kl} \right) \right.$$

$$\left. + C_{s3}^u \left(a_{ik} W_{kj} + a_{jk} W_{ki} \right) \right] e$$

$$+ C_b^u \left(\beta_i \langle u_j' \theta_v' \rangle + \beta_j \langle u_i' \theta_v' \rangle - \frac{2}{3} \delta_{ij} \beta_k \langle u_k' \theta_v' \rangle \right)$$

$$+ 2 C_c^u \left(\epsilon_{ilk} \Omega_l \langle u_k' u_j' \rangle + \epsilon_{jlk} \Omega_l \langle u_k' u_i' \rangle \right) + NLT, \tag{12}$$

$$\Pi_{\theta i} = C_t^\theta \frac{\langle u_i' \theta' \rangle}{\tau_\theta} + \left(C_{s1}^\theta S_{ij} + C_{s2}^\theta W_{ij} \right) \langle u_j' \theta' \rangle$$

$$+ C_b^\theta \beta_i \langle \theta' \theta_v' \rangle + 2 C_c^\theta \epsilon_{ijk} \Omega_j \langle u_k' \theta' \rangle + NLT. \tag{13}$$

Here, $a_{ij} = 2 \langle u_i' u_j' \rangle / \langle u_k' u_k' \rangle - \frac{2}{3} \delta_{ij}$ is the departure-from-isotropy tensor, $S_{ij} = \frac{1}{2} \left(\partial \langle u_i \rangle / \partial x_j + \partial \langle u_j \rangle / \partial x_i \right)$ and $W_{ij} = \frac{1}{2} \left(\partial \langle u_i \rangle / \partial x_j - \partial \langle u_j \rangle / \partial x_i \right)$ are the symmetric and the antisymmetric parts of the mean–velocity gradient tensor, respectively, and C_t^u, C_{s1}^u, C_{s2}^u, C_{s3}^u, C_b^u, C_c^u, C_t^θ, C_{s1}^θ, C_{s2}^θ, C_b^θ and C_c^θ are dimensionless coefficients. The return-to-isotropy time scales τ_u and τ_θ should not be confused with τ_{u*} and $\tau_{\theta *}$ in Eqs. (10) and (11).

Jones and Musogne [94] (see also [50]) added an additional term to the parameterization of $\Pi_{\theta i}$ that is proportional to the gradient of mean scalar concentration. It reads $C_m^\theta e a_{ij} \partial \langle \theta \rangle / \partial x_j$, where C_m^θ is a dimensionless coefficient. The Jones and Musogne approach was further developed by Craft et al. [46] who incorporated the mean–gradient term into the expression for the non-linear turbulence (return-to-isotropy) contribution to $\Pi_{\theta i}$.

The mean–shear, buoyancy and Coriolis terms on the r.h.s. of Eqs. (12) and (13) represent linear contributions to the so-called rapid parts of Π_{ij} and $\Pi_{\theta i}$ (the return-to-isotropy contributions to Π_{ij} and $\Pi_{\theta i}$ are referred to as slow parts of the pressure terms). Notice that these linear rapid terms have the same form as the mean–shear, buoyancy and Coriolis terms in Eq. (3)

for the Reynolds stress and Eq. (4) for the temperature flux. Therefore, the effect of linear rapid terms, sometimes referred to as the implicit mean–shear, buoyancy and Coriolis terms, is simply to partially offset the respective explicit terms already present in Eqs. (3) and (4).

Generally speaking, the pressure–velocity and the pressure–scalar covariances depend non-linearly on the departure-from-isotropy tensor and on the other tensors involved, e.g. on the rotation tensor $\epsilon_{ikj}\Omega_k$. The non-linear parts of Π_{ij} and $\Pi_{\theta i}$ are symbolized by NLT on the r.h.s. of Eqs. (12) and (13). Numerous elaborate non-linear formulations of the pressure terms have been proposed (e.g. [46, 85, 130, 144, 166, 213]). Particular emphasis is placed on the realizability of parameterized pressure terms, following different ways of imposing realizability constraints. The non-linear models perform better than the simplified linear models, particularly in flows with large departures from isotropy. However, the non-linear models are inevitably complex. They are often inconvenient to use and are computationally expensive. It is therefore common practice in geophysical applications to put up with the shortcomings of linear models and apply Eqs. (12) and (13) without $NLTs$. In doing so the TKE dissipation time scale $\tau_\epsilon = e/\epsilon$ is typically used instead of the return-to-isotropy time scales τ_u and τ_θ, assuming that all these time scales are proportional to each other. The dimensionless coefficients C_t^u through C_c^θ in Eqs. (12) and (13) are adjusted to provide a good fit of the model results to empirical data, i.e. these coefficients are treated as tuning model parameters. Some estimates of these dimensionless coefficients used in geophysical turbulence modelling are given in [202]. Several important points should be discussed in relation to the parameterization of the pressure terms.

First and foremost we recall how the parameterizations for the pressure terms are derived. Taking the divergence of the transport equation for the fluctuating velocity, a Poisson equation for the fluctuating pressure is obtained. Parameterizations for various contributions to Π_{ij} and $\Pi_{\theta i}$ are then developed on the basis of the Green's function solution to the Poisson equation [45]. That solution depends on the entire fluid domain considered. In practice, however, the two-point correlations are assumed to be different from zero only in the vicinity of the point where the pressure terms are evaluated. Then, the pressure–velocity and pressure–scalar covariances are modelled as if they were local (dependent on the flow variables at the same point), although they may actually be non-local (dependent on the flow variables in the entire domain). Therefore, formulations of the type given by Eqs. (12) and (13) have inherent limitations. These formulations may not perform well in situation where turbulence is essentially non-local, as is, for example, the case for atmospheric convection. Notice that *both* linear *and* non-linear one-point formulations for Π_{ij} and $\Pi_{\theta i}$ suffer from this shortcoming.

Although linear models of Π_{ij} and $\Pi_{\theta i}$ are attractive from the standpoint of practical applications, they may entirely fail in some situations of interest. An illustrative example is turbulent convection driven by the surface buoyancy flux and affected by rotation [144, 147]. In the seemingly simple case where the rotation axis is aligned with the vector of gravity, the linear model predicts a Coriolis contribution to $\Pi_{\theta i}$ that is identically zero, although the Coriolis contribution becomes one of the dominant parts of $\Pi_{\theta i}$ as the rotation rate increases. As shown in [144], a non-linear formulation is required which is at least quadratic in the rotation tensor $\epsilon_{ikj}\Omega_k$. Fortunately, the effect of rotation on turbulence is of little importance in most atmospheric flows.[2] Similar problems may, however, be encountered when the effects of buoyancy and shear are considered. Caution must be exercised when simplified parameterizations are applied.

To conclude this section, we remark that modelling the pressure transport term $\frac{2}{3}\delta_{ij}\partial\langle u_k'p'\rangle/\partial x_k$ in Eq. (3) represents a separate problem. Lumley [130] and Shih [173] discussed this problem in some detail. The pressure transport term is usually smaller than the third-order velocity correlation term, although this is not always the case. It is standard practice in applied turbulence modelling to neglect the pressure transport term entirely, or to incorporate it into a parameterization of the third-order velocity correlation.

Third-Order Moments

The second group of terms that require closure assumptions includes turbulence moments of the third order. These terms enter Eqs. (3, 4, 5, 6, 7, 8 and 9) in the divergence form. They describe the transport of the second-order moments by the fluctuating velocity. Numerous formulations have been proposed for the third-order transport terms (e.g. [43, 49, 83, 86, 117]), ranging from the simplest down-gradient approximations to very complex formulations based on a sophisticated treatment of transport equations for the third-order turbulence moments. Simple down-gradient approximations have been most popular in geophysical applications. They read

$$\langle u_i'u_j'u_k'\rangle = -K_{uu}\left(\frac{\partial\langle u_i'u_j'\rangle}{\partial x_k} + \frac{\partial\langle u_i'u_k'\rangle}{\partial x_j} + \frac{\partial\langle u_j'u_k'\rangle}{\partial x_i}\right), \qquad (14)$$

$$\langle u_i'u_j'\theta'\rangle = -K_{u\theta}\left(\frac{\partial\langle u_i'\theta'\rangle}{\partial x_j} + \frac{\partial\langle u_j'\theta'\rangle}{\partial x_i}\right), \qquad (15)$$

$$\langle u_i'\theta'^2\rangle = -K_{\theta\theta}\frac{\partial\langle\theta'^2\rangle}{\partial x_i}, \qquad (16)$$

---

[2] This holds for deep convective updraughts, boundary-layer plumes and rolls, and eddy motions on a smaller scale. Eddy motions of larger spatial scale, such as synoptic weather systems (e.g. cyclones and fronts) and regional circulations, do feel the earth's rotation. These motions are, however, resolved by the present-day NWP models so that there is no need for a parameterization.

where K_{uu}, $K_{u\theta}$ and $K_{\theta\theta}$ are the eddy diffusion coefficients. Other scalar quantities are treated in the same way as the potential temperature. The down-gradient approximation for the TKE transport term reads

$$\langle u_i' u_k'^2 \rangle = -K_e \frac{\partial \langle u_k'^2 \rangle}{\partial x_i}, \tag{17}$$

where K_e is the eddy diffusion coefficient with respect to the TKE.

The down-gradient formulations (14, 15, 16 and 17) are attractive for their simplicity. It has long since been recognized, however, that their performance in complex flows leaves very much to be desired and a more accurate treatment of the third-order moments is required. This is particularly true for convective flows (e.g. [147, 150]), but may also be the case for stably stratified flows (e.g. [46]). In an attempt to develop a physically plausible parameterization, the focus has been on buoyant convection where the third-order moments are largely responsible for non-local transport properties of turbulent motions.

A straightforward way is to derive expressions for the third-order moments from their budget equations. These equations require closure assumptions in much the same way as the second-moment equations. In particular, the fourth-order moments that describe the transport of the third-order quantities by the fluctuating velocity should be parameterized. The so-called Millionshchikov hypothesis [143] has been used for this purpose over several decades. It states that the fourth-order moments can be considered as quasi-Gaussian, even though the third-order moments are nonzero. That is, the following relation holds for any four fluctuating quantities a, b, c and d:

$$\langle a'b'c'd' \rangle = \langle a'b' \rangle \langle c'd' \rangle + \langle a'c' \rangle \langle b'd' \rangle + \langle a'd' \rangle \langle b'c' \rangle. \tag{18}$$

Using Eq. (18) for the fourth-order moments, the Rotta-type formulations for the pressure terms, and the relaxation-type formulations for the dissipation terms, then neglecting the advection and the time-rate-of-change of the third-order moments, a closed set of algebraic expressions for the third-order moments is derived. Canuto et al. [43] developed such expressions for the horizontally homogeneous convective boundary layer (CBL). The third-order moments appear to be linear combinations of the derivatives (in the x_3 vertical direction only) of all second-order moments involved multiplied by certain combinations of governing parameters with dimensions of eddy diffusivity. Some of those combinations explicitly depend on the buoyancy parameter. Canuto et al. [42] employed a modified quasi-normal approximation that basically amounts to multiplying the r.h.s. of Eq. (18) by a correction function of the dissipation time scale and of the buoyancy time scale (a reciprocal of the buoyancy frequency). These authors proposed modified (and slightly simplified) expressions for the third-order moments that show a better agreement with large-eddy simulation (LES) data from a shear-free CBL than the expressions given in [43].

An attractive way of looking at the problem of non-local convective transport is based on the observation that convective turbulence is skewed. For example, in the CBL driven by the surface buoyancy flux, the vertical transport in mid-CBL is dominated by quasi-organized motions, convective updraughts, whose size is of the order of the CBL depth. The updraughts are more localized (occupy a smaller area) than the compensating downward motions, downdraughts. A quantitative measure of this localization is the vertical-velocity skewness $S_w = \langle u_3'^3 \rangle / \langle u_3'^2 \rangle^{3/2}$. Likewise the potential–temperature skewness $S_\theta = \langle \theta'^3 \rangle / \langle \theta'^2 \rangle^{3/2}$ is a quantitative measure of the localization of potential–temperature anomalies (with respect to a horizontal mean). Guided by this view of convective circulation and of the bottom–up top–down transport asymmetry [209], Abdella and McFarlane [1], Canuto and Dubovikov [41] and Zilitinkevich et al. [225] proposed the following parameterization for the flux of potential–temperature flux:

$$\langle u_3'^2 \theta' \rangle = S_w \langle u_3'^2 \rangle^{1/2} \langle u_3' \theta' \rangle . \tag{19}$$

This expression has an advective rather than a down-gradient diffusive form, indicating that the temperature flux is transported by the CBL-scale quasi-organized eddies rather than diffused by small-scale random turbulence. The quantity $S_w \langle u_3'^2 \rangle^{1/2} = \langle u_3'^3 \rangle / \langle u_3'^2 \rangle$ was termed "large-eddy skewed-turbulence advection velocity" in [225]. Zilitinkevich et al. [225] (see also [83]) added a conventional down-gradient diffusion term $-K_{w\theta} \partial \langle u_3' \theta' \rangle / \partial x_3$, $K_{w\theta}$ being the turbulent diffusivity with respect to $\langle u_3' \theta' \rangle$, to the r.h.s. of Eq. (19) in order to arrive at an interpolation formula that should work in both well-mixed regions of the flow, where advective transport by the CBL-scale eddies dominates, and in stratified regions, where turbulent transport is primarily of diffusive character.

A skewness-dependent parameterization for the flux of potential–temperature variance was formulated by Mironov et al. [146], Abdella and McFarlane [2] and Abdella and Petersen [3]. It reads

$$\langle u_3' \theta'^2 \rangle = S_\theta \langle \theta'^2 \rangle^{1/2} \langle u_3' \theta' \rangle . \tag{20}$$

An interpolation formula for $\langle u_3' \theta'^2 \rangle$ that incorporates the down-gradient term $-K_{\theta\theta} \partial \langle \theta'^2 \rangle / \partial x_3$, $K_{\theta\theta}$ being the turbulent diffusivity with respect to $\langle \theta'^2 \rangle$, was presented in [83].

Equations (19) and (20) are consistent with the top-hat representation of fluctuating quantities. The top-hat representation is central to the mass–flux approach widely used to parameterize convection in numerical models of the atmosphere. The simplest top-hat mass–flux model can be formulated in terms of a probability distribution function (PDF) which consists of only two Dirac delta functions, i.e. the probabilities of motions to be either updraughts or downdraughts are P_u and P_d, respectively, and $P_u + P_d = 1$. A comprehensive account of the two-delta-function mass–flux framework is given by Randall et al. [162], Lappen and Randall [110, 111] and Gryanik and Hartmann [83]. In order to emphasize a different localization (different fractional area coverage of positive/negative anomalies with respect to a horizontal mean) of the vertical velocity and of the scalar quantities, as manifested by a difference between S_w and S_θ (see Fig. 1 in [146]), Gryanik and Hartmann [83] refer to their approach as to the two-scale mass–flux approach. Notice that different PDFs can be used to develop parameterizations of statistical moments of turbulence. For example, Larson and Golaz [116] developed parameterizations of various third-order and fourth-order moments, using a combination of two trivariate Gaussian functions. These authors considered moist CBL and presented their results in terms of vertical velocity, liquid water potential temperature and total water specific humidity. The formulations based on the two-Gaussian-function PDF revealed a somewhat improved fit to data for some moments as compared to the formulations based on the two-delta-function PDF.

Equations (19) and (20) require that S_w and S_θ be specified. If the budget equations for the third-order moments are used for this purpose as discussed above, formulations for the fourth-order moments are required. Taking the two-scale mass–flux approach, Gryanik and Hartmann [83] and Gryanik et al. [84] proposed (see also [2])

$$\langle u_3'^4 \rangle = 3 \left(1 + \frac{1}{3}S_w^2 \right) \langle u_3'^2 \rangle^2 , \qquad \langle \theta'^4 \rangle = 3 \left(1 + \frac{1}{3}S_\theta^2 \right) \langle \theta'^2 \rangle^2 , \qquad (21)$$

$$\langle u_3'^3 \theta' \rangle = 3 \left(1 + \frac{1}{3}S_w^2 \right) \langle u_3'^2 \rangle \langle u_3'\theta' \rangle , \qquad (22)$$

$$\langle u_3' \theta'^3 \rangle = 3 \left(1 + \frac{1}{3}S_\theta^2 \right) \langle \theta'^2 \rangle \langle u_3'\theta' \rangle , \qquad (23)$$

$$\langle u_3'^2 \theta'^2 \rangle = \langle u_3'^2 \rangle \langle \theta'^2 \rangle + 2 \langle u_3'\theta' \rangle^2 + S_w S_\theta \langle u_3'\theta' \rangle \langle u_3'^2 \rangle^{1/2} \langle \theta'^2 \rangle^{1/2} . \qquad (24)$$

Similar expressions for the fourth-order moments that incorporate horizontal velocity components u_1 and u_2 are presented in [84]. Both Eqs. (19) and (20) for the third-order moments and Eqs. (21, 22, 23 and 24) for the fourth-order moments were favourably tested against data from LES and from aircraft measurements in the atmospheric CBL [83, 84], from numerical simulation of open-ocean deep convection [128], and from numerical simulation of solar and stellar convection [108].

Equations (21, 22, 23 and 24) amount to a generalization of the Millionshchikov hypothesis. Indeed, in the case of isotropic turbulence, S_w and S_θ vanish and Eqs. (21, 22, 23 and 24) reduce to the form given by Eq. (18). In the other limiting case of very skewed turbulence, the terms with S_w and

S_θ dominate over the other terms and Eqs. (21, 22, 23 and 24) take on the form suggested by the top-hat mass–flux approach. Then, Eqs. (21, 22, 23 and 24) represent the simplest linear interpolation between the two limiting cases, where dimensionless coefficients on the r.h.s. (3, 1 and 1/3) are chosen in such a way that these limiting cases are satisfied exactly. It should be emphasized that Eqs. (19) and (20) for the third-order moments and Eqs. (21, 22, 23 and 24) for the fourth-order moments taken in the limit of large skewness are in essence the top-hat mass–flux parameterizations expressed in terms of the ensemble–mean quantities. Analogies between the ensemble–mean and the mass–flux modelling frameworks are discussed below in greater depth.

Noteworthy also is that the expressions (21, 22, 23 and 24) for the fourth-order moments satisfy the realizability constraints [5–7, 84] regardless of the magnitude of skewness. This is not the case for Eq. (18) that violates realizability if the magnitude of S_w or of S_θ exceeds $2^{1/2}$ [84].

Parameterizations (19, 20, 21, 22, 23 and 24) are developed for the temperature-stratified horizontally homogeneous CBL, where potential temperature is the only thermodynamic variable and all quantities of interest depend on the x_3 vertical co-ordinate only. Their extension to the three-dimensional case is by no means trivial but seems to be manageable (D. Mironov, A note on the parameterization of the third-order transport in skewed convective boundary-layer turbulence, unpublished manuscript; V. Gryanik, personal communication). Such an extension is highly desirable and should be developed. The same is true for the extension to the moist atmosphere, where, apart from potential temperature, water in its three phases should be considered.

Dissipation Rates

Finally, the rates of dissipation of the second-order turbulence moments should be parameterized. It is common practice to assume, following Kolmogorov [104], local isotropy at small scales, giving $\epsilon_{i\theta} = 0$, $\epsilon_{iq} = 0$, and $\epsilon_{ij} = \frac{2}{3}\delta_{ij}\epsilon$. In order to determine the TKE dissipation rate, a prognostic equation for ϵ has been used in engineering and geophysics over several decades (e.g. [10, 16, 37, 46, 49, 58, 117, 167, 201]). The dissipation rates of the scalar quantities, ϵ_θ, ϵ_q and $\epsilon_{\theta q}$, are either related to e and ϵ through $\epsilon_\theta \propto e^{-1}\epsilon \langle \theta'^2 \rangle$ (similarly for ϵ_q and $\epsilon_{\theta q}$), or computed from their own prognostic equations (e.g. [156]). Once the dissipation rates are determined, the various time scales, length scales and eddy diffusion coefficients are computed diagnostically through these dissipation rates and the corresponding variances. For example, the quantities with respect to the TKE are given by $l \propto \epsilon^{-1}e^{3/2}$, $\tau \propto \epsilon^{-1}e$ and $K \propto \epsilon^{-1}e^2$. In this way the system of the second-order equations is closed.

Apart from the ϵ equation, prognostic equations have been formulated for other quantities that determine the turbulence length/time scale. Prognostic equations for the product $e\,l$ of the TKE and the turbulence length scale [142], for the reciprocal of turbulence time scale $e^{-1}\epsilon$ [203, 205], and for the eddy diffusivity $\epsilon^{-1}e^2$ [212] are examples. These closure ideas were generalized by Umlauf and Burchard [201] who developed a generic equation for the quantity $e^m l^n$ that incorporates the equations mentioned above as particular cases. These authors proposed a rational way to calibrate their generalised model in the so-called two-equation second-order modelling framework, where only the TKE equation and the equation for $e^m l^n$ are carried as prognostic equations whereas the other second-order equations are reduced to the diagnostic algebraic expressions. The exponents m and n along with the other model parameters are evaluated by demanding consistency with a number of well-documented reference cases, such as the logarithmic boundary layer and the decay of homogeneous turbulence.

The prognostic equations for the dissipation rates of TKE and of scalar variances are very complex. They contain a number of terms whose physical nature is not satisfactorily understood. In fact, all terms in the dissipation-rate equations that describe production, destruction and turbulent transport of the dissipation rates should be parameterized, and the validity of those parameterizations is uncertain. It has often been questioned whether prognostic equations for the dissipation rates are really necessary, or diagnostic expressions may be sufficient, at least in case of a relatively simple flow geometry. The latter viewpoint is often held in geophysical applications. A simple and an economical way to determine the dissipation rates of the TKE and of the scalar variances is to compute them from the following expressions:

$$\epsilon = C_{\epsilon e}\frac{e^{3/2}}{l}, \qquad \epsilon_\theta = C_{\epsilon\theta}\frac{\langle\theta'^2\rangle e^{1/2}}{l}, \tag{25}$$

using one or the other formulation for the turbulence length scale l. Here, $C_{\epsilon e}$ and $C_{\epsilon\theta}$ are dimensionless coefficients. The dissipation rates ϵ_q and $\epsilon_{\theta q}$ of the humidity variance and of the potential temperature–humidity covariance, respectively, are computed similarly to ϵ_θ. The above expressions for the dissipation rates can be recast in terms of the turbulence time scale $\tau = e^{-1/2}l$.

There have been numerous proposals for expressions to compute the length scale l. The simplest of them seems to have been Blackadar's formula [28], $l^{-1} = l_{\text{sfc}}^{-1} + l_\infty^{-1}$, that interpolates between the two limits, namely, $l = l_{\text{sfc}} = \kappa x_3$, κ being the von Kármán constant, as $x_3 \to 0$, and $l = l_\infty$ as $x_3 \to \infty$. This yields the logarithmic profiles close to the underlying surface and prevents the turbulence length scale from growing without bound well above the surface. The free-flow length scale l_∞ is either set proportional to the PBL depth, or simply set to a constant value (typically from one hundred to a few hundred metres in the atmospheric models). One more formulation is due to Mellor and Yamada [141] who proposed $l_\infty = C_\infty \left(\int_0^\infty x_3 e^{1/2} dx_3\right)^{-1} \int_0^\infty e^{1/2} dx_3$, where C_∞ is a dimensionless coefficient of order 10^{-1}. Other limitations on l have also been used. The length scale is taken to be limited by the shear length scale $l_s = C_{ls} (S_{ij}S_{ij})^{-1/2} e^{1/2}$ (e.g. [44] and references therein), and, in case of stable density stratification, by the buoyancy length scale $l_b = C_{lb}N^{-1}e^{1/2}$, where N is the buoyancy frequency (e.g. [34, 215, 226]). In rotating flows, the length scale is taken to be limited by $l_r = C_{lr} \left(\Omega_i \Omega_i \right)^{-1/2} e^{1/2}$ (e.g. [87]). This limitation is of little importance in the atmosphere. It is important in many geophysical, astrophysical and technical applications. A prominent geophysical example is open-ocean deep convection [137]. Dimensionless constants C_{ls}, C_{lb} and C_{lr} are evaluated on the basis of empirical and numerical data. Readers are referred to the chapter "Turbulence in Astrophysical and Geophysical Flows" of this volume for further discussion of the dissipation rates and of the turbulence length scale.

2.3 Simplifications

The second-order equations closed as discussed above would probably do a fairly nice job of describing most salient features of turbulence in the lower troposphere. However, the full set of the (time-dependent, three-dimensional) second-order equations is still far too complex and expensive computationally. Further simplifications are necessary in order to obtain a reasonably simple turbulence parameterization scheme that can be accommodated by a numerical model of the atmosphere.[3]

Truncation

A family of second-order closures has been developed by Mellor and Yamada [141] (see also [142], a comprehensive discussion of the Mellor and Yamada closures and their numerous derivatives is given in [154]). They utilized the second invariant $A_2 = a_{ij} a_{ij}$ of the departure-from-isotropy tensor as the scaling parameter that measures the degree of flow anisotropy. Using the observation that A_2 is small and invoking additional arguments to scale the advection and the turbulent diffusion terms, they successively discarded terms of different order in A_2 in the second-moment equations. The result proved to be a hierarchy of truncated turbulence closure schemes, ranging from the complete second-order closure to a simple algebraic stress model, where all second-moment equations are reduced to algebraic expressions. That hierarchy of closure schemes has found a wide utility in geophysical applications and is often referred to collectively as the Mellor–Yamada closures ever since.

The scheme termed the level 2.5 Mellor–Yamada scheme has been most popular in practical applications. The only prognostic equation carried by that scheme is the TKE equation. The TKE diffusion is usually parameterized through the simplest down-gradient approximation. All other second-moment equations are reduced to algebraic expressions by neglecting the time-rate-of-change, the advection and the turbulent diffusion terms. The pressure–velocity

---

[3] The material in Sects. 2.3, 2.4 and 2.5 is somewhat more technical. Some readers may prefer to proceed directly to Sect. 3 for an outline of the mass–flux convection schemes.

and the pressure–scalar covariances are parameterized through Eqs. (12) and (13), typically without non-linear terms, or simply through the Rotta-type return-to-isotropy formulations (10) and (11). Notice that the use of Eqs. (12) and (13) without non-linear terms instead of Eqs. (10) and (11) does not radically change the result. Since the linear rapid terms on the r.h.s. of Eqs. (12) and (13) have the same form as the respective terms in Eqs. (3) and (4), the only (though not unimportant) effect of their inclusion is to modify dimensionless coefficients in front of various terms in the resulting expressions for the Reynolds stress and for the scalar fluxes. The dissipation rates of the TKE and of the scalar variances are parameterized through the algebraic relations (25). A turbulence model that carries only one prognostic equation, namely, the TKE equation, is referred to as the one-equation model. In case the transport equation is used for the TKE dissipation rate, or for any quantity $e^m l^n$ (see above), the resulting turbulence model is referred to as the two-equation model.

Boundary-Layer Approximation

Another simplification typically involved in geophysical applications is the so-called boundary-layer approximation where the flow is treated as horizontally homogeneous. This approximation is fairly accurate for large-scale and mesoscale NWP models, whose grid-box aspect ratio (the ratio of the horizontal grid size to the vertical grid size) is large. In the framework of the boundary-layer approximation, all derivatives in x_1 and x_2 horizontal directions in the second-moment equations are neglected and the grid-box mean vertical velocity $\langle u_3 \rangle$ is set to zero (in the second-moment equations, but not in the equations for the mean fields). The one-equation turbulence closure scheme in the boundary-layer approximation has been probably the most popular turbulence scheme in NWP. The scheme carries the prognostic TKE equation. All other second-moment equations are reduced to algebraic relations that constitute a system of linear equations for variances and fluxes. The solution to that system yields the expressions for the vertical fluxes of momentum and scalars in the following down-gradient form:

$$\langle u_3' f' \rangle = -S_f l e^{1/2} \frac{\partial \langle f \rangle}{\partial x_3}, \tag{26}$$

where a generic variable f stands for u_1, u_2, θ or q. The so-called stability functions S_f depend on the dimensionless buoyancy gradient $\epsilon^{-2} e^2 N^2$ and on the dimensionless shear $\epsilon^{-2} e^2 \left[(\partial \langle u_1 \rangle / \partial x_3)^2 + (\partial \langle u_2 \rangle / \partial x_3)^2 \right]$. They incorporate various combinations of dimensionless coefficients that stem from the parameterizations of the pressure–velocity and pressure–scalar covariances and of the dissipation rates. The turbulence length scale l is parameterized algebraically as discussed above. Examples of one-equation turbulence closure schemes for NWP purposes are the schemes used operationally in the limited-area model COSMO (formerly referred to as LM [163, 164, 186]) and HIRLAM [204].

At the next level of simplification, the time-rate-of-change, the advection and the turbulent diffusion of the TKE are neglected so that all second-moment equations are reduced to algebraic relations. The resulting expressions for fluxes are essentially of the down-gradient form $\langle u_3' f' \rangle = -K_f \partial \langle f \rangle / \partial x_3$, where the diffusion coefficients K_f are functions of the turbulence length scale and of the vertical gradients of velocity and buoyancy. These diffusion coefficients are often adjusted in a somewhat ad hoc manner in order to improve the overall performance of an NWP model (cf. the situation with mixing-length models in astrophysics discussed in the chapter "Turbulent Convection and Numerical Simulations in Solar and Stellar Astrophysics" of this volume). The algebraic turbulence closure schemes are used, for example, in the global NWP models GME [135] of the German Weather Service (DWD) and IFS (Integrated Forecasting System) [93] of the European Centre for Medium-Range Weather Forecasts (ECMWF).

Unfortunately, no simplification is possible without the sacrifice of relevant information and hence of accuracy, and this is particularly true of truncated second-order closures. Well-calibrated algebraic and one-equation turbulence closure schemes show a good performance in turbulent flows where the static stability is close to neutral. However, they are known to have serious problems in stratified flows, both stable and convective.

Performance of Simplified Closures in Stratified Flows

Turbulence in *stably stratified boundary layers* is weak and often intermittent in space and time [77]. The stable boundary layer (SBL) is exposed to various types of meso-scale motions, such as gravity waves and meanders of cold air, to horizontal inhomogeneity of the underlying surface, and to the radiation flux divergence. These and other effects significantly complicate the SBL structure [62, 133, 134, 182, 217]. Current turbulence schemes do not include many of these important effects in a physically meaningful way [132] and are not able to satisfactorily describe the SBL turbulence structure. Most current turbulence schemes tend to extinguish turbulence in case of strong static stability, when the gradient Richardson number $\mathrm{Ri} = \left[\left(\partial \langle u_1 \rangle / \partial x_3 \right)^2 + \left(\partial \langle u_2 \rangle / \partial x_3 \right)^2 \right]^{-1} N^2$ exceeds its critical value of order 0.2 (see the chapter "Turbulence in Astrophysical and Geophysical Flows" of this volume for further discussion of this issue). Then, the schemes are tuned in an ad hoc way to prevent turbulence from dying out entirely as the static stability increases. A simple device often applied in NWP models is a "minimum diffusion coefficient". That is, the eddy diffusivity for momentum and scalars is limited from below by a predefined constant value to provide "residual" mixing when the turbulence parameterization scheme predicts no turbulence at all. A tuning device of this sort may have a detrimental effect on the NWP model performance in some important situations. For example, it may destroy a delicate balance of physical processes (radiative and evaporative cooling, advection by mean vertical velocity, and turbulent entrainment) near the top of the stable or neutral PBL capped by stratocumulus clouds, leading to the disappearance of clouds where they should actually be maintained.

A physically plausible approach to the problem of maintaining turbulence in case of strong static stability was taken by Raschendorfer [163, 164]. He surmised that turbulence in the shear-driven SBL would not collapse entirely, if the underlying surface at the sub-grid scale is horizontally inhomogeneous with respect to the temperature. Spatial buoyancy differences due to this temperature inhomogeneity induce horizontal pressure gradients that in turn set the air in motion. Although these air motions experience friction at the underlying surface, they may not contribute to the grid-box mean momentum flux as the flow patterns in different directions may counteract each other (cf. cell-like motions in the shear-free CBL that efficiently transport heat but make no contribution to the grid-box momentum flux). However, they do contribute to the grid-scale mean TKE, preventing the SBL from collapsing entirely. Having assumed the above mechanism of maintenance of turbulence in stable stratification, Raschendorfer extended the one-equation turbulence closure scheme of the NWP model COSMO so that the scheme is guarded against sharp turbulence cut-off at a critical Richardson number.

Worthy of mention is an attempt to derive eddy viscosity in stably strati-fied turbulent flows from first principles made by Sukoriansky et al. [189, 191]. Their spectral model is free of the sharp cut-off critical Richardson number deficiency. It predicts turbulent eddy diffusivities for wind and scalar quanti-ties in good agreement with observations. The new theoretical findings have been used in the framework of the two-equation e-ϵ turbulence closure scheme to model atmospheric SBL over sea ice [190].

Difficulties of simplified turbulence closure schemes in *convective condi-tions* are associated first of all with their inability to adequately account for non-local transport properties of convective turbulence. This is not particu-larly surprising, however, considering that in the simplified truncated closures the third-order terms largely responsible for non-local transport of momen-tum and scalars are either entirely neglected or parameterized very crudely. Local turbulence schemes are typically unable to reproduce the well-mixed character of the CBL with counter–gradient fluxes of scalars often encoun-tered in the upper part of the boundary layer. They also fail to correctly represent entrainment at the CBL top, leading to erroneous prediction of the CBL temperature and humidity and of the CBL height. One way to cope with these difficulties is to introduce more of the essential physics into the second-order closure scheme, e.g. by using the skewness-dependent formulations of third-order transport terms discussed above.

Notice that simplified truncated second-order closure schemes are almost inevitably non-realisable. In order to prevent such schemes from producing unphysical solutions, a clipping operation is usually applied. Normal stresses and scalar variances are set to zero if they become negative, and Schwarz' inequalities for the third-order moments are strictly enforced. The clipping

operation has proven to be an effective tool [55] and is considered to be legiti-
mate in engineering and geophysical applications [173]. It should, however, be
avoided whenever possible; that is to say, effort should be mounted to develop
closure schemes where clipping is reduced to a minimum

The Similarity Approach

An alternative way to describe boundary-layer turbulence and shallow convection is through the use of the similarity theory for boundary-layer flows. The
approach basically amounts to representing the vertical profiles of turbulent
quantities through the shape functions, using the scales of variables pertinent
to the mixing regime in question. The scaling ideas should be consistent with
the budget equations for turbulence moments, at least in the integral sense (cf.
the surface-layer flux–profile relationships of the Monin–Obukhov similarity
theory considered in the next section).

Taking the similarity approach, shapes of the vertical profiles of the
turbulent diffusion coefficients are prescribed and the magnitudes of diffusion coefficients and of other turbulence characteristics, such as the fluxes
due to entrainment at the boundary layer top, are expressed through the
appropriate scales of length, velocity, temperature and humidity. To this
end, the now classical Deardorff convective scaling [53, 54] is widely used
with the CBL depth h as the bulk length scale, and $w_* = (hB_s)^{1/3}$ and
$\theta_* = \langle u'_3\theta'\rangle_s / w_*$ as the bulk velocity and potential-temperature scales, respectively. Here, $B_s = \beta_3 \langle u'_3\theta'_v\rangle_s$ is the surface buoyancy flux, and $\langle u'_3\theta'\rangle_s$
and $\langle u'_3\theta'_v\rangle_s$ are the surface fluxes of potential temperature and of vir-
tual potential temperature, respectively. The humidity scale is introduced
similarly to the potential-temperature scale. Power-law functions of dimensionless height x_3/h are commonly utilized for the vertical-profile shape functions. For example, turbulent temperature diffusivity in shear-free CBL is
expressed as $K_\theta/w_*h = C_{K\theta}(x_3/h)(1 - C_{\mathrm{entr}}x_3/h)^\alpha$, where $C_{K\theta}$, C_{entr} and
α are disposable parameters, and C_{entr} is chosen so as to provide the right
amount of entrainment at the CBL top. In order to account for the produc-
tion of turbulence energy due to mean velocity shear, a convective velocity
scale is modified through the incorporation of the surface friction velocity
$u_* = \left(\langle u'_3u'_1\rangle_s^2 + \langle u'_3u'_2\rangle_s^2\right)^{1/4}$, where the subscript "s" indicates the surface
values. Using the similarity approach, momentum and scalar fluxes are not
directly dependent on local gradients; rather they are functions of the integral scales that characterize the CBL as a whole and thus account (at least
implicitly) for the non-local effects. Such a "non-local" scheme is proposed in
[92], using earlier ideas presented in [200].

An advanced boundary layer mixing scheme based on the similarity approach was developed by Lock et al. ([127, 138], see also [125, 126]) for use in
the UK Met Office NWP and climate models [48]. The scheme incorporates an
entrainment parameterization based on a generalized turbulent velocity scale

that accounts for the generation of turbulence due to the mean velocity shear, due to the surface heating, due to the cloud-top radiative cooling, and due to the evaporative cooling of entrained air. The unstable layers are identified on the basis of the buoyancy of undilute parcels lifted from the surface and lowered from the cloud top with due regard for latent heat effects. The mixing regimes considered by the scheme range from dry SBL to a complex configuration, where a layer of stratocumulus clouds is separated from the unstable surface layer by a cumulus cloud layer. As different mixing parameterizations are used for different regimes, the scheme includes a sophisticated decision tree to discriminate between various boundary-layer mixing regimes. It should be noted that the scheme does not operate throughout the atmosphere. It is applied to about the lowest 2.5 km [127]. Mixing through the rest of the atmosphere, as well as through the cumulus cloud layers diagnosed within the area of operation of the turbulence scheme, is computed with the convection scheme [82]. Convection schemes currently used in NWP and climate models are developed on the basis of the mass–flux approach (Sect. 3).

2.4 The Surface Layer

The layer in the immediate vicinity of the underlying surface, the surface layer, deserves special attention. The surface layer looms large in meteorology as it is this layer where the interaction of the atmosphere with the underlying surface takes place. In NWP and climate models, the surface-layer resistance, heat and mass transfer laws are used to compute surface fluxes of momentum, heat, water vapour, and if necessary, of other scalar quantities, and are, therefore, the key components of the physical parameterization package.

The now classical Monin–Obukhov similarity theory [151, 155] has been commonly used for more than half a century to describe the vertical structure of the atmospheric surface layer. For lack of space, it is impossible to give an account of the Monin–Obukhov theory in the present paper. Readers are referred to [63, 67, 95, 152, 183, 216], where various aspects of the surface-layer similarity are discussed. Here we only present the Monin–Obukhov surface-layer flux–profile relationships for the wind velocity and for the potential temperature (the formulation for specific humidity is similar to that for potential temperature). They read

$$u(z) - u_s = \frac{u_*}{\kappa} \left[\ln \frac{z}{z_{0u}} + \psi_u(z/L) \right], \tag{27}$$

$$\theta(z) - \theta_s = -\mathrm{Pr_n} \frac{\langle w'\theta' \rangle_s}{\kappa u_*} \left[\ln \frac{z}{z_{0\theta}} + \psi_\theta(z/L) \right]. \tag{28}$$

Here, $z = x_3$ is the height above the underlying surface, $u = u_1$ is the component of the wind vector along the x_1 horizontal axis that is taken to be aligned with the surface stress (then the wind component along the x_2-axis is zero),

$w = u_3$ is the vertical component of the wind vector (this notation is used to stress the one-dimensionality of the approach), u_s and θ_s are the values of u and θ, respectively, at the underlying surface (u_s is zero at the rigid surface), z_{0u} and $z_{0\theta}$ are the roughness lengths with respect to wind velocity and potential temperature, respectively, and $\mathrm{Pr_n}$ is the turbulent Prandtl number at neutral static stability. The dimensionless functions of the Monin–Obukhov similarity theory, ψ_u and ψ_θ, account for the effect of static stability in the surface layer. The Obukhov length [155] is defined as $L = -u_*^3/(\kappa B_s)$. The von Kármán constant κ is traditionally included into the definition of L. At $z/L \ll 1$, i.e. in the lower part of the stratified surface layer, or throughout the surface layer in near-neutral conditions, Eqs. (27) and (28) reduce to the classical logarithmic profiles, where the roughness lengths are the principal parameters that describe the interaction of the flow with the underlying surface.

The surface-layer formulations are often presented in terms of the drag coefficient and the heat and mass transfer coefficients. One more alternative formulation is through the resistance of the surface layer to the transfer of momentum, heat and mass. The resistance is a more "physical" parameter, i.e. the parameter more directly related to the flux and the gradient of the quantity in question than the roughness length that should be viewed as a more "derived" parameter [136]. Nonetheless, the majority of the surface-layer formulations have been given in terms of the roughness lengths and the Monin–Obukhov similarity functions, perhaps due to their convenience in representing the profiles.

There is a substantial body of literature on the Monin–Obukhov surface-layer similarity functions. Readers are referred to the review articles [88, 89] and to the historical surveys [38, 64], where numerous further references can be found. As to the parameterization of roughness lengths with respect to wind and scalar quantities (more generally, the air–land and air–sea interaction), these are discussed in [8, 24, 33, 35, 67, 68, 99, 102, 105, 136, 165, 223], to mention a few.

It should be emphasized that the Monin–Obukhov flux–profile relationships are consistent with the budget equations for the second-order turbulence moments. In essence, they represent the second-moment budgets that are truncated under the surface-layer similarity-theory assumptions. These are that (i) turbulence is continuous, stationary and horizontally homogeneous, (ii) third-order turbulent transport is negligible, and (iii) the surface layer is a small portion of the PBL, so that the directional wind turning is negligible and turbulent fluxes can be considered approximately height-constant, equal to their surface values (in other words, changes of fluxes over the surface layer are small as compared to their changes over the entire PBL). For example, the logarithmic wind profile is readily obtained from the TKE budget equation where only the shear-production term and the dissipation term are retained. Using the surface-layer scaling relations, $e \propto u_*^2$ and $l \propto z$, to express the TKE dissipation rate through Eq. (25) along with the assumption of height-constant momentum flux, $\langle u'w' \rangle = -u_*^2$, yields the flux gradient-relationship $d \langle u \rangle /dz \propto u_*/z$ (the omitted proportionality constant is the reciprocal of the von Kármán constant). Its integration over z results in the logarithmic wind profile. The lower limit of integration is the height z_{0u} where the wind profile extrapolated logarithmically downward approaches its surface value (zero over the rigid surface). Since the surface-layer flux–profile relationships are consistent with the (truncated) second-moment budgets, they suffer from the same shortcoming as the (truncated) second-order closures. They are known to experience problems in strongly convective and in strongly stable flows.

In conditions of free convection, when the mean wind vanishes, the surface-layer flux–profile relationships predict zero fluxes. The failure is due to the neglect of the CBL-scale cell-like coherent motions. As the flow patterns in different directions effectively counteract each other, these motions make no contribution to the transfer of mean momentum (mean over a horizontal area that is large enough to embrace a multitude of convective cells). However, these motions efficiently transport heat and other scalar quantities. Businger [39] introduced the concept of "minimum friction velocity", that is the friction velocity due to the effect of the CBL-scale motions which do experience friction at the surface, although mean wind is zero. The minimum friction velocity was assumed to scale on the Deardorff convective velocity w_* and to additionally depend on z_{0u}. Using the above concept, a number of heat and mass transfer laws have been proposed that are suitable for calculation of surface fluxes in conditions of free convection [4, 22, 171, 188, 193, 221, 224]. Comprehensive summaries are given in [221, 224]. Some authors used the classical $Nu \propto Ra^{1/3}$ heat transfer law to estimate surface fluxes in free convection. Notice that the Nusselt number Nu and the Rayleigh number Ra explicitly depend on the molecular viscosity and on the molecular heat conductivity of the medium in question. A generalization of this law to the case of a two-component medium, e.g. moist air, was proposed in [72, 73]. Examples of its successful application to the computation of surface fluxes of sensible and latent heat are given in [9, 78].

Problems of the surface-layer similarity theory in conditions of strong static stability are associated with the intermittent nature of turbulence and with many other effects, such as internal gravity waves and horizontal inhomogeneity of the underlying surface, that complicate the surface-layer structure (see e.g. [62, 76, 77, 132–134]). Traditional log-linear flux–profile relationships of the Monin–Obukhov theory predict zero fluxes as the static stability increases and the gradient Richardson number approaches its critical value. This is in conflict with most of the observational data which indicate that turbulence very often survives well above the critical Richardson number threshold and the surface fluxes of momentum and scalars are weak but non-negligible. Recall that the Monin–Obukhov flux–profile relationships are derived under a number of simplifying assumptions which restrict their limits of applicability. Their failure to describe the real-world strongly stable surface layers is not particularly surprising. Taking a pragmatic approach, the flux–profile relationships are adjusted in a somewhat ad hoc manner to enable the fluxes to be nonzero at sufficiently strong static stability (e.g. [24]). The study of the stably stratified PBL, including the surface layer, is a very active research area in which progress is being made and more physically justified remedial measures are proposed. Mention should be made of a series of publications by Zilitinkevich and co-authors who examined the effects of static stability at the SBL outer edge, that is characterized by the buoyancy frequency N, on the SBL mean and turbulence structure. The exchange of energy, both kinetic and potential, between the SBL and the overlying stably stratified atmosphere due to the radiation of internal gravity waves was analysed in [182, 217]. Equations for the SBL depth, the SBL resistance and heat transfer laws, and the surface-layer flux–profile relationships were modified to incorporate the dependence on N [218–220, 222, 227]. The surface–flux calculation algorithms were modified with due regard for the effect of N and applied to determine surface fluxes of momentum and heat in numerical models of the atmosphere [157, 228, 229].

2.5 Extension to Saturated Air

Up to this point the atmospheric air has been treated as unsaturated, characterized by the two thermodynamic variables, θ and q. The thermodynamic structure of the real atmosphere is strongly complicated by the presence of clouds. Clouds produce precipitation. They strongly interact with atmospheric radiation, changing the atmosphere energy budget, the energy budget of the underlying surface and of the PBL in particular. They also change the buoyancy of air parcels, thus affecting the rate of production/destruction of TKE by the gravitational force. All these effects related to the presence of clouds should be accurately represented in numerical models of the atmosphere. As far as the parameteriations of turbulence and of shallow non-precipitating convection are concerned, the primary goal is to account for the effect of clouds on the buoyancy production/destruction of the Reynolds stress, including its trace – the TKE, and of the scalar fluxes. The key issue is an accurate representation of the horizontal fractional cloud coverage of a given numerical-model grid box and of the amount of cloud condensate it contains [199].

In order to account for the presence of cloud condensate, turbulence and shallow-convection parameterization schemes are formulated in terms of variables that are approximately conserved for phase changes in the absence of precipitation. Consider first warm clouds that only contain water in liquid form. Possible extension to the case of three phases including cloud ice is briefly discussed at the end of this section. One pair of moist quasi-conservative variables often used in models of non-precipitating clouds consists of the total water specific humidity q_t and the liquid water potential temperature θ_l defined as [25, 56]

$$q_t = q + q_l, \qquad \theta_l = \theta - \frac{\theta}{T}\frac{L_v}{c_p}q_l, \tag{29}$$

where q_l is the liquid water specific humidity, L_v is the latent heat of vapourization, c_p is the specific heat of air at constant pressure, and T is the absolute temperature related to the potential temperature through $T = \theta(P/P_0)^{R_d/c_p}$, P and P_0 being the atmospheric pressure and its reference value, respectively. No supersaturation is assumed, so that $q_l = q_t - q_s$ if $q_t > q_s$, where q_s is the saturation specific humidity, and $q_l = 0$ otherwise. Clearly, q_t and θ_l reduce to the dry variables q and θ, respectively, in unsaturated conditions.

Using the above moist quasi-conservative variables, the second-moment equations (3, 4, 5, 6, 7, 8 and 9) remain the same to within the substitution of θ_l and q_t for θ and q, respectively. However, the buoyancy terms (the terms with β_i) in Eqs. (3), (9), (4) and (5) should be modified with due regard for the presence of cloud condensate. This problem amounts to modelling the virtual potential temperature flux $\langle u_i'\theta_v'\rangle$ and the scalar–virtual potential temperature covariances $\langle \theta_l'\theta_v'\rangle$ and $\langle q_t'\theta_v'\rangle$ in terms of fluctuations of θ_l and q_t. Using Eq. (29) and a generalized virtual potential temperature that accounts for the water loading effect [15, 124],

$$\theta_v = \theta\left[1 + (R - 1)q - q_l\right], \tag{30}$$

the above covariances are given by

$$\langle f'\theta_v'\rangle = [1 + (R - 1)\langle q_t\rangle - R\langle q_l\rangle]\langle f'\theta_l'\rangle + (R - 1)\langle\theta\rangle\langle f'q_t'\rangle$$
$$+ \left\{\frac{\langle\theta\rangle}{\langle T\rangle}\frac{L_v}{c_p}[1 + (R - 1)\langle q_t\rangle - R\langle q_l\rangle] - R\langle\theta\rangle\right\}\langle f'q_l'\rangle, \tag{31}$$

where $R = R_v/R_d$, and a generic variable f stands for u_i, θ_l or q_t. In order to arrive at Eq. (31), the third-order covariances and the pressure fluctuations are neglected (the latter assumption yields $\theta_l = \theta - \frac{\langle\theta\rangle}{\langle T\rangle}\frac{L_v}{c_p}q_l$).

In the "dry" limit, where a given numerical-model grid box is cloud free, Eq. (31) reduces to

$$\langle f'\theta_v'\rangle_d = [1 + (R - 1)\langle q_t\rangle]\langle f'\theta_l'\rangle + (R - 1)\langle\theta\rangle\langle f'q_t'\rangle, \tag{32}$$

where θ_l and q_t coincide with θ and q, respectively, as $q_l = 0$. In the "wet" limit, where a given grid box is uniformly saturated, $\langle f'\theta_v'\rangle$ can be expressed, to a good approximation, in terms of $\langle f'\theta_l'\rangle$ and $\langle f'q_t'\rangle$ as follows:

$$\langle f'\theta_v'\rangle_w = \left[1 + (R - 1)\langle q_t\rangle - R\langle q_l\rangle - \frac{\mathcal{A}\mathcal{P}}{\mathcal{Q}}\right]\langle f'\theta_l'\rangle$$
$$+ \left[(R - 1)\langle\theta\rangle + \frac{\mathcal{A}}{\mathcal{Q}}\right]\langle f'q_t'\rangle, \tag{33}$$

where $\mathcal{A} = \frac{\langle\theta\rangle}{\langle T\rangle}\frac{L_v}{c_p}[1 + (R - 1)\langle q_t\rangle - R\langle q_l\rangle] - R\langle\theta\rangle$, $\mathcal{P} = \frac{\langle T\rangle}{\langle\theta\rangle}\langle q_{sl,T}\rangle$, $\mathcal{Q} = 1 + \frac{L_v}{c_p}\langle q_{sl,T}\rangle$, and $\langle q_{sl,T}\rangle \equiv \left.\frac{\partial q_s}{\partial T}\right|_{T=\langle T_l\rangle}$ is computed from the Clausius–Clapeyron equation, $\partial q_s/\partial T = L_v q_s/(R_v T^2)$. A first-order Taylor expansion of the saturation-specific humidity $q_s(T)$ about $T = \langle T_l \rangle$ is used to derive Eq. (33).

The simplest way to determine $\langle f' \theta'_v \rangle$ is to use either Eq. (32) or Eq. (33), assuming that a given numerical-model grid box is either all clear or all cloudy, respectively. This "all-or-nothing" approach may be used in cloud-resolving models (although some caution is still required). Since it essentially assumes no sub-grid scale fluctuations of cloud water related variables, it is not applicable in the framework of the NWP and climate models whose horizontal resolution is too coarse to resolve cloud-scale motions. As sizable SGS fluctuations of cloud water related variables exist, an expression is needed that is valid not only in the dry and wet limits, but also in the general case of fractional cloudiness. To this end, an interpolation formula is used,

$$\langle f' \theta'_v \rangle = (1 - \hat{\mathcal{R}}) \langle f' \theta'_v \rangle_d + \hat{\mathcal{R}} \langle f' \theta'_v \rangle_w , \qquad (34)$$

where $\hat{\mathcal{R}}$ is the interpolation variable satisfying $0 \leq \hat{\mathcal{R}} \leq 1$. In case the PDFs of SGS fluctuations of θ_l, q_t and u_3 (vertical velocity) are Gaussian and the fluctuations of u_3 and q_l are uncorrelated, $\hat{\mathcal{R}}$ is identical to the fractional cloud cover $\hat{\mathcal{C}}$. In case the fluctuations of u_3 and q_l are correlated, the PDFs can differ significantly from the Gaussians, and $\hat{\mathcal{R}}$ can deviate widely from $\hat{\mathcal{C}}$. This is the case for shallow cumuli, where the fractional cloud cover is small, u_3 and q_l are strongly correlated, the PDFs of cloud related variables are highly skewed, and $\hat{\mathcal{R}}$ can be several times larger than $\hat{\mathcal{C}}$ [121]. In order to account for both Gaussian and non-Gaussian cases, Eq. (34) can be conveniently recast as follows:

$$\langle f' \theta'_v \rangle = (1 - \hat{\mathcal{C}}) \langle f' \theta'_v \rangle_d + \hat{\mathcal{C}} \langle f' \theta'_v \rangle_w + F_{NG} \hat{\mathcal{C}} (1 - \hat{\mathcal{C}}) \langle f' \theta'_v \rangle_w , \qquad (35)$$

where a correct behaviour in the dry $\hat{\mathcal{C}} = 0$ and the wet $\hat{\mathcal{C}} = 1$ limits is ensured. The deviations from the Gaussian limit are accounted for through the correction function F_{NG}. It is a complicated function of various cloud related quantities, such as the mean saturation deficit, variances of θ_l, q_t and u_3, and their skewness. In practice, simplified formulations of F_{NG} are utilized that ignore the dependencies on some of these quantities. Usable formulations that provide a smooth transition between the Gaussian state and the non-Gaussian skewed state are presented in e.g. [19, 20, 47, 121, 122].

The fractional cloud cover $\hat{\mathcal{C}}$ should now be determined. Cloud-cover parameterization schemes proposed to date vary in terms of their complexity and physical realism. Comprehensive reviews are given in [198, 199], where further references can be found. Here, only the very basic ideas are briefly outlined.

The commonly used relative humidity schemes are termed so since they employ the grid-scale mean relative humidity $\langle RH \rangle$ as the chief predictor of the cloud cover. The sub-grid scale fluctuations of temperature and humidity enable clouds to form even though a numerical-model grid box in question is unsaturated on the average, $\langle RH \rangle < 1$. A critical relative humidity $\langle RH \rangle_{\text{cr}}$, below which $\hat{\mathcal{C}} = 0$, is introduced, and the fractional cloud cover $\hat{\mathcal{C}}$ is assumed to increase monotonically with increasing $\langle RH \rangle$ until $\hat{\mathcal{C}} = 1$ when $\langle RH \rangle = 1$. Additional predictors, such as the grid-scale mean vertical velocity, are used in some schemes. Although the SGS variability of temperature and humidity is implicit in the relative humidity schemes, the connection between the fractional cloud cover and the SGS dynamics is rather loose.

The schemes referred to as the SGS statistical cloud schemes, pioneered by Sommeria and Deardorff [181] and Mellor [140], make use of PDFs of the SGS humidity (and temperature) fluctuations. Once the PDF is specified, the fractional cloud cover is simply the integral over a saturated part of the PDF. Since clouds can result both from the humidity fluctuations and from the temperature fluctuations, the latter ones change the local saturation vapour pressure, it is convenient to introduce, following Mellor [140], a variable $s = \mathcal{Q}^{-1}(\langle q_t \rangle - \langle q_{sl} \rangle + q_t' - \mathcal{P}\theta_l')$, where $\langle q_{sl} \rangle = q_s(\langle T_l \rangle)$ (s here should not be confused with the dry static energy used in Sect. 3.1). The variable s represents the local value of the liquid water specific humidity computed with respect to the linearized saturation specific humidity curve. This quantity has already been used above to express $\langle f' q_l' \rangle$ through $\langle f' q_t' \rangle$ and $\langle f' \theta_l' \rangle$ in Eq. (31), leading to Eq. (33). Assuming that no supersaturation occurs, the fractional cloud cover and the grid-box mean liquid water-specific humidity are given by

$$\hat{\mathcal{C}} = \int_0^\infty G(s)\mathrm{d}s, \qquad \langle q_l \rangle = \int_0^\infty sG(s)\mathrm{d}s, \tag{36}$$

where $G(s)$ is the PDF of s. If a Gaussian PDF is assumed, then $\hat{\mathcal{C}} = \frac{1}{2}\left[1 + \mathrm{erf}\left(\frac{\langle s \rangle}{\sqrt{2}\sigma_s}\right)\right]$ and $\langle q_l \rangle = \hat{\mathcal{C}}\langle s \rangle + \frac{\sigma_s}{\sqrt{2\pi}}\exp\left(\frac{\langle s \rangle^2}{2\sigma_s^2}\right)$, where erf is the error function, $\langle s \rangle = \mathcal{Q}^{-1}(\langle q_t \rangle - \langle q_{sl} \rangle)$ is the mean value of s and $\sigma_s \equiv \langle s'^2 \rangle^{1/2} = \mathcal{Q}^{-1}\left[\langle q_t'^2 \rangle + \mathcal{P}^2\langle \theta_l'^2 \rangle - 2\mathcal{P}\langle q_t'\theta_l' \rangle\right]^{1/2}$ is its standard deviation. Notice that σ_s depends on the variances of q_t and of θ_l and on their covariance. This provides an important link between the cloud cover and the dynamics of SGS motions. Apart from the Gaussian PDFs, various other PDFs have been proposed. A number of them are non-symmetric. Besides the first and the second moments of the distribution (i.e. the mean and the variance), they require higher-order moments, e.g. skewness, as an input. Tompkins [198, 199] presented a comprehensive review of the PDFs proposed by various authors and discussed several consistency issues, such as the use of statistical cloud schemes in the atmospheric models that carry a prognostic equation for $\langle q_l \rangle$. He also showed that there is no clear distinction between the statistical schemes and the relative humidity schemes. If the PDF moments are kept constant in space and time, the statistical cloud-cover formulations can be recast in terms of relative humidity.

Although prognostic equations may be (and often are) used to compute the PDF moments, the fractional cloud cover is determined diagnostically in the framework of statistical cloud schemes. Some other schemes, exemplified by the Tiedtke scheme [197], take a different approach – they determine $\hat{\mathcal{C}}$ from its own prognostic equation. Merits and shortcomings of such schemes are discussed in [40, 81, 115, 199, 206].

The q_t-θ_l system outlined above can be extended to the case of three phases including cloud ice [56]. To this end, the total water-specific humidity is generalized to account for the presence of ice, $q_t = q + q_l + q_i$, where q_i is the "solid water specific humidity" (the mass of cloud ice per unit mass of moist air), and the ice–liquid water potential temperature is introduced, $\theta_{il} = \theta - \frac{\theta}{T}\frac{L_v}{c_p}q_l - \frac{\theta}{T}\frac{L_i}{c_p}q_i$, where L_i is the specific heat of sublimation. The saturation specific humidity requires a generalized definition, the simplest of which is $q_s = (1 - \mathcal{F}_i)q_{sl} + \mathcal{F}_i q_{si}$, where $\mathcal{F}_i = q_i/(q_l + q_i)$ is the cloud-ice fraction of the total cloud condensate, and q_{sl} and q_{si} are the saturation-specific humidity for the vapour–liquid equilibrium and for the vapour–ice equilibrium, respectively. If ice and liquid water are allowed to co-exist over a certain temperature range, a simple function of temperature can be used to determine $\mathcal{F}_i$. Alternatively, a rate equation for $\mathcal{F}_i$ can be employed [56]. The use of the q_t-θ_{il} system raises various issues, such as allowance for supersaturation and consistency with the prognostic equations for $\langle q_i \rangle$ and its precipitating components that are carried by many atmospheric models. These issues necessitate an extensive discussion that is beyond the scope of the present paper.

In closing this section it should be emphasized that an accurate prediction of the fractional cloud cover and of the amount of cloud condensate is of great importance for radiation calculations. The SGS cloud scheme is thus an essential component of the physical parameterization package of an atmospheric model that provides a tight coupling between various parameterization schemes.

3 Mass–Flux Convection Schemes

This section discusses the mass–flux modelling framework that is widely used to parameterize convection, both deep precipitating and shallow non-precipitating, in numerical models of the atmosphere. First, the most salient features of the mass–flux convection schemes, as they are currently used in NWP and related applications, are recollected. Then, the analogy between budget equations for the second-order moments of fluctuating fields derived within the mass–flux modelling framework and within the ensemble–mean second-order modelling framework are examined. These exercises help to elucidate the physical meaning of some closure assumptions and disposable parameters of mass–flux schemes. They also demonstrate the similarities and the differences between the two approaches and suggest how the mass flux parameterizations can be formulated in terms of the ensemble–mean second-order closures and vice versa. The analysis of the second-order moment budgets is performed using the simplest "two-delta-function" mass–flux framework. Most currently used mass–flux schemes are (formally) based on a slightly more complex "three-delta-function" framework. However, the use of a simplified two-delta-function framework does not affect the principal results from the analysis.

3.1 Outline of Basic Features

In this section, the basic features of the mass–flux convection schemes are outlined. Attention is focused on the scheme developed by Tiedtke [196] (hereafter T89). That scheme is taken as an example as it was the first comprehensive mass–flux scheme that found a wide utility in NWP and climate modelling. Other mass–flux convection schemes have been developed to date, as for instance, the Kain–Fritsch scheme [96–98], the Gregory and Rowntree scheme [82], and the scheme used in the IFS of ECMWF [18, 93]. Further examples are the schemes proposed by Emanuel [59] and by Bechtold et al. [17]. Although various mass–flux schemes differ from the T89 scheme in many details, they rest on the same basic assumptions. Early ideas regarding the parameterization of convection in atmospheric models, including the moisture convergence schemes (e.g. [106, 107]), convective adjustment schemes (e.g. [26, 27]), and mass–flux schemes (e.g. [13, 32]), are discussed in [12, 65, 195]. A comparative analysis of several cumulus parameterization schemes is given in [185].

The T89 scheme, as well as its derivatives, utilises a triple top-hat decomposition. A fluctuating quantity in question is represented as

$$\overline{f} = a_u f_u + a_d f_d + a_e f_e, \tag{37}$$

where a generic variable f refers to the vertical velocity w, the dry static energy per unit mass $s = c_p T + gz$, the specific humidity q, the specific cloud water content q_l, or to any other quantity treated by a parameterization scheme. We focus attention on the mass–flux parameterization of scalar transport. Momentum transport by convection and its parameterization through the mass–flux approach are considered in e.g. [80, 101]. The notation with $w = u_3$, $z = x_3$ and $g = -g_3$ is used in this section to stress the one-dimensionality of the approach, where only vertical convective transport is considered. An overbar denotes a horizontal mean, and the subscripts "u", "d" and "e" refer to the contribution from convective updraughts, from convective downdraughts and from the environmental air, respectively. The fractional areas of updraughts, a_u, of downdraughts, a_d, and of environmental air, a_e, satisfy $a_u + a_d + a_e = 1$.

Notice that a coherent top-hat part of the quantity in question does not contain residual "sub-plume" fluctuations. Hence, the moments of fluctuating fields in the mass–flux approximation do not generally coincide with the moments in the ensemble framework (although they may be close to each other if the coherent part dominates). An overbar is used to denote quantities in the mass–flux approximation.

Although the T89 scheme is formally based on the triple decomposition (37), the properties of environmental air do not appear in the governing equations. The reason is that a mean over the environment is taken to be equal to a horizontal mean. For scalar quantities, this means $s_e = \bar{s}$, $q_e = \bar{q}$ and $q_{le} = \bar{q_l}$. Assuming further that updraughts and downdraughts are in a steady state, the T89 scheme solves a number of ordinary differential equations (in the vertical co-ordinate) for the mass fluxes and for the fluxes of scalar quantities in convective updraughts and convective downdraughts. The equations for convective updraughts read

$$\frac{\partial M_u}{\partial z} = E_u - D_u, \tag{38}$$

$$\frac{\partial}{\partial z} M_u X_u = E_u \overline{X} - D_u X_u + \bar{\rho} a_u F_{xu}, \tag{39}$$

where M_u is the updraught mass flux defined as

$$M_u = \bar{\rho} a_u (w_u - \overline{w}), \tag{40}$$

ρ is the density, and E_u and D_u are the rates of mass entrainment and detrainment per unit length. In order to closely follow the nomenclature traditionally used in the description of mass–flux schemes, the density appears explicitly in the equations of this section. A scalar X stands for s, q or q_l, and F_x stands for the source of the scalar X due to condensation/evaporation and precipitation fall-out. Similar equations are formulated for convective downdraughts, except that liquid water flux is zero. The problem is closed through the use of several parameterization rules to specify the vertical extent of convection, the fluxes through the cloud base and the cloud top, the type of convection (penetrative, mid-level or shallow), and the rates of entrainment and detrainment.

The vertical extent of convection is specified using the parcel method (see e.g. [174]). The fluxes through the cloud base are related to the moisture convergence in the sub-cloud layer (convergence of moisture fluxes due to both resolved-scale and sub-grid scale motions), as is the case in the original T89 scheme. An alternative formulation is through the convective available potential energy (CAPE), as, for example, in the ECMWF IFS convection scheme [93] in case of deep convection. The entrainment and detrainment rates are split into two parts, turbulent entrainment/detrainment through the cloud edges (lateral boundaries) and organized entrainment/detrainment through the cloud edges and through the cloud base and the cloud top. The organized entrainment through the updraught edges is set proportional to the large-scale moisture convergence. It is only considered for penetrative and mid-level convection and for the layer from the cloud base up to the level of strongest vertical ascent. The organized detrainment of updraughts occurs at the cloud top. The organized detrainment of downdraughts occurs in the sub-cloud layer. The rates of turbulent entrainment and detrainment are set proportional to the mass flux, that is $E_u = \epsilon_u M_u$ and $D_u = \delta_u M_u$ for the updraughts (similar for the downdraughts), where ϵ_u and δ_u are simply taken to be constant (different for different types of convection).

The simplifying assumptions of the T89 scheme, as well as of other mass–flux convection schemes, are many and varied. The two crucial assumptions have already been mentioned above: (i) convection is in a quasi-steady state and (ii) the mean over the environment is equal to the mean over a grid box. Several other points should be emphasized at once. First, although convective motions are driven by buoyancy, the buoyancy term is not explicitly present in Eq. (38). Second, neither Eq. (38) for the mass–flux nor Eq. (39) for the flux of a scalar contains pressure terms. In ensemble–mean second-order closure models, the pressure–velocity and pressure–scalar correlations appear explicitly (see Sect. 2). The fact that these terms do not appear explicitly in mass–flux equations suggests that the other terms serve to perform their function. This issue is discussed below. The third point to note is that the updraught fraction a_u and the updraught vertical velocity w_u are not estimated separately – only their combination, namely, the updraught mass flux M_u, is computed through Eqs. (38) and (40).

Some mass–flux convection schemes make use of an equation for the updraught kinetic energy. It reads (see e.g. [93])

$$\frac{1}{2}\frac{\partial w_u^2}{\partial z} = \frac{1}{C_{k1}(1+C_{k2})}g\frac{T_{vu}-\overline{T}}{\overline{T}} - \frac{\mu_u}{M_u}(1+C_{k3}C_{k4})w_u^2, \qquad (41)$$

where T_v is the virtual temperature, and C_{k1}, C_{k2}, C_{k3} and C_{k4} are empirical dimensionless constants. The "mixing coefficient" μ_u is set equal to either E_u or D_u, whichever is larger. The vertical acceleration on the left-hand side of Eq. (41) is a difference between a buoyancy force and a drag force represented by the first term and the second term on the r.h.s. of Eq. (41), respectively. Equation (41) originates from the work of Simpson et al. ([177], see also [79, 178]). It is meant to describe the vertical acceleration of a cumulus tower that is treated as an idealized jet, a buoyant rising thermal, or a "thermal" with vortical internal circulation. The tower kinetic energy is given by $\frac{1}{2}w_u^2$. In this respect, it is not clear if Eq. (41) is a good approximation to the budget equation for the kinetic energy of convective motions averaged over a grid box of a numerical model.

3.2 The Two-Delta-Function Mass–Flux Framework

In this section, the conventional two-delta-function mass–flux framework that has been widely used to parameterize atmospheric convection, shallow cumulus convection in particular, is briefly described. A more detailed account of the mass–flux framework is given in [52, 83, 110, 111, 162].

In the two-delta-function mass–flux framework, a fluctuating quantity in question is represented as

$$\overline{f} = a_u f_u + a_d f_d, \tag{42}$$

where the subscripts "u" and "d" refer to contributions from convective up-draughts and convective downdraughts, respectively The fractional areas of updraughts, a_u, and downdraughts, a_d, satisfy $a_u + a_d = 1$. The decomposition (42) can be formulated in terms of the probabilities of convective motions to be either updraughts, P_u, or downdraughts, P_d, so that $P_u + P_d = 1$ (e.g. [146, 209, 225]). The downdraught in the two-delta-function mass–flux framework should not be confused with the downdraught in the three-delta-function framework. Equation (37) reduces to Eq. (42) if the downdraught and the environment in the triple decomposition are treated together.

All moments of fluctuating fields in the mass–flux approximation are computed through the following averaging rule:

$$\begin{aligned}
\overline{w'^n X'^m} &= a \left(w_u - \overline{w}\right)^n \left(X_u - \overline{X}\right)^m + (1-a)\left(w_d - \overline{w}\right)^n \left(X_d - \overline{X}\right)^m \\
&= a(1-a)\left[(1-a)^{m+n-1} - (-a)^{m+n-1}\right]\left(w_u - w_d\right)^n \left(X_u - X_d\right)^m,
\end{aligned} \tag{43}$$

where $a = a_u$ and $1 - a = a_d$. According to Eq. (43), the flux of a quantity X is given by

$$\begin{aligned}
\overline{w'X'} &= a(w_u - \overline{w})(X_u - \overline{X}) + (1-a)\left(w_d - \overline{w}\right)\left(X_d - \overline{X}\right) \\
&= a(1-a)\left(w_u - w_d\right)\left(X_u - X_d\right) = \frac{M_c}{\overline{\rho}}(X_u - X_d), \tag{44}
\end{aligned}$$

where M_c is the convective mass flux introduced in [162],

$$M_c = \overline{\rho}a(1-a)(w_u - w_d). \tag{45}$$

The vertical-velocity variance and the scalar variance are given by

$$\overline{w'^2} = a(1-a)(w_u - w_d)^2, \qquad \overline{X'^2} = a(1-a)(X_u - X_d)^2, \tag{46}$$

and the third-order moments are given by

$$\overline{w'^3} = a(1-a)(1-2a)(w_u - w_d)^3, \tag{47}$$

$$\overline{X'^3} = a(1-a)(1-2a)(X_u - X_d)^3, \tag{48}$$

$$\overline{w'^2 X'} = a(1-a)(1-2a)(w_u - w_d)^2(X_u - X_d), \tag{49}$$

$$\overline{w'X'^2} = a(1-a)(1-2a)(w_u - w_d)(X_u - X_d)^2. \tag{50}$$

The following distinctive features of Eqs. (44, 45, 46, 47, 48, 49 and 50) should be emphasized. All second-order and third-order moments vanish in the limiting cases of $a = 0$ and of $a = 1$. The former case is merely the case of no convection. The case $a = 1$ corresponds to a convective updraught that covers the entire horizontal area in question, e.g. the entire grid box of a numerical model. Then, the updraught is no longer a sub-grid scale feature. With $a = 1$ it becomes a grid scale feature that should be described by the evolution equations for the resolved fields. Notice that this is not the case for the T89 and similar mass–flux convection schemes. Due to their implicit assumption that $a \ll 1$, those convection schemes remain active no matter how large/small the horizontal size of a grid box of a numerical model as compared to the size of an updraught. The lack of sensitivity to the grid size becomes a serious problem as the resolution of numerical models is increased.

The r.h.s. of Eqs. (47, 48, 49 and 50) contain a factor $1 - 2a$, i.e. the third-order moments vanish as $a = 1/2$. The value of $a = 1/2$ corresponds to zero skewness. It is readily shown using Eqs. (46) and (47) that the vertical-velocity skewness is given by $S_w = [a(1 - a)]^{-1/2}(1 - 2a)$. Notice that in the framework of the simplest mass–flux approach considered here the magnitude of the skewness S_x of a scalar field X is the same as the magnitude of S_w. An extended two-scale mass–flux framework [83] enables S_x to be different in magnitude from S_w. The fact that the third-order moments vanish if the sub-grid scale velocity and scalar fields are not skewed is accounted for by the expressions (19) and (20) discussed in Sect. 2.2. In this regard, Eqs. (19) and (20) are nothing but the mass–flux Eqs. (49) and (50) recast in terms of the ensemble–mean quantities used in the second-order closure approach.

The budget equations for the updraught and for the downdraught are (see e.g. [52, 110])

$$\frac{\partial}{\partial t}\overline{\rho}aX_u + \frac{\partial}{\partial z}\overline{\rho}aw_uX_u = EX_d - DX_u + \overline{\rho}aF_{xu}, \tag{51}$$

$$\frac{\partial}{\partial t}\overline{\rho}(1-a)X_d + \frac{\partial}{\partial z}\overline{\rho}(1-a)w_dX_d = DX_u - EX_d + \overline{\rho}(1-a)F_{xd}, \tag{52}$$

where E (D) is the lateral mass exchange rate from the sinking (rising) fluid into the rising (sinking) fluid.

Setting $X_u = X_d = 1$ and $F_{xu} = F_{xd} = 0$ in Eqs. (51) and (52) yields

$$\frac{\partial}{\partial t}\overline{\rho}a + \frac{\partial}{\partial z}\overline{\rho}aw_u = E - D, \tag{53}$$

$$\frac{\partial}{\partial t}\overline{\rho}(1-a) + \frac{\partial}{\partial z}\overline{\rho}(1-a)w_d = D - E. \tag{54}$$

Adding Eqs. (53) and (54) gives the continuity equation for the mean flow,

$$\frac{\partial\overline{\rho}}{\partial t} + \frac{\partial}{\partial z}\overline{\rho}\,\overline{w} = 0. \tag{55}$$

Multiplying Eq. (54) by a and subtracting the result from Eq. (53) times $1 - a$ yields the equation that relates the fractional area of the updraught with the mass–flux divergence and with the entrainment and detrainment rates. It reads

$$\overline{\rho}\left(\frac{\partial}{\partial t} + \overline{w}\frac{\partial}{\partial z}\right)a = -\frac{\partial}{\partial z}\overline{\rho}a(1-a)(w_u - w_d) + E - D$$

$$= -\frac{\partial M_c}{\partial z} + E - D. \tag{56}$$

Notice an essential difference to Eq. (38) where the substantial derivative of a is neglected. This neglect deprives the mass–flux convection schemes of memory, making the vertical profile of mass flux to adjust instantaneously to the current state of the atmosphere.

Adding Eqs. (51) and (52) and rearranging gives the equation for $\overline{X}$ in the mass–flux approximation,

$$\overline{\rho}\left(\frac{\partial}{\partial t} + \overline{w}\frac{\partial}{\partial z}\right)\overline{X} = -\frac{\partial}{\partial z}M_c(X_u - X_d) + \overline{\rho F_x}. \tag{57}$$

A direct analogy to the ensemble–mean equation for mean scalar concentration is immediately recognized. The first term on the r.h.s. of Eq. (57) is the mass–flux analogue of the turbulent scalar flux divergence term in the ensemble–mean equation.

3.3 Analogies Between the Mass–Flux and the Ensemble–Mean Second-Moment Budgets

In this section, analogies between the mass–flux and the ensemble–mean budget equations for the second-order moments are examined. The budgets of the scalar variance, of the vertical-velocity variance and of the vertical scalar flux are considered. It should be mentioned that these budget equations are not *explicitly* carried by most of the mass–flux models developed to date. Their consideration is, however, required in order to elucidate the physical meaning of the various terms in the mass–flux model equations. An analysis of the scalar–variance equations has been previously performed by de Roode et al. [52] and Lappen and Randall [110]. They found, among other things, that the sum of the lateral entrainment and detrainment rates in the mass–flux equation corresponds to the inverse scalar–variance dissipation time scale in the ensemble–mean equation. For the sake of clarity and completeness, the treatment of the scalar–variance budget is repeated here. We then extend the analysis of de Rode et al. and of Lappen and Randall to examine the budgets of the vertical-velocity variance and of the vertical scalar flux, giving particular attention to the role of the pressure–velocity and pressure–scalar covariances. The two-delta-function framework is used for the analysis. The use of the three-delta-function framework would make derivations more cumbersome, but would not affect the results in a principal way.

Scalar Variance

Subtracting Eq. (52) times a from Eq. (51) times $1 - a$, then multiplying the result by $X_u - X_d$ and rearranging yields the budget equation for the scalar variance in the mass–flux approximation. Omitting algebraic manipulations, we obtain

$$\frac{1}{2}\left(\frac{\partial}{\partial t}+\overline{w}\frac{\partial}{\partial z}\right)a(1-a)(X_u-X_d)^2 =$$

$$-\frac{M_c}{\overline{\rho}}(X_u-X_d)\frac{\partial\overline{X}}{\partial z}-\frac{1}{2\overline{\rho}}\frac{\partial}{\partial z}(1-2a)M_c(X_u-X_d)^2$$

$$-\frac{E+D}{2\overline{\rho}}(X_u-X_d)^2+a(1-a)(X_u-X_d)(F_{xu}-F_{xd}). \qquad (58)$$

This equation should be compared with the ensemble–mean budget equation for the scalar variance,

$$\frac{1}{2}\left(\frac{\partial}{\partial t}+\langle w\rangle\frac{\partial}{\partial z}\right)\langle X'^2\rangle =$$

$$-\langle w'X'\rangle\frac{\partial\langle X\rangle}{\partial z}-\frac{1}{2}\frac{\partial}{\partial z}\langle w'X'^2\rangle-\epsilon_x+\langle X'F'_x\rangle. \qquad (59)$$

The terms on the r.h.s. of Eq. (59) represent the mean–gradient production/destruction, turbulent transport, the dissipation rate of the scalar variance, and the source term. Except for the source term, Eq. (59) is simply a one-dimensional form of Eqs. (6) and (7).

A comparison shows that there is a direct analogy between the first, the second and the fourth terms on the r.h.s. of the mass–flux and of the ensemble–mean equations. The third term on the r.h.s. of the mass–flux equation (58) is negative definite. It acts to decrease the scalar difference between the updraught and the downdraught and can, by analogy with the third term on the r.h.s. of the ensemble–mean equation (59), be interpreted as the scalar-variance dissipation. Then, the quantity $2a(1-a)\overline{\rho}/(E+D)$ in the mass–flux equation corresponds to the scalar-variance dissipation time scale $\tau_{\epsilon x}\equiv\langle X'^2\rangle/\epsilon_x$ in the ensemble–mean equation. It is worth noting [52, 110] that the mass–flux "dissipation" term does not originate directly from the molecular diffusion term in the scalar equation. This term originates from the lateral exchange terms in the updraught–downdraught model and from their parameterization through the entrainment–detrainment concept. The above analogy is useful as it guides the way to set the rates of entrainment and detrainment in the mass-flux models. These quantities should be parameterised so as to provide the most realistic scalar–variance dissipation rate. Further requirements are imposed by the budgets of the vertical-velocity variance and of the vertical scalar flux.

Vertical-Velocity Variance

The budget equations for the vertical velocity in the updraught and in the downdraught are

$$\frac{\partial}{\partial t}\overline{\rho}aw_u + \frac{\partial}{\partial z}\overline{\rho}aw_u w_u =$$
$$- a(\partial p/\partial z)_u - \overline{\rho}a g\frac{\theta_r - \theta_u}{\theta_r} + Ew_d - Dw_u, \qquad (60)$$

$$\frac{\partial}{\partial t}\overline{\rho}(1-a)w_d + \frac{\partial}{\partial z}\overline{\rho}(1-a)w_d w_d =$$
$$- (1-a)(\partial p/\partial z)_d - \overline{\rho}(1-a)g\frac{\theta_r - \theta_d}{\theta_r} + Dw_u - Ew_d, \qquad (61)$$

where p is the deviation of pressure (here not divided by density) from its reference value in hydrostatic equilibrium. To simplify notation, θ is used in the buoyancy terms instead of θ_v.

Adding Eqs. (60) and (61), using Eqs. (42), (55) and (56) and rearranging gives the equation for $\overline{w}$,

$$\overline{\rho}\left(\frac{\partial}{\partial t} + \overline{w}\frac{\partial}{\partial z}\right)\overline{w} = -\overline{\partial p/\partial z} - \overline{\rho}g\frac{\theta_r - \overline{\theta}}{\theta_r} - \frac{\partial}{\partial z}M_c(w_u - w_d), \qquad (62)$$

where the third term on the r.h.s. is the mass–flux analogue of the Reynolds stress divergence term in the ensemble–mean equation.

Equations (60), (61) and (62) for the vertical velocity are similar to Eqs. (51), (52) and (57) for a scalar except that the equations for w contain the pressure–gradient and the buoyancy terms, the first and the second terms on the r.h.s. of Eqs. (60), (61) and (62), respectively.[4] The updraught–downdraught decomposition of the buoyancy term presents no difficulties. The treatment of the pressure terms in the mass–flux framework is tricky and requires special consideration.

First and foremost we emphasize that the updraught–downdraught decomposition cannot be applied to the pressure itself. A straightforward decomposition of $\overline{p}$ through Eq. (42) assumes a zero-order pressure jump across the updraught–downdraught interface that would result in a spurious source term in the equation for the vertical-velocity variance. A rigorous way to go is to take the divergence of the momentum equation and to solve the resulting Poisson equation for the fluctuating pressure in terms of the Green's function (see Sect. 2.2). The pressure field so obtained would be consistent with the governing momentum and scalar equations.

We take a different approach. It is based on the observation that the pressure terms are *actually not explicitly considered at all* in the mass–flux models.[5] For example, the updraught mass flux and the updraught fluxes of

---

[4] In our formulation, there is no source term in the vertical momentum equation. A more rigorous formulation, including momentum changes due to the presence of hydrometeors (e.g. rain and snow), is given in [14].

[5] Strictly speaking, the pressure terms are not present in the overwhelming majority of the mass–flux models developed to date. An exception is the mass–flux

scalar quantities in the T89 scheme are computed on the basis of Eqs. (38) and (39), where the scalar source terms account for the effects of condensation/evaporation and of precipitation fall-out. In this way no pressure effects are explicitly accounted for. This is apparently because the mass continuity within the mass–flux framework is assumed to be satisfied exactly from the very outset. Even so, the pressure effects should be implicitly accounted for in the mass–flux second-moment budgets, and it remains to be seen which terms in the budgets serve this function. From the above line of reasoning, the pressure terms in Eqs. (60), (61) and (62) should be set to zero so that they do not appear in their explicit form in the mass–flux second-moment budgets.

The equation for the vertical-velocity variance is derived in the same way as the equation for the scalar variance. Subtracting Eq. (61) times a from Eq. (60) times $1 - a$, then multiplying the result by $w_u - w_d$ and rearranging, we obtain the budget equation for the vertical-velocity variance in the mass–flux approximation. It reads

$$
\frac{1}{2}\left(\frac{\partial}{\partial t} + \overline{w}\frac{\partial}{\partial z}\right)a(1-a)(w_u - w_d)^2 = -\frac{M_c}{\overline{\rho}}(w_u - w_d)\frac{\partial \overline{w}}{\partial z} + \frac{g}{\theta_r}\frac{M_c}{\overline{\rho}}(\theta_u - \theta_d)
$$

$$
- \frac{1}{2\overline{\rho}}\frac{\partial}{\partial z}(1-2a)M_c(w_u - w_d)^2 - \frac{E+D}{2\overline{\rho}}(w_u - w_d)^2. \tag{63}
$$

We have omitted the algebraic manipulations leading to Eq. (63) as they are fairly straightforward. Equation (63) should be compared with the ensemble–mean budget equation for the vertical-velocity variance [cf. Eq. (3)],

$$
\frac{1}{2}\left(\frac{\partial}{\partial t} + \langle w\rangle\frac{\partial}{\partial z}\right)\langle w'^2\rangle = -\langle w'^2\rangle\frac{\partial\langle w\rangle}{\partial z} + \frac{g}{\theta_r}\langle w'\theta'\rangle
$$

$$
- \frac{1}{2}\frac{\partial}{\partial z}\langle w'^3\rangle - \frac{1}{\overline{\rho}}\langle w'\partial p'/\partial z\rangle - \epsilon_w, \tag{64}
$$

where the terms on the r.h.s. represent the mean–gradient production/destruction, the buoyancy production/destruction, turbulent transport, the vertical velocity-pressure gradient covariance, and the dissipation rate of the vertical-velocity variance.

Comparing Eqs. (63) and (64), we conclude that there is a direct analogy between the first, the second and the third terms on the r.h.s. of the mass–flux and the ensemble–mean equations. As to the last term on the r.h.s. of Eq. (63), two interpretations can be suggested.

---

model ADHOC developed by Lappen and Randall [110–112]. However, the parameterization of the pressure terms in ADHOC is based on the ensemble–mean second-order closure ideas. A new version of ADHOC, ADHOC2 [113], incorporates a representation of the pressure terms that is consistent with the mass–flux framework [114]. Earlier attempts to account for the effect of perturbation pressure on cumulus convection are reported in e.g. [90, 207, 211].

We recall (see Sect. 2.1) that it is common practice within the ensemble–mean second-order modelling framework to separate out the pressure transport from the pressure gradient–velocity correlation [represented by the fourth term on the r.h.s. of Eq. (64)] and to model pressure transport together with the turbulent transport [the divergence of the third-order velocity correlation represented by the third term on the r.h.s. of Eq. (64)]. The rest of the pressure term (pressure redistribution) is modelled separately. It is usually decomposed into the rapid part and the slow part, where the slow part is believed to return turbulence towards isotropy (Sect. 2.2). Numerous studies have revealed the importance of pressure terms in maintaining the second-moment budgets in turbulent flows. Inadequate modelling of pressure terms, the pressure redistribution in particular, most often results in inaccurate prediction of fluxes and variances and consequently of the mean fields.

Assume that no account whatsoever is taken of the pressure effects in the mass–flux framework. Then the last term on the r.h.s. of the mass–flux equation (63) can be interpreted as the dissipation of the vertical-velocity variance. This term acts to decrease the vertical-velocity difference between the updraught and the downdraught and is negative definite. With no pressure terms in the mass–flux budget, the transport of variance is solely due to the third-order velocity correlation, and the pressure redistribution is not accounted for. In convective flows, both pressure transport and pressure redistribution are known to be substantial. Their neglect results in a deficient vertical-velocity variance budget.

Another possible interpretation of the last term on the r.h.s. of Eq. (63) can be offered by assuming that, although the pressure term is not *explicitly* considered in the mass–flux budget, the pressure effects are *implicitly* accounted for. Then, the last term on the r.h.s. of (63) should describe the *combined* effect of the dissipation and of the pressure redistribution. In most convective flows, the pressure redistribution acts to reduce the vertical-velocity variance. This is explained from the following simple reasoning. The major source of energy in convection is the buoyancy that directly feeds the vertical component of the fluctuating velocity. The horizontal components of the fluctuating velocity are not fed directly. They grow at the expense of the vertical velocity component. By this means turbulence is driven toward isotropy. It is the traceless part of the pressure gradient–velocity correlation, i.e. the pressure redistribution term, that accounts for the inter-component energy exchange. Since both the dissipation and the pressure redistribution tend to reduce the vertical-velocity variance, it is reasonable to assume that the last term on the r.h.s. of Eq. (63) accounts for their combined effect. A problem, however, arises if we attempt to reconcile the last term in the vertical-velocity variance budget (63) with a similar term in the scalar variance budget (58). The term with $E + D$ in the scalar variance budget describes the scalar variance dissipation, whereas a similar term in the vertical-velocity variance budget describes the combined effect of the dissipation and of the pressure redistribution. Putting it differently, the same quantity $2a(1-a)\bar{\rho}/(E+D)$ should characterize both the scalar-variance relaxation time scale due to dissipation and the velocity-variance relaxation time scale due to both dissipation and pressure redistribution. Further difficulties are encountered with the mass–flux budget equation for the vertical scalar flux.

Scalar Flux

Subtracting Eq. (52) times a from Eq. (51) times $1 - a$ and rearranging gives the equation for the updraught–downdraught scalar difference $X_u - X_d$. Similar manipulations with Eq. (61) and (60) (recall that there are no explicit pressure terms in the mass–flux equations) gives the equation for the updraught–downdraught vertical-velocity difference $w_u - w_d$. Then, adding the equation for $X_u - X_d$ multiplied by $w_u - w_d$ and the equation for $w_u - w_d$ multiplied by $X_u - X_d$ and rearranging, we obtain the budget equation for the vertical scalar flux in the mass–flux approximation. Omitting algebraic manipulations, we obtain

$$
\left(\frac{\partial}{\partial t} + \overline{w} \frac{\partial}{\partial z} \right) a(1-a)(w_u - w_d)(X_u - X_d) =
$$

$$
- \frac{M_c}{\overline{\rho}} (w_u - w_d) \frac{\partial \overline{X}}{\partial z} - \frac{M_c}{\overline{\rho}} (X_u - X_d) \frac{\partial \overline{w}}{\partial z}
$$

$$
+ a(1-a) \frac{g}{\theta_r} (\theta_u - \theta_d)(X_u - X_d) - \frac{1}{\overline{\rho}} \frac{\partial}{\partial z} (1 - 2a) M_c (w_u - w_d)(X_u - X_d)
$$

$$
- \frac{E+D}{\overline{\rho}} (w_u - w_d)(X_u - X_d) + a(1-a)(w_u - w_d)(F_{xu} - F_{xd}). \qquad (65)
$$

Equation (65) should be compared with the ensemble–mean budget equation for the vertical scalar flux [cf. Eqs. (4) and (5)],

$$
\left(\frac{\partial}{\partial t} + \langle w \rangle \frac{\partial}{\partial z} \right) \langle w'X' \rangle = - \langle w'^2 \rangle \frac{\partial \langle X \rangle}{\partial z} - \langle w'X' \rangle \frac{\partial \langle w \rangle}{\partial z}
$$

$$
+ \frac{g}{\theta_r} \langle \theta'X' \rangle - \frac{\partial}{\partial z} \langle w'^2 X' \rangle - \frac{1}{\overline{\rho}} \langle X' \partial p'/\partial z \rangle + \langle w'F_x' \rangle, \qquad (66)
$$

where the terms on the r.h.s. represent the mean–gradient production/destruction (the first two terms on the r.h.s.), the buoyancy production/destruction, turbulent transport, the pressure gradient–scalar covariance, and the source term.

There is a direct analogy between all the terms in the mass–flux and the ensemble–mean budgets, except the previous last terms. The previous last term on the r.h.s. of the mass–flux equation (65) is purely destructive. It acts to decrease the magnitude of the vertical scalar flux. Its counterpart in the ensemble–mean equation (66) is, however, not the dissipation term, but the pressure gradient–scalar covariance. In high Reynolds number flows (with local isotropy at small scales, see Sect. 2.2), molecular dissipation of scalar fluxes is negligible, and it is the pressure gradient–scalar covariance that destroys the scalar flux. The term with $E + D$ in the mass–flux budget (65) should, therefore, be interpreted as the term that describes the pressure effects. This is in apparent contradiction with the interpretation of similar terms in the scalar variance budget (58) and in the vertical-velocity variance budget (63), where the terms with $E + D$ describe the dissipation and the combined effect of dissipation and pressure redistribution, respectively. Since the scalar–variance dissipation, the velocity-variance dissipation, the pressure redistribution and the pressure gradient–scalar covariance depend on the flow variables in different ways, it is not easy to describe all the above effects in terms of only two quantities, viz., the rates of lateral entrainment and detrainment (cf. [51, 192]).

A positive outcome of the above analysis of the second-moment budgets is that it suggests an extended formulation for the rates of turbulent entrainment E and detrainment D. Recall that the traditional formulation sets E and D proportional to the mass flux M_c through the constant fractional entrainment and detrainment rates, ϵ and δ, respectively (their dimensions is m^{-1}; ϵ here should not be confused with the TKE dissipation rate). An extended formulation is proposed in [148]. It reads

$$(E, D) = M_c \left[(\epsilon, \delta) + C_B a^2 (1-a)^2 \frac{g}{\theta_r} \frac{\theta_u - \theta_d}{(M_c/\bar{\rho})^2} \right]$$
$$= M_c \left[(\epsilon, \delta) + C_B \frac{g}{\theta_r} \frac{\theta_u - \theta_d}{(w_u - w_d)^2} \right], \qquad (67)$$

where C_B is a dimensionless constant.

The flow of arguments leading to Eq. (67) is as follows. First, the pressure redistribution term in Eq. (64) and the pressure gradient–scalar covariance term in Eq. (66) are parameterized through Eqs. (12) and (13), respectively, where only the first and the third terms on the r.h.s. are retained. These are return-to-isotropy and the buoyancy parts of the pressure terms that are typically the dominant contributions in convective flows [144, 149]. Next, it is assumed that all pressure relaxation (return-to-isotropy) time scales and the dissipation time scales are proportional to each other and to the "master" relaxation time scale $\tau_m = e^{-1/2} l_m$, l_m being the "master" length scale, and that the vertical-velocity variance $\langle w'^2 \rangle$ is proportional to the TKE e. Then, the above analogies between the mass–flux and the ensemble–mean second-moment budgets are exploited to infer that $(E + D)/\bar{\rho} \propto a(1-a)/\tau_m = a(1-a)e^{1/2}/l_m$. Finally, using the definition of the convective mass flux (45) and recalling that $(w_u - w_d)$ is the mass–flux analogue of $\langle w'^2 \rangle^{1/2}$ which is assumed to scale on e, we obtain $(E, D) = (\epsilon, \delta) M_c$ with $(\epsilon, \delta) \propto l_m^{-1}$. Then, setting $l_m = const$ yields $(\epsilon, \delta) = const$. It is easy to verify that the second term in brackets on the r.h.s. of Eq. (67) stems from the buoyancy contributions to the pressure terms in the budgets of the vertical-velocity variance and of the scalar flux.

Notice that the traditional formulation for E and D, i.e. Eq. (67) without the second term in brackets on the r.h.s., can be obtained from the above reasoning if the simplest Rotta-type formulations (10) and (11) are used instead of (12) and (13), respectively, to parameterize the pressure terms in the second-moment budgets. It should also be mentioned that the above analysis of the second-moment budgets does not allow to discriminate between E and D. That is, it provides no guidance as to whether the buoyancy correction term in Eq. (67) should be applied to E, to D, or to both E and D.

A formulation for the fractional entrainment rate ϵ that is very similar to the second term in brackets on the r.h.s. of Eq. (67) was proposed by Gregory [79] from different physical considerations.

4 Towards a Unified Description of Boundary-Layer Turbulence and Shallow Convection

Having discussed the second-order closure and the mass–flux modelling frameworks in some detail, it is appropriate to return to the question raised in the Introduction. That is, whether regime-dependent parameterization schemes should be developed to describe various types of fluctuating motions, or some unification of different parameterization frameworks could be achieved. Although a definitive answer to this question does not seem to exist at present, there is a growing interest in unifying various parameterization ideas (see discussions in [12, 145, 187]). Considering the cumulus parameterization problem, Arakawa [12] states:

> It is rather obvious that for future climate models the scope of the problem must be drastically expanded from "cumulus parameterization" to "unified cloud parameterization" or even to "unified model physics". This is an extremely challenging task, both intellectually and computationally, and the use of multiple approaches is crucial even for a moderate success.

The tasks of developing a "unified cloud parameterization" and eventually a "unified model physics" are very ambitious. Most NWP and climate models will unlikely enjoy the use of such general parameterization frameworks for some, perhaps many, years to come. However, a less ambitious task, namely, a unified description of boundary-layer turbulence and shallow convection, seems to be feasible. There are several ways to do so, but it is not a priory clear which way should be preferred. A number of attempts have been made to develop a more unified turbulence-shallow convection parameterization schemes. They can be classified, rather loosely, into three groups.

Extended mass–flux schemes are built around the top-hat updraught–downdraught representation of fluctuating quantities. As discussed above, the simplest top-hat mass–flux representation is equivalent to assuming a two-delta-function PDF, where the motions can be either updraughts or

downdraughts and the sum of the probabilities of the two admissible states
is one. Since the variety of motions is not exhausted by quasi-organized updraughts and downdraughts, the mass–flux equations are extended by adding
the "sub-plume scale" motions. These motions are thought to be small-scale
and chaotic, so that they can be parameterized on the basis of the second-
order closure ideas the simplest of which is the down-gradient approximation
of fluxes. Lappen and Randall [110–112] developed an extended mass–flux
scheme termed Assumed-Distribution Higher-Order Closure (ADHOC) that
parameterizes boundary-layer turbulence and shallow convection in a unified
framework. As the heart of the scheme is the two-delta-function mass–flux representation, ADHOC is attractive for describing non-local convective transport. Missing components, namely, parameterizations of the subplume scale
fluxes, of the pressure terms, and, to some extent, of the dissipation terms, are
borrowed from the ensemble–mean second-order modelling framework. An up-
dated version of ADHOC, ADHOC2 [113, 114], includes parameterizations of
pressure terms and of momentum fluxes consistent with the mass–flux framework (more specifically, with the assumed spatial distribution based on the
two types of idealized coherent structures – plumes and rolls).

Parameterization schemes where the mass–flux closure ideas and the
ensemble–mean second-order closure ideas have roughly equal standing can
be labelled as *hybrid schemes*. These are exemplified by the Eddy-Diffusivity/
Mass–Flux (EDMF) scheme proposed by Soares et al. [180] based on earlier work of Siebesma and Teixeira [176]. In the framework of the EDMF
scheme, the vertical flux of a fluctuating quantity f is represented as a
sum of two contributions [175], one is assumed to stem from the small-scale
chaotic eddies and is described with the eddy-diffusivity down-gradient formulation, and the other is assumed to stem from the convective-layer-scale
quasi-organized plumes and is described with the mass–flux formulation. That
is, $\langle w'f' \rangle = -K_f \partial \langle f \rangle / \partial z + (M_u/\overline{\rho})(f_u - \langle f \rangle)$, where M_u is the convective updraught mass flux, f_u is the value of f in the updraught, and $\langle f \rangle$ is the value
of f averaged over a grid box of a host numerical model. The eddy diffusivity
K_f is estimated on the basis of the TKE, for which a prognostic equation is
carried, and of a diagnostic formulation for the turbulence length scale [180].
A simple entraining parcel model is used to determine M_u and f_u. The EDMF
scheme operates throughout the convective layer, from the (near vicinity of
the) surface up to the top of shallow cumuli, and does not require switching
between turbulence and convection schemes (until deep convection is triggered
that still requires a separate parameterization scheme). The scheme is formulated in terms of moist quasi-conservative variables, the liquid water potential
temperature θ_l and the total water specific humidity q_t, and an SGS statistical cloud scheme (Sect. 2.5) is used to predict fractional cloud cover and the
amount of cloud condensate.

Some features of the EDMF scheme deserve critical consideration. A parcel
model used to determine M_u and f_u assumes that the updraughts fractional
area coverage a_u is small as compared to the horizontal grid size of a host

atmospheric model. Then, the mass–flux component of the EDMF scheme inherits all shortcomings of the "traditional" mass–flux schemes (Sect. 3). In particular, there is no resolution dependency – the mass–flux component remains active irrespective of the ratio of the horizontal size of numerical grid to the size of the updraught. Furthermore, there is no dependency on the skewness of fluctuating fields – the skewness is always large by virtue of a small a_u. This is an important difference to the ADHOC scheme which guarantees that a_u approaches $1/2$ as skewness approaches zero.

Variances $\langle f'^2 \rangle$ of scalar quantities in the EDMF scheme are diagnosed on the basis of a truncated scalar–variance equation, see Eqs. (6) and (7), where only the dissipation term and the mean-gradient term are retained, that is $\epsilon_f = -\langle w'f' \rangle \partial \langle f \rangle /\partial z$. Then, parameterizing the dissipation rate through the dissipation time scale, $\epsilon_f = \langle f'^2 \rangle /\tau_f$, and using the above EDMF formulation for the flux $\langle w'f' \rangle$, yields the expression for $\langle f'^2 \rangle$. Notice that using the mass–flux formulation for the flux and at the same time neglecting the third-order transport term in the scalar–variance equation is not quite consistent. Numerous analyses of observational and LES data (e.g. [118, 147, 150]) indicate the importance of the turbulent transport term $-\frac{1}{2}\partial \langle w'f'^2 \rangle /\partial z$ in maintaining the scalar-variance budget in the well-mixed CBL core, where the mean-gradient term is small. Furthermore, as discussed in Sects. 2.2 and 3.2, it is the third-order transport term in the scalar-variance equation that accounts for the non-local transport by skewed convective turbulence. Neglecting this term is inconsistent with the assumption of large skewness (small a_u). Notice also that the formulation for $\langle f'^2 \rangle$ neglecting the third-order transport term should not allow for the counter-gradient scalar flux (when $\langle w'f' \rangle$ and $\partial \langle f \rangle /\partial z$ have the same sign) that is known to often occur in convective flows. A counter-gradient scalar flux would lead to totally spurious negative values of the scalar–variance dissipation and hence of the scalar variance. In the EDMF scheme of Soares et al. [180], this situation is avoided by applying a clipping operation – the mass–flux contribution to the scalar–variance is set to zero whenever it becomes negative. As discussed in Sect. 2.5, variances of θ_l and of q_t are the key input parameters for statistical parameterizations of fractional cloudiness. They should be accurately predicted. This may not be achieved without an accurate treatment of the third-order transport terms in the scalar–variance equations.

Notwithstanding their shortcomings, the EDMF-like hybrid schemes are attractive for they are simple and computationally efficient. They enjoy growing popularity in atmospheric modelling (e.g. [11, 103]).

Non-local second-order closure schemes represent one more alternative to describe boundary-layer turbulence and shallow convection in a unified framework. In pursuing this aim, a number of schemes based on the ensemble–mean equations for the statistical moments of fluctuating fields have been developed. These range from low-order turbulence closures, where the only prognostic equation is the TKE equation (e.g. [19, 21]), to high-order closures, where transport equations are carried for all second-order and third-order moments involved (e.g. [29–31]). These schemes proved to do a fair job of describing turbulence and shallow convection. Using moist quasi-conservative variables and well-tuned statistical parameterizations of fractional cloudiness, these schemes appeared to be capable of describing cumuliform and stratiform boundary-layer clouds in a unified framework.

Turbulence closure schemes based on the ensemble–mean equations are often blamed for their inability to describe non-local transport due to quasi-organized convective motions. Both heavily truncated second-order closures and sophisticated high-order closures suffer from this drawback. The incorporation of additional transport equations for third-order and possibly higher-order moments makes the schemes very complex and computationally expensive. The gain in terms of accuracy of their performance is, however, not as tangible as one would expect in the hope that making crude assumptions on the high-order moments would still yield an accurate prediction of low-order moments of interest, viz., of fluxes and variances and of the mean fields. What is most likely to be at fault is the assumption, which is either explicit or implicit in most turbulence closures based on the ensemble–mean equations, that the PDF of fluctuating fields is approximately Gaussian. For example, a third-order closure that makes use of the Millionshchikov hypothesis (quasi-Gaussian approximation, Sect. 2.2) to parameterise fourth-order moments is fairly sophisticated, and yet it fails to properly account for non-local nature of convective turbulence.

As the analysis in Sects. 2.2 and 3.2 suggests, the inability of traditional ensemble–mean closures to describe non-local convective transport may well be apparent rather than real. In the second-order modelling framework, one of the key points is the parameterization of the third-order transport moments in the second-moment equations. It is these terms that are largely responsible for non-local transport properties of convective motions. Recall that the third-order terms are usually parameterized through the simple down-gradient diffusive approximations, or through the use of their own transport equations where the fourth-order moments are taken to be Gaussian. In order to account for non-local transport properties of convection, additional terms should be added to the formulations for the third-order moments, namely, the terms dependent on the skewness of fluctuating fields. Such additional terms for the third-order moments in the temperature–flux equation and in the temperature–variance equation are given by Eqs. (19) and (20) respectively. It must be stressed that Eqs. (19) and (20) are simply the mass–flux Eqs. (49) and (50) recast in terms of the ensemble–mean quantities. Since the mass–flux formulations are advective rather than diffusive in character and are intrinsically non-local [110], their ensemble–mean counterparts should also be able to properly account for non-local convective transport. The skewness of fluctuating fields (may be different for different quantities, e.g. [83, 146]) should be determined from transport equations for the third-order moments, where the fourth-order moments are represented through skewness-dependent formulations (21, 22, 23 and 24) (the generalized Millionshchikov hypothesis) which again are consistent with the mass–flux formulations in the non-Gaussian limit. In some situations, simplified algebraic formulations for skewness may appear to be sufficient [19] (see also the chapter "Turbulence in Astrophysical and Geophysical Flows" of this volume).

Notice that extended non-local second-order closure schemes with skewness-dependent formulations for the third-order transport terms are likely to be more stable numerically as they would not violate realisability in case of large skewness. They are also more consistent with statistical parameterizations of fractional cloudiness many of which utilize a skewed PDF to describe cumuliform clouds (see Sect. 2.5).

The above classification of unified turbulence-shallow convection schemes is rather arbitrary. For example, the "assumed PDF scheme" of Golaz et al. [70, 71] can be viewed as an extension of the ADHOC scheme of Lappen and Randall, where a two-delta-function PDF is replaced with a two-Gaussian-function PDF. Alternatively, it can be viewed as an extended ensemble–mean high-order closure based on a rather flexible two-Gaussian PDF. Generally speaking, any of the three approaches outlined above should yield the same result if parameterizations are formulated and implemented clearly and consistently. Putting it differently, it should not matter much which conceptual framework is used as a basis, i.e. whether a unified scheme is built within the mass–flux modelling framework and the missing components (e.g. parameterization of the sub-plume scale fluxes) are borrowed from the ensemble–mean framework, or whether it is built within the ensemble–mean modelling framework and the missing components (e.g. parameterization of the third-order transport) are borrowed from the mass–flux framework. However, a clear and consistent formulation requires certain level of complexity, and that level is likely to be higher than most NWP and climate models can afford. In view of stringent requirements of computational economy, simpler schemes are called for that are based on a (heavily) truncated set of equations for statistical moments of fluctuating fields. Mass–flux schemes (or hybrid schemes) are likely to be preferred for some years to come in situations where non-local transport properties of fluctuating fields is a major concern. In a long-term perspective, however, unified schemes built around the ensemble–mean second-order closures seem to be more appealing. The following arguments in favour of this viewpoint can be adduced. Due to a rapid development of computers, the resolution of numerical models of the atmosphere is continuously refined. As the mesh size becomes small, increasingly more quasi-organized flow structures, that are chiefly responsible for non-local transport, are resolved. Then, the focus of SGS parameterizations is shifted towards motions at smaller scales, which are (presumably) more chaotic, and towards other issues, such as the anisotropy of turbulence near the surface and in stably stratified regions of the flow and an accurate parameterization of pressure redistribution and pressure transport. The second-order closures are attractive for describing these very features.

Now we outline the next step that, in the author's opinion, should be made to go beyond the level of one-equation closure schemes (the level 2.5 schemes in the Mellor–Yamada nomenclature) that have been and still are the draft horses of atmospheric turbulence modelling in NWP, climate studies, and related applications. Closure schemes that presently carry only one prognostic equation, viz., the TKE equation, should be extended to incorporate prognostic equations for the scalar variances. This suggestion is almost trivial as may be inferred from the following arguments. The key to successful modelling of any turbulent flow is an adequate description of the flow energy. In neutrally stratified flows, the kinetic energy of turbulence is a major (or the only) concern. This explains why the one-equation closure schemes have been used to advantage in simulating neutral flows. The situation is essentially different in flows where the density (buoyancy) stratification is different from neutral. In such flows, the turbulence potential energy (TPE) plays an important part along with the TKE. The TKE is spent to work against the gravity and is converted into the TPE in stably stratified flows. In convective flows, the TKE grows at the expense of the TPE. The rate of TKE↔TPE conversion is represented by the buoyancy–flux term $\beta_i \langle u_i' \theta_v' \rangle$ that enters the TKE equation (9) as a source (sink) term. Since the atmospheric flows are virtually never hydrostatically neutral, and the TKE and the TPE in stratified flows are equally important, it is difficult to adduce plausible arguments in favour of one form of energy over the other. Both energies should be treated in a similar way. In the dry atmospheric CBL, the potential temperature is the only thermodynamic variable that affects the distribution of buoyancy. The TPE is proportional to the temperature variance [172] that should be determined from Eq. (6), where the representation of the third-order transport term should account for the non-local character of skewed convective motions. In case of moist atmosphere, the TPE depends on the variances $\langle \theta_l'^2 \rangle$ and $\langle q_t'^2 \rangle$ of moist quasi-conservative variables and on their correlation $\langle \theta_l' q_t' \rangle$. A parameterisation of fractional cloudiness is additionally required to determine the buoyancy terms.

Closure schemes for atmospheric applications, that carry transport equations for both the TKE and for the scalar variances, have been developed by e.g. Kenjereš and Hanjalić [100] and Nakanishi and Niino [153]. Curiously, it was noticed already by Mellor and Yamada in their classical 1974 paper [141] that the scheme (level 3 in their nomenclature) that carries two prognostic equations, for the TKE and for the potential–temperature variance, is particularly attractive. Three closure schemes of various complexity were applied to simulate a PBL subject to a diurnally varying surface heat flux. The level 3 scheme proved to outperform an algebraic closure scheme. Little was gained if the most complex of the three schemes, that carries transport equations for all second-order moments involved, was used. Thus, the level 3 scheme was found to be the best compromise between physical realism and computational economy. The message does not seem to have been got by the geophysical turbulence-modelling community. Most users of the Mellor–Yamada closures gave preference to the level 2.5 scheme in spite of its obvious shortcomings.

In closing this section, yet another way of representing convection and turbulence in numerical models of the atmosphere should be mentioned. Two- or even three-dimensional models capable of resolving cloud scales are embedded into grid-boxes of coarse-resolution atmospheric models. This way to tackle the sub-grid scale parameterization problem was unthinkable a decade ago, but a drastically increased computer power has made it possible nowadays. Consideration of this innovative approach called "cloud-resolving convective parameterization" [74, 75] or "super-parameterization" [161] is beyond the scope of the present chapter.

5 Conclusions

Modelling (parameterizing) turbulence and shallow convection in the lower troposphere as it is practised in NWP and related applications is discussed. Although turbulence and convection are both unresolved, sub-grid scale motions and a distinction between the two is quite ambiguous, different concepts are typically used to parameterize them in numerical models of the atmosphere. The ensemble–mean second-order closure approach is taken to describe turbulence, deemed to represent quasi-random small-scale motions, whereas the mass–flux closure approach is taken to describe convection, deemed to represent quasi-organized motions of larger scales.

The ensemble–mean second-order closure framework is outlined with the emphasis on the parameterization of the pressure redistribution and of the third-order transport. A rather lengthy treatment appeared to be necessary to demonstrate how simplified turbulence parameterization schemes are obtained and what is lost on the way. As we have seen, only a small fraction of what is available nowadays is actually used in applications. This "keep it simple" strategy is justified in view of stringent requirements of computational economy that parameterization schemes for NWP, climate modelling and similar applications should necessarily meet. Nonetheless, incorporating more of the essential physics into the existing turbulence schemes is highly desirable.

The mass–flux closure framework is outlined with reference to its simplifying assumptions and to its limits of applicability. The analogies between the second-moment budgets derived in the mass–flux and in the ensemble-mean second-order modelling frameworks are analysed. The analysis shows that the two modelling frameworks have very much in common and that the parameterization ideas developed in one framework can be translated into the language of the other. Further outcome of the analysis is an extended formulation for the rates of turbulent entrainment and detrainment, Eq. (67), the key parameters in the mass–flux convection schemes.

As the artificial separation of processes and scales in numerical models of the atmosphere causes many conceptual and practical problems [12] (the turbulence–convection separation being an example), there is a growing need for a more consistent description of turbulence and shallow non-precipitating convection within a unified parameterization framework. Several alternative ways to achieve such a description are considered and their pros and cons are discussed. A non-local second-order closure scheme, that carries prognostic equations for both kinetic energy and potential energy of sub-grid scale fluctuating motions and incorporates skewness-dependent formulations for the third-order moments, seems to be an attractive alternative.

In a long-term perspective, deep precipitating convection should also be incorporated into a unified turbulence–convection scheme. This task is very difficult and intellectually challenging, and quick success is by no means guaranteed. Except for very high-resolution atmospheric models capable of resolving deep convective motions, separate parameterization schemes for deep precipitating convection will be used over some, perhaps many, years to come. Then, these schemes should be adjusted to adequately respond to an increasing resolution of host atmospheric models and to work in harmony with improved turbulence-shallow convection schemes. To this end, some restrictive assumptions of deep convection schemes may need to be relaxed, e.g. the assumptions of steady-state and of small fractional area coverage of convective updraughts. Steps forward in this direction are described in [69, 158].

Acknowledgements

The author is grateful to Wolfgang Hillebrandt and Friedrich Kupka for the invitation to prepare this article, and to Peter Bechtold, Michael Buchhold, Vittorio Canuto, Sergey Danilov, Stephan de Roode, Jean-François Geleyn, Andrey Grachev, Erdmann Heise, Martin Köhler, Friedrich Kupka, Cara-Lyn Lappen, Vasily Lykossov, Detlev Majewski, Pedro Miranda, Veniamin Perov, Jean-Marcel Piriou, David Randall, Matthias Raschendorfer, Bodo Ritter, Pier Siebesma, Pedro Soares, Joao Teixeira, Jun-Ichi Yano, Sergej Zilitinkevich, and particularly to Evgeni Fedorovich and Vladimir Gryanik for useful discussions and helpful suggestions at various stages of the preparation of this article.

References

1. Abdella, K., McFarlane, N.: A new second-order turbulence closure scheme for the planetary boundary layer. J. Atmos. Sci. **54**, 1850–1867 (1997)
2. Abdella, K., McFarlane, N.: Reply. J. Atmos. Sci. **56**, 3482–3483 (1999)
3. Abdella, K., Petersen, A.C.: Third-order moment closure through the mass-flux approach. Boundary-Layer Meteorol. **95**, 303–318 (2000)
4. Akylas, E., Tombrou, M., Lalas, D., Zilitinkevich, S.: Surface fluxes under shear-free convection. Quart. J. Roy. Meteorol. Soc. **127**, 1–15 (2001)

5. André, J.C., De Moor, G., Lacarrère, P., du Vachat, R.: Turbulence approximation for inhomogeneous flows. Part I: The clipping approximation. J. Atmos. Sci. **33**, 476–481 (1976)
6. André, J.C., De Moor, G., Lacarrère, P., du Vachat, R.: Turbulence approximation for inhomogeneous flows. Part II: The numerical simulation of a penetrative convection experiment. J. Atmos. Sci. **33**, 482–491 (1976)
7. André, J.C., De Moor, G., Lacarrère, P., Therry, G., du Vachat, R.: Modelling the 24-hour evolution of the mean and turbulent structures of the planetary boundary layer. J. Atmos. Sci. **35**, 1861–1883 (1978)
8. Andreas, E.L.: Parameterizing scalar transfer over snow and ice: A review. J. Hydrometeor. **3**, 417–432 (2002)
9. Andreas, E.L., Cash, B.A.: Convective heat transfer over wintertime leads and polynyas. J. Geophys. Res. **104**, 25721–25734 (1999)
10. Andrén, A.: A TKE-dissipation model for the atmospheric boundary layer. Boundary-Layer Meteorol. **56**, 207–221 (1991)
11. Angevine, W.M.: An integrated turbulence scheme for boundary layers with shallow cumulus applied to pollutant transport. J. Appl. Meteorol. **44**, 1436–1452 (2005)
12. Arakawa, A.: The cumulus parameterization problem: past, present, and future. J. Climate **17**, 2493–2525 (2004)
13. Arakawa, A., Schubert, W.H.: Interaction of a cumulus cloud ensemble with the large-scale environment, Part I. J. Atmos. Sci. **31**, 674–701 (1974)
14. Bannon, P.R.: Theoretical foundations for models of moist convection. J. Atmos. Sci. **59**, 1967–1982 (2002)
15. Bannon, P.R.: Virtualization. J. Atmos. Sci. **64**, 1405–1409 (2007)
16. Baumert, H.Z., Simpson, J.H., Sündermann, J. (eds.): Marine Turbulence. Theories, Observations, and Models, 630 pp. Cambridge Univ. Press, Cambridge (2005)
17. Bechtold, P., Bazile, E., Guichard, F., Mascart, P., Richard, E.: A mass-flux convection scheme for regional and global models. Quart. J. Roy. Meteorol. Soc. **127**, 869–886 (2001)
18. Bechtold, P., Chaboureau, J.-P., Beljaars, A., Betts, A.K., Köhler, M., Miller, M., Redelsperger, J.-L.: The simulation of the diurnal cycle of convective precipitation over land in a global model. Quart. J. Roy. Meteorol. Soc. **130**, 3119–3137 (2004)
19. Bechtold, P., Cuijpers, J.W.M., Mascart, P., Trouilhet, P.: Modeling of trade wind cumuli with a low-order turbulence model: Toward a unified description of Cu and Sc clouds in meteorological models. J. Atmos. Sci. **52**, 455–463 (1995)
20. Bechtold, P., Siebesma, P.: Organization and representation of boundary layer clouds. J. Atmos. Sci. **55**, 888–895 (1998)
21. Bechtold, P., Fravalo, C., Pinty, J.-P.: A model of marine boundary-layer cloudiness for mesoscale applications. J. Atmos. Sci. **49**, 1723–1744 (1992)
22. Beljaars, A.C.M.: The parameterization of surface fluxes in large-scale models under free convection. Quart. J. Roy. Meteorol. Soc. **121**, 255–270 (1995)
23. Beljaars, A.C.M., Brown, A.R., Wood, N.: A new parametrization of turbulent orographic form drag. Quart. J. Roy. Meteorol. Soc. **130**, 1327–1347 (2004)
24. Beljaars, A.C.M., Holtslag, A.A.M.: Flux parameterization over land surfaces for atmospheric models. J. Appl. Meteorol. **30**, 327–341 (1991)

25. Betts, A.K.: Non-precipitating cumulus convection and its parameterization. Quart. J. Roy. Meteorol. Soc. **99**, 178–196 (1973)

26. Betts, A.K.: A new convective adjustment scheme. Part I: Observational and theoretical basis. Quart. J. Roy. Meteorol. Soc. **112**, 677–691 (1986)

27. Betts, A.K., Miller, M.J.: A new convective adjustment scheme. Part II: Single column tests using GATE wave, BOMEX, ATEX and arctic air-mass data sets. Quart. J. Roy. Meteorol. Soc. **112**, 693–709 (1986)

28. Blackadar, A.K.: The vertical distribution of wind and turbulent exchange in neutral atmosphere. J. Geophys. Res. **67**, 3095–3102 (1962)

29. Bougeault, P.: Modeling the trade-wind cumulus boundary-layer. Part I: Testing the ensemble cloud relations against numerical data. J. Atmos. Sci. **38**, 2414–2428 (1981)

30. Bougeault, P.: Modeling the trade-wind cumulus boundary-layer. Part II: A high-order one-dimensional model. J. Atmos. Sci. **39**, 2429–2439 (1981)

31. Bougeault, P.: Cloud-ensemble relations based on the Gamma probability distribution for the higher-order models of the planetary boundary layer. J. Atmos. Sci. **39**, 2691–2700 (1982)

32. Bougeault, P.: A simple parameterization of the large-scale effects of cumulus convection. Mon. Weather Rev. **113**, 2108–2121 (1985)

33. Bradley, F., Fairall, C.: A Guide to Making Climate Quality Meteorological and Flux Measurements at Sea. NOAA Technical Memorandum OAR PSD-311, 109 pp. Earth System Research Laboratory, Boulder, Colorado, USA (2006)

34. Brost, R.A., Wyngaard, J. C.: A model study of the stably stratified planetary boundary layer. J. Atmos. Sci. **35**, 1427–1440 (1978)

35. Brutsaert, W.: Evaporation into the Atmosphere, 299 pp. D. Reidel, Dordrecht, etc. (1982)

36. Bretherton, F.P.: Momentum transport by gravity waves. Quart. J. Roy. Meteorol. Soc. **95**, 213–243 (1969)

37. Burchard, H.: Applied Turbulence Modelling in Marine Waters, 215 pp. Springer, Berlin, etc. (2002)

38. Businger, J.A.: Reflections on boundary-layer problems of the last 50 years. Boundary-Layer Meteorol. **4**, 323–326 (1973)

39. Businger, J.A.: A note on free convection. Boundary-Layer Meteorol. **116**, 161–173 (2005)

40. Bushell, A.C., Wilson, D.R., Gregory, D.: A description of cloud production by non-uniformly distributed processes. Quart. J. Roy. Meteorol. Soc. **129**, 1435–1455 (2003)

41. Canuto, V.M., Dubovikov, M.: Stellar turbulent convection. I. Theory. Astrophys. J. **493**, 834–847 (1998)

42. Canuto, V.M., Cheng, Y., Howard, A.: New third-order moments for the convective boundary layer. J. Atmos. Sci. **58**, 1169–1172 (2001)

43. Canuto, V.M., Minotti, F., Ronchi, C., Ypma, R.M., Zeman, O.: Second-order closure PBL model with new third-order moments: comparison with LES data. J. Atmos. Sci. **51**, 1605–1618 (1994)

44. Cheng, Y., Canuto, V.M.: Stably stratified shear turbulence: A new model for the energy dissipation length scale. J. Atmos. Sci. **51**, 2384–2396 (1994)

45. Chou, P.-Y.: On velocity correlations and the solutions of the equations of turbulent fluctuation. Quart. Appl. Math. **3**, 38–54 (1945)

46. Craft, T.J., Ince, N.Z., Launder, B.E.: Recent developments in second-moment closure for buoyancy-affected flows. Dyn. Atmos. Oceans **23**, 99–114 (1996)
47. Cuijpers, J.W.M., Bechtold, P.: A simple parameterization of cloud water related variables for use in boundary layer models. J. Atmos. Sci. **52**, 2486–2490 (1995)
48. Cullen, M.J.P.: The unified forecast/climate model. Meteor. Mag. **122**, 81–94 (1993)
49. Daly, B. J., Harlow, F.H.: Transport equations in turbulence. Phys. Fluids **13**, 2634–2649 (1970)
50. Dakos, T., Gibson, M.M.: On modelling the pressure terms of the scalar flux equations. In: Turbulent Shear Flows 5, pp. 7–18 Durst, F., et al. (eds.), Springer, Berlin (1987)
51. de Roode, S.R., Bretherton, C.S.: Mass-flux budgets of shallow cumulus clouds. J. Atmos. Sci. **60**, 137–151 (2003)
52. de Roode, S.R., Duynkerke, P.G., Siebesma, A.P.: Analogies between mass-flux and Reynolds-averaged equations. J. Atmos. Sci. **57**, 1585–1598 (2000)
53. Deardorff, J.W.: Preliminary results from numerical integrations of the unstable planetary boundary layer. J. Atmos. Sci. **27**, 1209–1211 (1970)
54. Deardorff, J.W.: Convective velocity and temperature scales for the unstable planetary boundary layer and for Rayleigh convection. J. Atmos. Sci. **27**, 1211–1213 (1970)
55. Deardorff, J.W.: The use of subgrid transport equations in a three-dimensional model of atmospheric turbulence. J. Fluids Eng. **95**, 429–438 (1973)
56. Deardorff, J.W.: Usefulness of liquid-water potential temperature in a shallow-cloud model. J. Appl. Meteorol. **15**, 98–102 (1976)
57. du Vachat, R.: Realizability inequalities in turbulent flows. Phys. Fluids **20**, 551–556 (1977)
58. Duynkerke, P.G.: Application of the $E - \epsilon$ turbulence closure model to the neutral and stable atmospheric boundary layer. J. Atmos. Sci. **45**, 865–880 (1988)
59. Emanuel, K.A.: A scheme for representing cumulus convection in large-scale models. J. Atmos. Sci. **48**, 2313–2335 (2001)
60. Emanuel, K.A.: Atmospheric Convection, 580 pp. Oxford Univ. Press, Oxford (1994)
61. Fedorovich, E., Rotunno, R., Stevens, B. (eds.): Atmospheric Turbulence and Mesoscale Meteorology, 280 pp. Cambridge Univ. Press, Cambridge (2004)
62. Finnigan, J.J.: A note on wave-turbulence interaction and the possibility of scaling the very stable boundary layer. Boundary-Layer Meteorol. **90**, 529–539 (1999)
63. Fleagle, R.G., Businger, J.A.: An Introduction to Atmospheric Physics, 346 pp. Academic Press, New York (1963)
64. Foken, T.: 50 years of the Monin-Obukhov similarity theory. Boundary-Layer Meteorol. **119**, 431–447 (2006)
65. Frank, W.M.: The cumulus parameterization problem. Mon. Weather Rev. **111**, 1859–1871 (1983)
66. Galperin, B., Orszag, S.A. (eds.): Large Eddy Simulation of Complex Engineering and Geophysical Flows, 622 pp. Cambridge Univ. Press, Cambridge (1993)

67. Garratt, J.R.: The Atmospheric Boundary Layer, 316 pp. Cambridge Univ. Press, Cambridge (1992)

68. Geernaert, G.L. (ed.): Air-Sea Exchange: Physics, Chemistry and Dynamics, 578 pp. Kluwer Acad. Publ., Dordrecht, etc. (1999)

69. Gerard, L., Geleyn, J.-F.: Evolution of a subgrid convection parameterization in a limited-area model with increasing resolution. Quart. J. Roy. Meteorol. Soc. **131**, 2293–2312 (2005)

70. Golaz, J.-C., Larson, V.E., Cotton, W.R.: A PDF-based model for boundary layer clouds. Part I: Method and model description. J. Atmos. Sci. **59**, 3540–3551 (2002)

71. Golaz, J.-C., Larson, V.E., Cotton, W.R.: A PDF-based model for boundary layer clouds. Part II: Model results. J. Atmos. Sci. **59**, 3552–3571 (2002)

72. Golitsyn, G.S., Grachev, A.A.: Velocities and heat and mass transfer during convection in a two-component medium. Doklady Acad. Nauk SSSR **255**, 548–552 (1980)

73. Golitsyn, G.S., Grachev, A.A.: Free convection of multi-component media and parameterization of air-sea interaction at light winds. Ocean-Air Interactions **1**, 57–78 (1986)

74. Grabowski, W.W.: MJO-like coherent structures: sensitivity simulations using cloud-resolving convection parameterization (CRCP). J. Atmos. Sci. **60**, 847–876 (2003)

75. Grabowski, W.W., Smolarkiewicz, P.K.: CRCP: a cloud resolving convection parameterization for modeling the tropical convective atmosphere. Physica D **133**, 171–178 (1999)

76. Grachev, A.A., Andreas, E.L., Fairall, Ch.W., Guest, P.S., Persson, P.O.G.: Turbulent measurements in the stable atmospheric boundary layer during SHEBA: ten years after. Acta Geophys. **56**, 142–166 (2008)

77. Grachev, A.A., Fairall, C.W., Persson, P.O.G., Andreas, E.L., Guest, P.S.: Stable boundary-layer scaling regimes: the SHEBA data. Boundary-Layer Meteorol. **116**, 201–235 (2005)

78. Grachev, A.A., Panin, G.N.: Parameterization of the sensible and latent heat fluxes above the water surface in calm weather under natural conditions. Izvestija AN SSSR. Fizika Atmosfery i Okeana **20**, 364–371 (1984)

79. Gregory, D.: Estimation of entrainment rate in simple models of convective clouds. Quart. J. Roy. Meteorol. Soc. **127**, 53–72 (2001)

80. Gregory, D., Kershaw, R., Inness, P.M.: Parametrization of momentum transport by convection. II: Tests in single-column and general circulation models. Quart. J. Roy. Meteorol. Soc. **123**, 1153–1183 (1997)

81. Gregory, D., Wilson, D., Bushell, A.: Insights into cloud parametrization provided by a prognostic approach. Quart. J. Roy. Meteorol. Soc. **128**, 1485–1504 (2002)

82. Gregory, D., Rowntree, P.R.: A mass flux convection scheme with representation of cloud ensemble characteristics and stability-dependent closure. Mon. Weather Rev. **118**, 1483–1506 (1990)

83. Gryanik, V.M., Hartmann, J.: A turbulence closure for the convective boundary layer based on a two-scale mass-flux approach. J. Atmos. Sci. **59**, 2729–2744 (2002)

84. Gryanik, V. M., Hartmann, J., Raasch, S., Schröter, M.: A refinement of the Millionshchikov quasi-normality hypothesis for convective boundary layer turbulence. J. Atmos. Sci. **62**, 2632–2638 (2005)

85. Hallbäck, M., Henningson, D.S., Johansson, A.V., Alfredsson, P.H. (eds.): Turbulence and Transition Modelling, 369 pp. Kluwer Acad. Publ., Dordrecht, etc. (1996)

86. Hanjalić, K., Launder, B.E.: A Reynolds stress model of turbulence and its application to thin shear flows. J. Fluid Mech. **52**, 609–638 (1972)

87. Hassid, S., Galperin, B.: Modeling rotating flows with neutral and unstable stratification. J. Geophys. Res. **99**, 12,533–12,548 (1994)

88. Högström, U.: Non-dimensional wind and temperature profiles in the atmospheric surface layer: A re-evaluation. Boundary-Layer Meteorol. **42**, 55–78 (1988)

89. Högström, U.: Review of some basic characteristics of the atmospheric surface layer. Boundary-Layer Meteorol. **78**, 215–246 (1996)

90. Holton, J.R.: A one-dimensional cumulus model including pressure perturbations. Mon. Weather Rev. **101**, 201–205 (1973)

91. Houze, R.A.: Cloud Dynamics, 573 pp. Academic Press, San Diego, etc. (1993)

92. Holtslag, A.A.M., Boville, B.: Local versus nonlocal boundary-layer diffusion in a global climate model. J. Climate **6**, 1825–1842 (1993)

93. IFS Documentation, Cycle 28r1. Available from the web site of ECMWF, http://www.ecmwf.int (2004)

94. Jones, W.P., Musogne, P.: Modelling of scalar transport in homogeneous turbulent flows. In: Proc. of the 4th Turbulent Shear Flows Symposium, pp. 17.18–17.24 Karlsruhe, Germany (1983)

95. Kaimal, J.C., Finnigan, J.J.: Atmospheric Boundary Layer Flows. Their Structure and Measurements, 289 pp. Oxford Univ. Press, New York (1994)

96. Kain, J.S.: The Kain-Fritsch convection parameterization: an update. J. Appl. Meteorol. **43**, 170–181 (2004)

97. Kain, J.S., Fritsch, J.M.: A one-dimensional entraining/detraining plume model and its application in convective parameterization. J. Atmos. Sci. **47**, 2784–2802 (1990)

98. Kain, J.S., Fritsch; J.M.: Convective parameterization for mesoscale models: the Kain-Fritsch scheme. In: The Representation of Cumulus Convection in Numerical Models, pp. 165–170 Meteorol. Monogr. No. 24, Amer. Meteor. Soc., (1993)

99. Kantha, L.H., Clayson, C.A.: Small Scale Processes in Geophysical Fluid Flows, 888 pp. Academic Press, San Diego, etc. (2000)

100. Kenjereš, S., Hanjalić, K.: Combined effects of terrain orography and thermal stratification on pollutant dispersion in a town valley: a T-RANS simulation. J. Turbulence **3**, 1–25, doi:10.1088/1468-5248/3/1/026 (2002)

101. Kershaw, R., Gregory, D.: Parametrization of momentum transport by convection. I: Theory and cloud modelling results. Quart. J. Roy. Meteorol. Soc. **123**, 1133–1151 (1997)

102. Kitaigorodskii, S.A.: The Physics of Air-Sea Interaction, 284 pp. Gidrometeoizdat, Leningrad (1970) (English translation by A. Baruch, Israel Program for Scientific Translation, Jerusalem, 1973, 273 pp.)

103. Köhler, M.: Improved prediction of boundary layer clouds. ECMWF Newsletter, No. 104, 18–22 (2005)

104. Kolmogorov, A.N.: Local structure of turbulence in an incompressible fluid at very high Reynolds numbers. Doklady Acad. Nauk SSSR **30**, 299–303 (1941)

105. Komen, G.J., Cavalery, L., Donelan, M., Hasselmann, K., Hasselmann, S., Janssen, P.A.E.M.: Dynamics and Modelling of Ocean Waves, 532 pp. Cambridge Univ. Press, Cambridge (1994)
106. Kuo, H.L.: On formation and intensification of tropical cyclones through latent heat release by cumulus convection. J. Atmos. Sci. **22**, 40–63 (1965)
107. Kuo, H.L.: Further studies of the parameterization of the influence of cumulus convection on large-scale flow. J. Atmos. Sci. **31**, 1232–1240 (1974)
108. Kupka, F., Robinson, F.J.: On the effects of coherent structures on higher order moments in models of solar and stellar surface convection. Mon. Not. R. Astron. Soc. **374**, 305–322 (2007)
109. Lange, H.-J.: Die Physik des Wetters und des Klimas. Ein Grundkurs zur Theorie des Systems Atmosphäre. 625 pp. Dietrich Reimer Verlag, Berlin (2002)
110. Lappen, C.-L., Randall, D.A.: Toward a unified parameterization of the boundary layer and moist convection. Part I: A new type of mass-flux model. J. Atmos. Sci. **58**, 2021–2036 (2001)
111. Lappen, C.-L., Randall, D.A.: Toward a unified parameterization of the boundary layer and moist convection. Part II: Lateral mass exchanges and subplume-scale fluxes. J. Atmos. Sci. **58**, 2037–2051 (2001)
112. Lappen, C.-L., Randall, D.A.: Toward a unified parameterization of the boundary layer and moist convection. Part III: Simulations of clear and cloudy convection. J. Atmos. Sci. **58**, 2052–2072 (2001)
113. Lappen, C.-L., Randall, D.A.: Using idealized coherent structures to parameterize momentum fluxes in a PBL mass-flux model. J. Atmos. Sci. **62**, 2829–2846 (2005)
114. Lappen, C.-L., Randall, D.A.: Parameterization of pressure perturbations in a PBL mass-flux model. J. Atmos. Sci. **63**, 1726–1751 (2006)
115. Larson, V.E.: Prognostic equations for cloud fraction and liquid water, and their relation to filtered density functions. J. Atmos. Sci. **61**, 338–351 (2004)
116. Larson, V.E., Golaz, J.-C.: Using probability density functions to derive consistent closure relationships among higher-order moments. Mon. Weather Rev. **133**, 1023–1042 (2005)
117. Launder, B.E., Reece, G.J., Rodi, W.: Progress in the development of a Reynolds-stress turbulence closure. J. Fluid Mech. **68**, 537–566 (1975)
118. Lenschow, D.H., Wyngaard, J.C., Pennell, W.T.: Mean-field and second-moment budgets in a baroclinic, convective boundary layer. J. Atmos. Sci. **37**, 1313–1326 (1980)
119. Lenschow, D.H., Mann, J., Kristensen, L.: How long is long enough when measuring fluxes and other turbulence statistics? J. Atmos. Oceanic Technol. **11**, 661–673 (1994)
120. Lesieur, M.: Turbulence in Fluids, 515 pp. Kluwer Acad. Publ., Dordrecht, etc. (1997)
121. Lewellen, D.C., Lewellen, W.S.: Buoyancy flux modeling for cloudy boundary layers. J. Atmos. Sci. **61**, 1147–1160 (2004)
122. Lewellen, W.C., Yoh, S.: Binormal model of ensemble partial cloudiness. J. Atmos. Sci. **50**, 1228–1237 (1993)
123. Lilly, D.K.: The representation of small-scale turbulence in numerical simulation experiments. In: Proc. of the IBM Scientific Computing Symposium on Environmental Sciences, pp. 195–210 Yorktown Heights, New York, Thomas J. Watson Research Center, IBM (1967)

124. Lilly, D.K.: Models of cloud-topped mixed layers under a strong inversion. Quart. J. Roy. Meteorol. Soc. **94**, 292–309 (1968)

125. Lock, A.P.: The parameterization of entrainment in cloudy boundary layers. Quart. J. Roy. Meteorol. Soc. **124**, 2729–2753 (1998)

126. Lock, A.P.: The numerical representation of entrainment in parameterizations of boundary layer turbulent mixing. Mon. Weather Rev. **129**, 3187–3199 (2001)

127. Lock, A.P., Brown, A.R., Bush, M.R., Martin, G.M., Smith, R.N.: A new boundary layer mixing scheme. Part I: Scheme description and single-column model tests. Mon. Weather Rev. **128**, 3187–3199 (2000)

128. Losch, M.: On the validity of the Millionshchikov quasi-normality hypothesis for open-ocean deep convection. Geophys. Res. Lett. **31**, L23301, doi:10.1029/2004GL021412 (2004)

129. Lott, F., Miller, M.J.: A new subgrid-scale orographic drag parametrization: its formulation and testing. Quart. J. Roy. Meteorol. Soc. **123**, 101–127 (1997)

130. Lumley, J.L.: Computational modeling of turbulent flows. Adv. Appl. Mech. **18**, 123–176 (1978)

131. Lumley, J.L., Panofsky, H.: The Structure of Atmospheric Turbulence, 239 pp. John Wiley and Sons, New York (1964)

132. Mahrt, L.: Stratified atmospheric boundary layers and breakdown of models. J. Theor. Comp. Flud Dyn. **11**, 263–279 (1998)

133. Mahrt, L.: Stratified atmospheric boundary layers. Boundary-Layer Meteorol. **90**, 375–396 (1999)

134. Mahrt, L., Sun, J., Blumen, W., Delany, T., Oncley, S.: Nocturnal boundary-layer regimes. Boundary-Layer Meteorol. **88**, 375–396 (1998)

135. Majewski, D., Liermann, D., Prohl, P., Ritter, B., Buchhold, M., Hanisch, T., Paul, G., Wergen, W., Baumgardner, J.: Icosahedral-hexagonal gridpoint model GME: description and high-resolution tests. Mon. Weather Rev. **130**, 319–338 (2002)

136. Malhi, Y.: The behaviour of the roughness length for temperature over heterogeneous surfaces. Quart. J. Roy. Meteorol. Soc. **122**, 1095–1125 (1996)

137. Marshall, J., Schott, F.: Open-ocean convection: Observations, theory, and models. Rev. Geophys. **37**, 1–64 (1999)

138. Martin, G.M., Bush, M.R., Brown, A.R., Lock, A.P., Smith, R.N.: A new boundary layer mixing scheme. Part II: Tests in climate and mesoscale models. Mon. Weather Rev. **128**, 3200–3217 (2000)

139. McFarlane, N.A.: The effect of orographically excited gravity wave drag on the circulation of the lower stratosphere and troposphere. J. Atmos. Sci. **44**, 1775–1800 (1987)

140. Mellor, G.L.: The Gaussian cloud model relations. J. Atmos. Sci. **34**, 356–358 [Corrigendum 1483–1484] (1977)

141. Mellor, G.L., Yamada, T.: A hierarchy of turbulence closure models for planetary boundary layers. J. Atmos. Sci. **31**, 1791–1806 (1974)

142. Mellor, G.L., Yamada, T.: Development of a turbulence closure model for geophysical fluid problems. Rev. Geophys. **20**, 851–875 (1982)

143. Millionshchikov, M.D.: On the theory of homogeneous isotropic turbulence. Doklady Acad. Nauk SSSR **32**, 611–614 (1941)

144. Mironov, D.V.: Pressure-potential-temperature covariance in convection with rotation. Quart. J. Roy. Meteorol. Soc. **127**, 89–100 (2001)

145. Mironov, D., Jones, C.: Summary of the working group discussion on the representation of convection in high resolution numerical models. In: Proc. of the HIRLAM/NetFAM Workshop on Convection and Clouds, pp. 113–116 24–26 January 2005, Tartu, Estonia (2005)

146. Mironov, D.V., Gryanik, V.M., Lykossov, V.N., Zilitinkevich, S.S.: Comments on "A New Second-Order Turbulence Closure Scheme for the Planetary Boundary Layer" by K. Abdella and N. McFarlane. J. Atmos. Sci. **56**, 3478–3481 (1999)

147. Mironov, D.V., Gryanik, V.M., Moeng, C.-H., Olbers, D.J., Warncke, T.H.: Vertical turbulence structure and second-moment budgets in convection with rotation: a large-eddy simulation study. Quart. J. Roy. Meteorol. Soc. **126**, 477–515 (2000)

148. Mironov, D., Ritter, B.: Parameterization of convection in the global NWP system GME of the German Weather Service. In: Proc. of the HIRLAM/NetFAM Workshop on Convection and Clouds, pp. 68–72 24–26 January 2005, Tartu, Estonia (2005)

149. Moeng, C.-H., Wyngaard, J.C.: An analysis of closures for pressure-scalar covariances in the convective boundary layer. J. Atmos. Sci. **43**, 2499–2513 (1986)

150. Moeng, C.-H., Wyngaard, J.C.: Evaluation of turbulent transport and dissipation closure in second order modeling. J. Atmos. Sci. **46**, 2311–2330 (1989)

151. Monin, A.S., Obukhov, A.M.: Basic laws of turbulent mixing in the atmospheric surface layer. Trudy Geofiz. Inst. Akad. Nauk SSSR No. 24 (151), 163–187 (1954)

152. Monin, A.S., Yaglom, A.M.: Statistical Fluid Mechanics. Vol. 1, 769 pp. MIT Press, Cambridge, Massachusetts (1971)

153. Nakanishi, M., Niino, H.: An improved Mellor-Yamada level-3 model with condensation physics: its design and verification. Boundary-Layer Meteorol. **112**, 1–31 (2004)

154. Nurser, A.J.G.: A Review of Models and Observations of the Oceanic Mixed Layer. Southampton Oceanographic Centre, Internal document No. 14, 247 pp. Southampton, U.K. (1996)

155. Obukhov, A.M.: Turbulence in an atmosphere with a non-uniform temperature. Trudy Inst. Teor. Geofiz. Akad. Nauk SSSR **1**, 95–115 (1946) (English translation: Boundary-Layer Meteorol. **2**, 7–29, 1971)

156. Otić, I., Grötzbach, G., Wörner, M.: Analysis and modelling of the temperature variance equation in turbulent natural convection for low-Prandtl-number fluids. J. Fluid Mech. **525**, 237–261 (2005)

157. Perov, V., Zilitinkevich, S.: Application of an extended similarity theory for the stably stratified atmospheric surface layer to the HIRLAM. HIRLAM Newsletter, No. 35, 137–142 (2000)

158. Piriou, J.-M., Redelsperger, J.-L., Geleyn, J.-F., Lafore, J.-P., Guichard, F.: An approach for convective parameterization with memory: separating microphysics and transport in grid-scale equations. J. Atmos. Sci. **64**, 4127–4139 (2007)

159. Plate, E.J., Fedorovich, E.E., Viegas, D.X., Wyngaard, J.C. (eds.): Buoyant Convection in Geophysical Flows, 504 pp. Kluwer Acad. Publ., Dordrecht, etc. (1998)

160. Pope, S.B.: Turbulent Flows, 771 pp. Cambridge Univ. Press, Cambridge (2000)

161. Randall, D., Khairoutdinov, M., Arakawa, A., Grabowski, W.: Breaking the cloud parameterization deadlock. Bull. Amer. Met. Soc. **84**, 1547–1564 (2003)
162. Randall, D. A., Shao, Q., Moeng, C.-H.: A second-order bulk boundary-layer model. J. Atmos. Sci. **49**, 1903–1923 (1992)
163. Raschendorfer, M.: Special topic: The new turbulence parameterization of LM. Quarterly Report of the Operational NWP-Models of the Deutscher Wetterdienst, No. 19, 3–12 (1999)
164. Raschendorfer, M.: The new turbulence parameterization of LM. COSMO Newsletter, No. 1, 89–97 (2001) (available from www.cosmo-model.org)
165. Raupach, M. R.: Drag and drag partition on rough surfaces. Boundary-Layer Meteorol. **60**, 375–395 (1992)
166. Ristorcelli, J.R., Lumley, J.L., Abid, R.: A rapid-pressure covariance representation consistent with the Taylor-Proudman theorem materially frame indifferent in the two-dimensional limit. J. Fluid Mech. **292**, 111–152 (1995)
167. Rodi, W.: Examples of calculation methods for flow and mixing in stratified fluids. J. Geophys. Res. **92**, 5305–5328 (1987)
168. Rotta, J. C.: Statistische Theorie nichthomogener Turbulenz. 1. Zs. Phys. **129**, 547–572 (1951)
169. Sander, J.: Dynamical equations and turbulent closures in geophysics. Continuum Mech. Therm. **10**, 1–28 (1998)
170. Schumann, U.: Realizability of Reynolds-stress turbulence models. Phys. Fluids **20**, 721–725 (1977)
171. Schumann, U.: Minimum friction velocity and heat transfer in the rough surface layer of a convective boundary layer. Boundary-Layer Meteorol. **44**, 311–326 (1988)
172. Schumann, U., Gerz, T.: Turbulent mixing in stably stratified shear flows. J. Appl. Meteorol. **34**, 33–48 (1995)
173. Shih, T.-H.: Constitutive relations and realizability of single-point turbulence closures. In: Hallbäck, M., Henningson, D.S., Johansson, A.V., Alfredsson, P.H. (eds.) Turbulence and Transition Modelling, pp. 155–192 Hallbäck, M., et al. (eds.) Kluwer Acad. Publ., Dordrecht, etc. (1996)
174. Siebesma, A.P.: Shallow cumulus convection. In: Buoyant Convection in Geophysical Flows, pp. 441–486 Plate, E., et al. (eds.) Kluwer Acad. Publ., Dordrecht, etc., (1998)
175. Siebesma, A.P., Cuijpers, J.W.M.: Evaluation of parametric assumptions for shallow cumulus convection. J. Atmos. Sci. **55**, 650–666 (1995)
176. Siebesma, A.P. Teixeira, J.: An advection-diffusion scheme for the convective boundary layer: Description and 1D results. In: Proc. of the 14th AMS Symposium on Boundary Layers and Turbulence, pp. 133–136 American Meteorol. Soc, Boston, USA (2000)
177. Simpson, J., Simpson, R.H., Andrews, D.A., Eaton, M.A.: Experimental cumulus dynamics. Rev. Geophys. **3**, 387–431 (1965)
178. Simpson, J., Wiggert, V.: Models of precipitating cumulus towers. Mon. Weather Rev. **97**, 471–489 (1969)
179. Smith, R.K.: The role of cumulus convection in hurricanes and its representation in hurricane models. Rev. Geophys. **38**, 465–489 (2000)
180. Soares, P.M.M., Miranda, P.M.A., Siebesma, A.P., Teixeira, J.: An eddy-diffusivity/mass-flux parameterization for dry and shallow cumulus convection. Quart. J. Roy. Meteorol. Soc. **130**, 3365–3383 (2004)

181. Sommeria, G., Deardorff, J.W.: Subgrid-scale condensation in models of non-precipitating clouds. J. Atmos. Sci. **34**, 344–355 (1977)
182. Soomere, T., Zilitinkevich, S S.: Supplement to 'Third-order transport due to internal waves and non-local turbulence in the stably stratified surface layer'. Quart. J. Roy. Meteorol. Soc. **128**, 1029–1031 (2002)
183. Sorbjan, Z.: Structure of the Atmospheric Boundary Layer, 317 pp. Prentice Hall, Englewood Cliffs, New Jersey 07632 (1989)
184. Speziale, C.G.: Modeling the pressure gradient-velocity correlation of turbulence. Phys. Fluids **28**, 69–71 (1985)
185. Stensrud, D.J.: Parameterization Schemes: Keys to Understanding Numerical Weather Prediction Models, 478 pp. Cambridge Univ. Press, Cambridge (2007)
186. Steppeler, J., Doms, G., Schättler, U., Bitzer, H.W., Gassmann, A., Damrath, U., Gregoric, G.: Meso-gamma scale forecasts using the non-hydrostatic model LM. Meteorol. Atmos. Phys. **82**, 75–96 (2003)
187. Stevens, B.: Atmospheric moist convection. Ann. Rev. Earth Planet. Sci. **33**, 605–643 (2005)
188. Stull, R.B.: A convective transport theory for surface fluxes. J. Atmos. Sci. **51**, 3–22 (1994)
189. Sukoriansky, S., Galperin, B.: A spectral closure model for turbulent flows with stable stratification. In: Marine Turbulence. Theories, Observations, and Models, pp. 53–65 Baumert, H.Z., et al. (eds.) Cambridge Univ. Press, Cambridge (2005)
190. Sukoriansky, S., Galperin, B., Perov, V.: Application of a new spectral theory of stably stratified turbulence to the atmospheric boundary layer over sea ice. Boundary-Layer Meteorol. **117**, 231–257 (2005)
191. Sukoriansky, S., Galperin, B., Staroselsky, I.: A quasinormal scale elimination model of turbulent flows with stable stratification. Phys. Fluids **17**, 085107–085107–28, doi:10.1063/1.2009010 (2005)
192. Swann, H.: Evaluation of the mass-flux approach to parametrizing deep convection. Quart. J. Roy. Meteorol. Soc. **127**, 1239–1260 (2001)
193. Sykes, R.I., Henn, D.S., Lewellen, W.S.: Surface-layer description under free-convection conditions. Quart. J. Roy. Meteorol. Soc. **119**, 409–421 (1993)
194. Tennekes, H., Lumley, J.L.: A First Course in Turbulence, 300 pp. MIT Press, Cambridge, Massachusetts (1972)
195. Tiedtke, M.: The Parameterization of Moist Processes. Part 2: Parameterization of Cumulus Convection. Meteorological Training Course, Lecture Series, 78 pp. European Centre for Medium-Range Weather Forecasts, Reading, U.K. (1988)
196. Tiedtke, M.: A comprehensive mass flux scheme for cumulus parameterization in large-scale models. Mon. Weather Rev. **117**, 1779–1800 (1989)
197. Tiedtke, M.: Representation of clouds in large-scale models. Mon. Weather Rev. **121**, 3040–3061 (1993)
198. Tompkins, A.M.: A prognostic parameterization for the subgrid-scale variability of water vapor and clouds in large-scale models and its use to diagnose cloud cover. J. Atmos. Sci. **59**, 1917–1942 (2002)
199. Tompkins, A.M.: The parameterization of cloud cover. Technical Memorandum, 23 pp. European Centre for Medium-Range Weather Forecasts, Reading, U.K. (2005)
200. Troen, I., Mahrt, L.: A simple model of the atmospheric boundary layer; sensitivity to surface evaporation. Boundary-Layer Meteorol. **37**, 129–148 (1986)

201. Umlauf, L., Burchard, H.: A generic length-scale equation for geophysical turbulence models. J. Marine Res. **61**, 235–265 (2003)
202. Umlauf, L., Burchard, H.: Second-order turbulence closure models for geophysical boundary layers. A review of recent work. Cont. Shelf Res. **25**, 795–827 (2005)
203. Umlauf, L., Burchard, H., Hutter, K.: Extending the $k - \omega$ turbulence model towards oceanic applications. Ocean Modelling **5**, 195–218 (2003)
204. Unden, P., et al.: HIRLAM-5 Scientific Documentation. HIRLAM Report, 144 pp. Swedish Meteorological and Hydrological Institute (2002)
205. Wilcox, D.C.: Reassessment of the scale-determining equation for advanced turbulence models. AIAA J. **26**, 1299–1310 (1988)
206. Wilson, D., Gregory, D.: The behaviour of large-scale model cloud schemes under idealized forcing scenarios. Quart. J. Roy. Meteorol. Soc. **129**, 967–986 (2003)
207. Wu, X., Yanai, M.: Effects of vertical wind shear on the cumulus transport of momentum: Observations and parameterizations. J. Atmos. Sci. **51**, 1640–1660 (1994)
208. Wyngaard, J.: Atmospheric turbulence. Ann. Rev. Fluid Mech. **24**, 205–233 (1992)
209. Wyngaard, J.: A physical mechanism for the asymmetry in top-down and bottom-up diffusion. J. Atmos. Sci. **44**, 1083–1087 (1987)
210. Wyngaard, J.: Toward numerical modeling in the "Terra Incognita". J. Atmos. Sci. **61**, 1816–1826 (2004)
211. Yau, M.K.: Perturbation pressure and cumulus convection. J. Atmos. Sci. **36**, 690–694 (1979)
212. Zeierman, S., Wolfstein, M.: Turbulent time scale for turbulent-flow calculations. AIAA J. **24**, 1606–1610 (1986)
213. Zeman, O.: Progress in the modelling of planetary boundary layers. Ann. Rev. Fluid Mech. **13**, 253–272 (1981)
214. Zeman, O., Lumley, J.L.: Modeling buoyancy driven mixed layers. J. Atmos. Sci. **33**, 1974–1988 (1976)
215. Zeman, O., Tennekes, H.: Parameterization of the turbulent energy budget at the top of the daytime atmospheric boundary layer. J. Atmos. Sci. **34**, 111–123 (1977)
216. Zilitinkevich, S.S.: Dynamics of the Atmospheric Boundary Layer, 292 pp. Gidrometeoizdat, Leningrad (1970)
217. Zilitinkevich, S.: Third-order transport due to internal waves and non-local turbulence in the stably stratified surface layer. Quart. J. Roy. Meteorol. Soc. **128**, 913–925 (2002)
218. Zilitinkevich, S., Baklanov, A., Rost, J., Smedman, A.-S., Lykosov, V., Calanca, P.: Diagnostic and prognostic equations for the depth of the stably stratified Ekman boundary layer. Quart. J. Roy. Meteorol. Soc. **128**, 25–46 (2002)
219. Zilitinkevich, S., Calanca, P.: An extended similarity-theory for the stably stratified atmospheric surface layer. Quart. J. Roy. Meteorol. Soc. **126**, 1913–1923 (2000)
220. Zilitinkevich, S. S., Esau, I.N.: Resistance and heat-transfer laws for neutral and stable planetary boundary layers: Old theory advanced and reevaluated. Quart. J. Roy. Meteorol. Soc. **131**, 1863–1892 (2005)

221. Zilitinkevich, S.S., Hunt, J.C.R., Esau, I.N., Grachev, A.A., Lalas, D.P., Akylas, E., Tombrou, M., Fairall, C.W., Fernando, H.J.S., Baklanov, A.A., Joffre, S.M.: The influence of large convective eddies on the surface-layer turbulence. Quart. J. Roy. Meteorol. Soc. **132**, 1423–1456 (2006)

222. Zilitinkevich, S., Johansson, P.-E., Mironov, D.V., Baklanov, A.: A similarity-theory model for wind profile and resistance law in stably stratified planetary boundary layers. J. Wind Eng. Indust. Aerodyn. **74–76**, 209–218 (1998)

223. Zilitinkevich, S.S., Grachev, A.A., Fairall, C.W.: Scaling reasoning and field data on the sea surface roughness lengths for scalars. J. Atmos. Sci. **58**, 320–325 (2001)

224. Zilitinkevich, S., Grachev, A.A., Hunt, J.C.R.: Surface frictional processes and non-local heat/mass transfer in the shear-free convective boundary layer. In: Buoyant Convection in Geophysical Flows, pp. 83–113 Plate, E.J., et al. (eds.) Kluwer Acad. Publ., Dordrecht, etc. (1998)

225. Zilitinkevich, S., Gryanik, V.M., Lykossov, V.N., Mironov, D.V.: Third-order transport and non-local turbulence closures for convective boundary layers. J. Atmos. Sci. **56**, 3463–3477 (1999)

226. Zilitinkevich, S.S., Mironov, D.V.: Theoretical model of thermocline in a freshwater basin. J. Phys. Oceanogr. **22**, 988–996 (1992)

227. Zilitinkevich, S., Mironov, D.V.: A multi-limit formulation for the equilibrium depth of a stably stratified boundary layer. Boundary-Layer Meteorol. **81**, 325–351 (1996)

228. Zilitinkevich, S.S., Perov, V.L., King, J.C.: Calculation of turbulent fluxes in stable stratification in numerical weather prediction. HIRLAM Newsletter, No. 37, 83–92 (2001)

229. Zilitinkevich, S.S., Perov, V.L., King, J.C.: Near-surface turbulent fluxes in stable stratification: calculation techniques for use in general circulation models. Quart. J. Roy. Meteorol. Soc. **128**, 1571–1587 (2002)

Magnetohydrodynamic Turbulence

W.-C. Müller

Max-Planck-Institut für Plasmaphysik, 85748 Garching, Germany
http://www.ipp.mpg.de/~wcm

1 Introduction

Turbulence remains one of the last unresolved problems of classical physics. Turbulent flows in electrically conducting media represent an important aspect of this problem, because of their general importance for the evolution of astro- and geophysical plasmas [10, 87]. Turbulence in plasmas, i.e. ionized gases, also offers valuable insights into the not yet fully understood nonlinear dynamics of spectral cascades and structure formation due to the presence or generation of magnetic fields (see, e.g. [73, 95, 96]). These allow additional diagnostic access to the underlying nonlinear interaction of turbulent fluctuations.

In experimental devices for thermonuclear fusion the magnetically confined hot plasma is basically collisionless and requires kinetic treatment. Exceptions are the thin and comparably cool edge layer near the vessel boundaries [94] and plasmas in reversed-field pinch configurations [71]. In contrast turbulent plasmas in or beyond the earth often allow a fluid description due to the immense size of the dynamical regions and associated time-scales of interest compared to the effective mean-free-path and the frequencies related to the plasma particles [3].

Since plasma turbulence is a fully nonlinear problem comprising the dynamics of many interacting degrees of freedom, the relatively simple single fluid description of magnetohydrodynamics (MHD) represents a sensible starting point for theoretical and numerical investigations. While it is often desirable to include additional and more complex physical components in the model of the turbulent medium, we will refrain from doing so in this chapter not to obfuscate the inherent properties of MHD turbulence. The interest in these properties lies mainly in their potential universality, that is to say the inherent properties of turbulence might well be important for the dynamics of systems involving additional physics, e.g. gravity, radiation, rotation, or convection.

For additional simplicity of the MHD description, the mass density of the magnetofluid is assumed to be constant in time and spatially uniform, $\rho = \rho_0 = 1$. In addition relativistic effects are neglected and fluid velocities are

Müller, W.-C.: *Magnetohydrodynamic Turbulence*. Lect. Notes Phys. **756**, 223–254 (2009)
DOI 10.1007/978-3-540-78961-1_6　　　　　　　　© Springer-Verlag Berlin Heidelberg 2009

assumed to be much smaller than the magnetosonic speeds in the plasma. The flow can therefore be regarded as incompressible, $d\rho/dt = 0$ (cf. for example, [92]). This condition though rarely fulfilled in realistic plasma flows yields another simplification of the problem by reducing the continuity equation,

$$\frac{d}{dt}\rho + \rho\nabla \cdot \boldsymbol{v} = 0$$

to a simple solenoidality constraint on the velocity field, $\nabla \cdot \boldsymbol{v} = 0$.

The dimensionless incompressible MHD equations governing the motions of an electrically conducting fluid on large space- and timescales on which fluctuations of the electrical charge density are levelled out quasi-instantaneously and the kinetic nature of the medium becomes invisible then are

$$\partial_t \boldsymbol{v} = -\boldsymbol{v} \cdot \nabla\boldsymbol{v} - \nabla p - \boldsymbol{b} \times (\nabla \times \boldsymbol{b}) + \hat{\mu}\Delta\boldsymbol{v}\,, \tag{1}$$

$$\partial_t \boldsymbol{b} = \nabla \times (\boldsymbol{v} \times \boldsymbol{b}) + \hat{\eta}\Delta\boldsymbol{b}\,, \tag{2}$$

$$\nabla \cdot \boldsymbol{v} = \nabla \cdot \boldsymbol{b} = 0\,. \tag{3}$$

The magnetic field, $\boldsymbol{b}$, is given in Alfvén-speed units with the Alfvén speed defined as $v_A = B/\sqrt{\mu_0\rho}$, the vacuum permeability μ_0 being set to unity and the mass density $\rho = \rho_0 = 1$. In addition the dimensionless dissipation coefficients $\hat{\mu}$ and $\hat{\eta}$ are introduced.

The nondimensional diffusivities are related to the kinematic viscosity, μ, and magnetic diffusivity, η, by the relations $\hat{\mu} = \mu/(L_0V_0) = \mathrm{Re}^{-1}$ and $\hat{\eta} = \eta/L_0V_0 = \mathrm{Rm}^{-1}$ where L_0 and V_0 are a characteristic length and velocity of the configuration considered. Additionally, $\hat{\mu}^{-1}$ and $\hat{\eta}^{-1}$ correspond formally to the kinetic Reynolds number, Re and the magnetic Reynolds number, Rm, which roughly quantify the ratio of nonlinear to dissipative terms in (1) and (2). The magnetic Prandtl number, $\mathrm{Pr_m} = \mathrm{Rm}/\mathrm{Re} = \mu/\eta$, is another significant parameter of incompressible MHD and, for example, of importance in studies of magnetic field generation by MHD turbulence (turbulent dynamo effect, see e.g. [9, 17]). Throughout this chapter $\mathrm{Pr_m}$ is set to unity to achieve a formally symmetric configuration with regard to $\boldsymbol{v}$ and $\boldsymbol{b}$.

Smoothly evolving observables in time and space can be extracted from turbulent flows by statistical treatment of its quasi-stochastic fluctuations. A straightforward approach is to use statistical averages taken at one specific point in time or space [65] which however does not suffice to fully characterize the quadratic nonlinearities underlying turbulent dynamics. Two-point statistics which involve averaged field differences between two points, $\boldsymbol{r}$ and $\boldsymbol{r}'$, [66] are more appropriate for the problem since they are linked to experimentally measurable scale-dependent quantities like the energy spectrum or the spatial structure of the fluctuations [79]. In fact the useful concept of fluctuations on a spatial scale $\ell = |\boldsymbol{r} - \boldsymbol{r}'|$ naturally follows from two-point statistics (see below).

The ensemble average of a two-point observable, $\langle f(\boldsymbol{r}, \boldsymbol{r}')\rangle$, can be defined as the mean of this quantity over a large number of systems which have developed over the same period of time starting from a macroscopically identical state. For practical purposes it is however more convenient to use spatial or temporal averages in one realization with sufficient spatial or temporal extent, V and T respectively, instead of the unwieldy and sometimes impossible preparation of a large ensemble suitable for averaging. For convergence towards ensemble averages when $V, T \to \infty$ quasi-ergodicity [6] is presupposed, i.e. the trajectory of the system in phase-space comes arbitrarily close to any theoretically reachable state of this configuration. Quasi-ergodicity is a plausible assumption for turbulent systems but has not been proven rigorously, yet.

When investigating the intrinsic properties of turbulence it is common to regard idealized systems neglecting the influence of a particular geometry, boundaries as well as other large-scale constraints on the flow. This approach is motivated by the observation that small-scale fluctuations of turbulence appear to be independent of the specific setup on large scales. The accompanying algebraic simplification of statistical turbulence theory is of course gratefully accepted.

A general and widely applied assumption is statistical homogeneity, i.e. invariance of statistical quantities under translation in space or time, e.g. $\langle f(r, r') \rangle = \langle f(r - r') \rangle$. Sometimes the more restrictive symmetry of statistical isotropy, which includes homogeneity, is chosen expressing additional invariance under rotation, $\langle f(r, r') \rangle = \langle f(|r - r'|) \rangle = \langle f(\ell) \rangle$.

The latter assumptions simplify theoretical considerations though in real turbulence which is usually driven by a large-scale gradient, instabilities or other energy-sustaining processes statistical homogeneity or isotropy are only found in a limited range of scales if at all. Nevertheless, due to their utility, theory is usually dealing with idealized turbulence having the aforementioned statistical symmetries. Consequently, their validity in direct numerical simulations of turbulence has to be checked before comparison of numerical results with theory can be undertaken.

2 Macroscopic Properties

The Reynolds numbers parameterize a laminar flow a priori since normally L_0 and V_0 of such a configuration can be unambiguously identified. In contrast statistically homogeneous turbulence only permits an a posteriori estimate of L_0 and V_0 and consequently the associated Reynolds numbers are merely diagnostic in character.

The definitions of Re and Rm can be expressed with the help of macroscopic quantities characteristic of the flow, the total energy per unit mass, $E = E^K + E^M = \frac{1}{2} \int_V dV (v^2 + b^2)$, and the energy dissipation rate, $\varepsilon = -\dot{E}$. These allow, using dimensional analysis, to estimate a characteristic length scale $L_0 = E^{3/2}/\varepsilon$ and large-scale velocity $V_0 = (E^K)^{1/2}$ giving

$$\mathrm{Re} \simeq \frac{\left(E^{\mathrm{K}}\right)^{1/2} E^{3/2}}{\mu\varepsilon} \quad \text{and} \quad \mathrm{Rm} \simeq \frac{\left(E^{\mathrm{K}}\right)^{1/2} E^{3/2}}{\eta\varepsilon} \; .$$

2.1 Ideal Invariants

Special quantities, in 3D hydrodynamics total energy $E = \frac{1}{2}\int_V \mathrm{d}V v^2$ and kinetic helicity $H^{\mathrm{K}} = \frac{1}{2}\int_V \mathrm{d}V \boldsymbol{v}\cdot\boldsymbol{\omega}$ where $\boldsymbol{\omega} = \nabla\times\boldsymbol{v}$ is the vorticity, in 3D magnetohydrodynamics total energy $E = \frac{1}{2}\int_V \mathrm{d}V(v^2+b^2)$, cross helicity, $H^{\mathrm{C}} = \frac{1}{2}\int_V \mathrm{d}V \boldsymbol{v}\cdot\boldsymbol{b}$ and magnetic helicity, $H^{\mathrm{M}} = \frac{1}{2}\int_V \mathrm{d}V \boldsymbol{b}\cdot\boldsymbol{a}$, defined with the dimensionless magnetic vector potential $\boldsymbol{a}$, $\boldsymbol{b} = \nabla\times\boldsymbol{a}$, are important for characterizing the macroscopic state of a turbulent flow.

The utility of these variables stems from their conservation in closed ideal systems, a property that even survives in representations using finite sets of Fourier modes. In MHD this ruggedness of the ideal invariants E, H^{C} and H^{M} is caused by their detailed conservation in the nonlinear interaction of modes associated with three arbitrary wave-vectors $\boldsymbol{k}$, $\boldsymbol{p}$ and $\boldsymbol{q}$ forming a triangle, $\boldsymbol{k} = \boldsymbol{p} + \boldsymbol{q}$ (see, for example, [56]). Triad interactions are a mathematical property of convolution integrals which are Fourier-space representations of the quadratic nonlinearities occurring in (1) and (2) [58]. Detailed conservation also is a prerequisite for turbulent cascade processes. Consequently, only ideal invariants are usually subject to this special kind of spectral transport that will be discussed later in this chapter.

Cross helicity and magnetic helicity of the turbulent fields quantify topological properties that are conserved if $\mu = \eta = 0$. Ideal invariance of H^{C} expresses the fact that the mean orientation of velocity and magnetic field cannot arbitrarily be changed due to the magnetic field being frozen into the fluid. A change of magnetic helicity, H^{M}, which measures the linkage and 'knottedness' of magnetic field lines [64] is suppressed by the lack of magnetic reconnection under ideal conditions. The kinetic helicity, H^{K}, which owes its conservation in ideal hydrodynamics to Kelvin's circulation theorem [2] has some importance in the context of dynamo theory [9] but will not play a role in this chapter.

2.2 Selective Decay

Turbulence occurs only in nonideal, dissipative systems like Eqs. (1, 2 and 3). If the turbulent flow is not sustained by some driving mechanism constantly replacing energy lost through dissipation the ideal invariants decay. Their decay rates can be straightforwardly calculated using the MHD equations as

$$\dot{E} = -\varepsilon = -\int_V \mathrm{d}V\left(\mu\omega^2 + \eta j^2\right), \tag{4}$$

$$\dot{H}^{\mathrm{C}} = -\varepsilon^{\mathrm{C}} = -\frac{\mu+\eta}{2}\int_V \mathrm{d}V(\boldsymbol{\omega}\cdot\boldsymbol{j}), \tag{5}$$

$$\dot{H}^{\mathrm{M}} = -\varepsilon^{\mathrm{M}} = -\eta \int_V \mathrm{d}V(\boldsymbol{j} \cdot \boldsymbol{b}), \tag{6}$$

with the electric current density $\boldsymbol{j} = \nabla \times \boldsymbol{b}$.

Since the Fourier-transformed integrands in (4), (5) and (6) are of different order in the spatial wavenumber k with (5) and (6) not being positive-definite, the associated decay rates differ systematically with $\varepsilon \gg \varepsilon^{\mathrm{C}} \gtrsim \varepsilon^{\mathrm{M}}$. Direct numerical simulations (DNS) confirm this behaviour known as *selective decay* [91].

Maximum cross and magnetic helicity states correspond to minimal-energy configurations. For maximal H^{C} there is $\boldsymbol{v} \parallel \boldsymbol{b}$ everywhere and the interaction of magnetic and velocity field becomes minimal while a maximum of H^{M} corresponds to a force-free magnetic field topology with $\boldsymbol{b} \parallel \boldsymbol{j}$. The development of decaying MHD turbulence is therefore determined by the initial values of the ideal invariants. For example, if the normalized cross helicity, $\sigma = 2H^{\mathrm{C}}/\sqrt{E^{\mathrm{K}}E^{\mathrm{M}}}$, which is also called *alignment*, stays within a few percent it will continue to fluctuate around zero while $\sigma \simeq 20\%$ and above gives rise to a *dynamic alignment* process [24, 39] increasing σ and eventually switching off turbulent dynamics.

2.3 Energy Decay

While the energy of a freely decaying laminar flow diminishes exponentially, $E(t) \sim \exp(-2\hat{\mu}k^2 t)$, the decay of turbulence is known to exhibit a period of self-similar power-law behaviour $E(t) \sim t^{-\beta}$ with constant exponent β. For the hydrodynamic case, $\beta = 10/7$ was derived by Kolmogorov [47] based on the questionable invariance of the Loitsianskii integral $\sim \int \mathrm{d}r r^4 \langle v(\boldsymbol{x}+\boldsymbol{r})v(\boldsymbol{x}) \rangle$ (see, e.g. [44]) which is used to estimate the integral length scale of the flow ℓ_0. A different approach [57] which utilizes a statistical closure theory known as eddy-damped quasi-normal Markovian approximation (EDQNM) [70] in combination with the postulated 'permanence of the big eddies' (invariance of the low wavenumber part of the energy spectrum (see (12)), $E_k \sim k^s \simeq$const) gives the expression $\beta = 2(s + 1)/(s + 3)$ with $1 \simeq s < 4$.

For $s = 4$ the latter expression would coincide with the value obtained by Kolmogorov, but the theoretical limit $s < 4$ excludes this value. Experiments (cf. for example [23, 44, 86]), do not give a clear-cut picture with β ranging between 1 and 2. Therefore neither of the phenomenologies can be verified at present.

The temporal evolution of total energy in three-dimensional decaying MHD turbulence has been subject to various numerical investigations (see, e.g. [11, 60, 90]. The interest in this problem is mainly driven by an evident discrepancy between the observed life times of molecular clouds in the interstellar medium and the free-fall time associated with their gravitational collapse, see, e.g. [10]. MHD turbulence was suggested to be a process delaying gravitational collapse by acting as an effective outward-directed pressure.

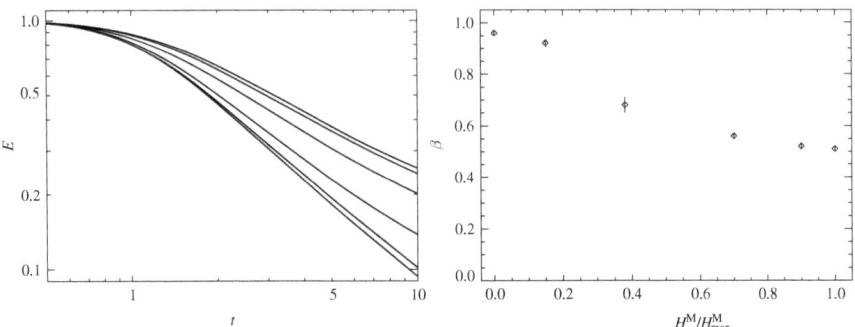

Fig. 1. *Left*: energy evolution for varying magnetic helicity $H^M = 0, 0.15, 0.38,$ $0.7, 0.92, 1[H^M_{max}]$ (bottom to top) in DNS of decaying MHD turbulence. *Right*: corresponding values of β as estimated from logarithmic derivatives of $E(t)$

As can be inferred from Fig. 1, the decay exponent in direct numerical simulations strongly depends on the level of magnetic helicity H^M being the lower the larger $|H^M|$, i.e. the higher the topological complexity of $\boldsymbol{b}$. A quasi-linear tuning of β can be achieved via the alignment σ. Since, however, an increase of σ simply depletes the nonlinear coupling of velocity and magnetic field (cf. Eq. (2)) the influence of this quantity on energy decay is rather uninteresting.

A model of energy decay in MHD turbulence that follows essentially the argumentation of Kolmogorov's hydrodynamic approach assumes invariance of an MHD-generalization of the Loitsianskii integral with Elsässer variables, $\boldsymbol{z}^{\pm} = \boldsymbol{v} \pm \boldsymbol{b}$ [27], in the involved correlation function [33]. By additionally postulating nonlinear energy-transfer driven by mutual Alfvén-wave collisions as in the Iroshnikov–Kraichnan picture of MHD turbulence [46, 52] (cf. Sect. 3.2), $\beta = 5/6$ is obtained. However, an inconsistency in the derivation [12] puts the result in question.

Finite magnetic helicity is characteristic of many astrophysical plasmas, since plasma turbulence usually occurs in rotating systems with mean gradients of temperature, so that the combined action of Coriolis and buoyancy forces naturally leads to twisted field lines. For finite H^M and if the turbulence is not continuously driven, selective decay, i.e. a much slower decay of H^M compared to that of the energy, will dominate the development of energy.

The basic framework utilizing this idea has been developed for the decay of enstrophy, $\Omega = \frac{1}{2} \int \omega^2 dS$ with dS denoting a surface element, in 2D hydrodynamic turbulence [1] and was later applied to MHD turbulence [41, 91] (and references therein). For high Reynolds number H^M can be considered invariant during energy decay. This property which is more robust than the questionable invariance of the Loitsianskii integral allows the construction of a phenomenological model for the energy decay.

Defining the characteristic length scale ℓ_0 by $E^M \ell_0 = H^M$ and applying the relations $H^M =$const, $\ell_0 \sim E^{3/2}/\varepsilon$ and $E \sim E^M$ one finds

$$\frac{dE}{dt} \sim \frac{E^{5/?}}{H^M} \tag{7}$$

which has the asymptotic solution $E \sim t^{-2/3}$ [7].

DNS reveal significant deviations from $\beta = 2/3$ (see Fig. 1), which can be attributed to a departure from self-similarity in the energy decay. In particular the ratio $\Gamma = E^K/E^M$ is not constant as in 2D MHD turbulence simulations [14] and as implicitly assumed in the derivation of (7), but decreases at a rate comparable to that of the energy, $\Gamma \simeq 0.1 \times E/H^M$ [11].

To account explicitly for the variation of Γ, it is assumed that the most important nonlinearities in the MHD equations arise from the advective terms, giving

$$-\frac{dE}{dt} = \varepsilon \sim \boldsymbol{v} \cdot \nabla E \sim (E^K)^{1/2} \frac{E}{\ell_0}.$$

Substitution of the integral scale ℓ_0 introduced above gives

$$\frac{E^{5/2}}{\varepsilon H^M} \frac{\Gamma^{1/2}}{(1+\Gamma)^{3/2}} = \text{const.} \tag{8}$$

which together with $\Gamma \sim E/H^M$ describes the energy decay.

In the limit $\Gamma \ll 1$, which is the asymptotic state decaying MHD turbulence, one finds the similarity solution $E \sim t^{-1/2}$. For larger Γ the decay is somewhat steeper, flattening to $t^{-1/2}$ as Γ becomes small [12].

The approximate constancy of expression (8) over the period of fully developed turbulence ($t > 3$) shown in Fig. 2, which depicts data gained with DNS of varying resolution (512^3, 256^3) and at different values of H^M [11], validates the associated simple phenomenology. The verification by use of a differential relation like (2) is easier and more accurate than extracting a power-law exponent from the curve $E(t)$ since the decay law has the general form $E(t) \sim (t - t_0)^{-\beta}$. The function $\sim t^{-\beta}$ appears only asymptotically for $t \gg t_0$ where the offset t_0 is undetermined and can be of the order of a large-eddy turnover time.

In the case of vanishing H^M, one finds a different decay law [12],

$$\frac{dE}{dt} \sim E^2, \tag{9}$$

which corresponds to the solution $E \sim t^{-1}$ in agreement with DNS of three-dimensional compressible and incompressible MHD turbulence [11, 60, 90]. The decay is similar to the one observed in two-dimensional MHD turbulence [14], though, contrary to the latter, there is no obvious selective decay process for $H^M = 0$ in the 3D case.

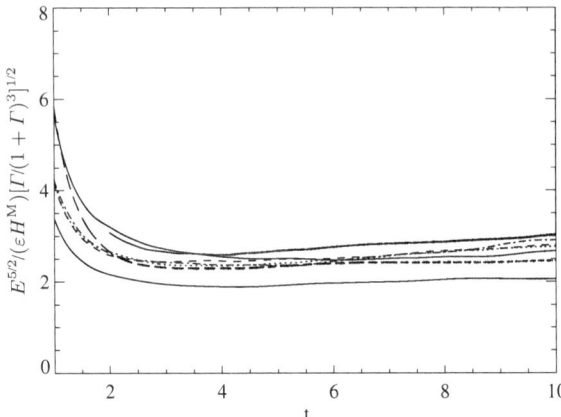

Fig. 2. Relation (8) for various DNS runs of decaying MHD turbulence with different values of H^M (cf. [11])

Because for $H^M = 0$ there is no clearly conserved quantity, it is difficult to interpret the decay law (9), a situation familiar from hydrodynamic turbulence. If the decay depends sensitively on the small-k spectrum, the t^{-1}-law may not be universal. The fact that the energy decay is more rapid for $H^M \simeq 0$ than for finite H^M, is plausible, since a nonhelical field is less constrained. The transition between the t^{-1} and the slower $t^{-1/2}$ behaviour is not abrupt, but occurs over a certain range $H^M/H^M_{\mathrm{max}} \simeq 0.2$–$0.6$ (cf. Fig. 1).

3 Small-Scale Dynamics

As mentioned before, the detailed conservation of ideal invariants in turbulent triad interactions [82] leads to nondissipative redistribution of ideally conserved quantities between different scales of fluid motion. While one nonlinear interaction results in transport quasi-randomly directed towards smaller or larger scales, on average the transfer has a preferred direction which depends on the kind and number of ideal invariants as well as on the dimensionality of the system under consideration.

Nonlinear spectral transport in fully developed turbulence is usually dominated by local triad interactions where for the three involved wave vectors $\boldsymbol{k}$, $\boldsymbol{p}$, $\boldsymbol{q}$ the relation $k \simeq p \simeq q$ holds. Therefore spectral nonlinear transfer proceeds in small steps, motivating the name *turbulent cascade*. Depending on its direction which can be inferred by study of absolute equilibrium ensembles (see, e.g. [31]) or direct numerical simulations a cascade is termed *direct* (towards small scales) or *inverse* (towards large scales) (cf. Table 1). While a direct cascade is characterized by the breakup of larger-scale fluctuations in smaller-scale ones which are eventually annihilated at the smallest-scales by dissipation processes, an inverse cascade leads to the formation of large-scale

Table 1. Cascading quantities in hydro- and magnetohydrodynamics, ↘: direct cascade, ↖: inverse cascade. The letter A signifies the mean square out-of-plane component of the magnetic vector potential

	2D		3D	
Navier–Stokes	E	↘	E	↘
	Ω	↘	H^{K}	↘
MHD	E	↘	E	↘
	H^{C}	↘	H^{C}	↘
	A	↖	H^{M}	↖

structures via subsequent merging of smaller-scale fluctuations. The largest scales of such self-organization are typically determined by the spatial exten- sions of the flow. In the case of magnetic helicity in MHD turbulence, however, the associated largest scales of the magnetic field associated with the inversely cascading H^{M} can be much larger than the volume occupied by the turbulent flow. Hence, it is possible and generally believed that the origin of dynamic magnetic fields accompanying many celestial bodies is plasma turbulence in their interior [19, 87] giving rise to an associated large-scale dynamo [80].

The fundament of the phenomenological understanding of turbulent flows is the hydrodynamic K41 picture put forward by Kolmogorov in the 1940s [49, 50] which also underlies all current phenomenologies of MHD turbulence [15, 36, 46, 52]. Turbulence is regarded as a superposition of structures or 'eddies' characterized by a spatial scale, ℓ and the associated velocity fluctu- ation, e.g. $\delta v_\ell = [\boldsymbol{v}(\boldsymbol{r} + \boldsymbol{\ell}) - \boldsymbol{v}(\boldsymbol{r})] \cdot \boldsymbol{\ell}/\ell$. For simplicity, the field is assumed to be statistically isotropic with the fluctuation amplitude depending on ℓ only. The characteristic velocity at scale ℓ can be defined via $v_\ell = \langle \delta v_\ell^2 \rangle^{1/2}$. As illustrated in Fig. 3, the K41 picture distinguishes different scales of motion: energy-containing scales driving the flow, the dissipation range at smallest scales, and the inertial range where nonlinear interactions dominate the dy- namics and the influence of driving and dissipation is negligible. It is within the latter region that spatial self-similarity is observed experimentally in the two-point structure functions of order p, $S_p^v(\ell) = \langle \delta v_\ell^p \rangle \sim \ell^{\zeta_p}$, with constant, p-dependent scaling exponents, ζ_p.

The K41 theory predicts values for these scaling exponents in the case of spatially self-similar turbulence at very large Reynolds number. However, the assumption of spatial self-similarity is not justified in reality due to the intermittent character of turbulent fields (cf. Sect. 4). Since the lowest or- der structure functions display only small deviations from nonintermittent behaviour (which however become important for $p \gtrsim 4$), see e.g. [29] and ref- erences therein, the K41-description is still a robust phenomenological starting point.

The structure functions are statistical moments of the two-point prob- ability distribution of the turbulent field. Their scaling exponents provide

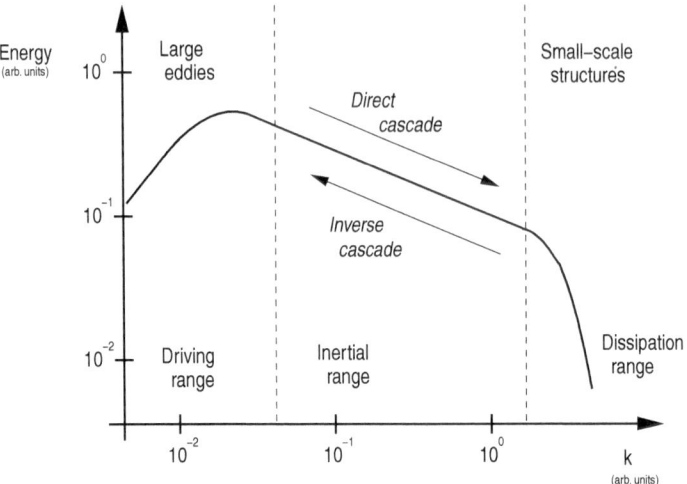

Fig. 3. Schematic view of the Kolmogorov (K41) picture of turbulence using the energy spectrum as an example

fundamental statistical information, e.g. $S_2^v(\ell)$ is linked to the one-dimensional energy spectrum while the p-dependence of the ζ_p characterizes the intermittency of flow structures.

3.1 Kolmogorov–Richardson Phenomenology

This model with the abbreviated name K41 successfully describes a wide range of hydrodynamical turbulent flows in laboratory, oceans and atmosphere. The inertial-range dynamics of the spectral energy flow is pictured as a Richardson cascade [81]. Turbulent eddies are forming a spatial hierarchy with energy being transferred to smaller scales by eddies becoming unstable and breaking up into smaller fluctuations.

Under quasi-stationary conditions, the resulting spectral energy flux Π in the inertial range is scale-independent and equal to the rate of energy dissipation, $\Pi = \varepsilon$ (cf. Kolmogorov's second similarity hypothesis [50]). In particular the flux can be approximated as $\Pi = \varepsilon \sim v_\ell^2/\tau_{\mathrm{NL}}$ with the scale-dependent nonlinear eddy turnover time $\tau_{\mathrm{NL}} = \ell/v_\ell$ implying a spectrally local transfer mechanism. Hence, $\varepsilon \sim v_\ell^2/\tau_{\mathrm{NL}} = v_\ell^3/\ell$ yielding the velocity scaling

$$v_\ell \sim (\varepsilon\ell)^{1/3} . \tag{10}$$

One of the few exact results in turbulence theory, the four-fifth law (see, e.g. [29]),

$$S_3^v(\ell) = -\frac{4}{5}\varepsilon\ell ,$$

yields $v_\ell^3 \sim \ell$ and motivates the generalization of (10),

$$S_p^v(\ell) \sim (\varepsilon\ell)^{p/3} \ . \tag{11}$$

Based on the Fourier-transformed turbulent velocity $\boldsymbol{v_k}$ the angle-integrated energy spectrum E_k with $k \sim \ell^{-1}$ usually considered in isotropic turbulence is defined as

$$E_k = \frac{1}{2}\int d^3 k' \, \delta(|\boldsymbol{k'}| - k)|v_{\boldsymbol{k'}}|^2 \ . \tag{12}$$

The relation $v_\ell^2 \simeq kE_k$ (see, e.g. [82]) links the scaling exponent ζ_2 of $S_2^v(\ell)$ with the inertial range scaling of $E_k \sim k^{-\alpha}$ yielding $\alpha = -(1+\zeta_2)$. Hence, the experimentally well-supported K41-spectrum of incompressible hydrodynamic turbulence follows as

$$E(k) = C_K \varepsilon^{2/3} k^{-5/3} \tag{13}$$

with the Kolmogorov constant $C_K \approx 1.6$.

The Kolmogorov dissipation scale ℓ_D gives an order of magnitude estimate of the spatial scales where dissipation $\sim v_\ell^2 \hat{\mu}/\ell^2$ starts to dominate over nonlinear transfer, $\sim v_\ell^2/\tau_{NL}$, marking the beginning of the dissipation range. This gives

$$\frac{\hat{\mu}}{\ell^2} \sim \tau_{NL}^{-1} = v_\ell/\ell \sim (\varepsilon\ell)^{1/3}/\ell$$

resulting in

$$\ell_D = \left(\frac{\hat{\mu}^3}{\varepsilon}\right)^{1/4} . \tag{14}$$

3.2 Iroshnikov–Kraichnan Phenomenology

The IK picture [46, 52] tries to capture the effect of magnetic fields present in MHD turbulence on the energy cascade by introducing a different model of nonlinear transfer. The model is meanwhile thought to be incorrect because it stays in the isotropic frame of the K41-phenomenology though the presence of magnetic fields is known to generate anisotropy with respect to the local field direction (see Sect. 3.3). Nevertheless it contains concepts probably important for further development of MHD turbulence phenomenology.

Eddy scrambling and breakup which underlie the K41-cascade are replaced by an energy transfer driven by shear Alfvén waves. The energy is spectrally redistributed between different length scales by nonlinear scattering of colliding Alfvén-wave packets counter-propagating along a magnetic field line.

In the following, the Elsässer quantity z_ℓ is used which is defined analogously to v_ℓ in the hydrodynamic case. Elsässer variables have the special property that $\boldsymbol{z}^\pm = 0$ are exact nonlinear solutions of the ideal incompressible MHD equations representing Alfvén wave pulses on a mean magnetic field. By restricting consideration to MHD turbulence with small mean $\boldsymbol{v}$-$\boldsymbol{b}$-alignment,

σ (see Sect. 2.3), it is not necessary to distinguish between z^+ and z^-. However, for finite σ the respective energy spectra can differ considerably [78], demanding a more complex theoretical approach [38].

The coupling of magnetic and velocity fields through the generation and attenuation of Alfvén waves, termed Alfvén effect, leads to approximate equipartition of magnetic and kinetic energy at small-scales where $\tau_A \ll \tau_{NL}$ making the effect dynamically dominant. The process is based on the interaction of eddies of size ℓ with the magnetic guide field b_0 which is generated by the largest energy-containing swirls or imposed externally. As a consequence the associated velocity perturbations v_ℓ are triggering Alfvén wave pulses by locally deforming b_0. The incompressible deformations then travel along the field lines. If the involved perturbations $\delta v, \delta B$ are small compared to b_0, one has $\delta v \simeq \pm \delta B$.

For two colliding Alfvén wave pulses of extent ℓ the interaction time for nonlinear energy transfer is given by $\tau_A = \ell/b_0$ (in the chosen nondimensional representation b_0 represents the Alfvén speed). Due to $b_0 \gg b_\ell$ this time is much shorter than the corresponding K41 transfer time, τ_{NL} and consequently only a fraction $\sim (\tau_A/\tau_{NL})$ of the energy exchanged in one K41 interaction is redistributed between the colliding wave packets. Due to the random character of the interactions, the cascade transfer time τ_* entering the IK energy flux z_ℓ^2/τ_* is enlarged by the factor (τ_{NL}/τ_A) compared to the K41 case [24].

Accordingly, the IK phenomenology follows along the same lines as the K41 model when replacing v_ℓ with z_ℓ and rewriting the characteristic cascade time as $\tau_* \sim (\tau_{NL}/\tau_A)\tau_{NL}$. This gives the following nonintermittent inertial range scaling

$$S_p^z(\ell) \sim (\varepsilon b_0 \ell)^{p/4} .$$

Analogously the energy spectrum is obtained as

$$E(k) = C_{IK}(\varepsilon b_0)^{1/2} k^{-3/2}$$

with the IK dissipation length

$$\ell_{IK} = \left(\frac{b_0 \hat{\eta}^2}{\varepsilon} \right)^{1/3} .$$

There, however, exists an exact relation corresponding to the four-fifth law for incompressible three-dimensional MHD [75, 76],

$$\sum_{i=1}^{3} \langle \delta z_\ell^{\mp} (\delta_i z_\ell^{\pm})^2 \rangle = -\frac{4}{3} \varepsilon^{\pm} \ell , \qquad (15)$$

with $\varepsilon^{\pm} = \frac{1}{2} \int_V dV [\mu \omega^2 + \eta j^2 \pm (\mu + \eta) \boldsymbol{\omega} \cdot \boldsymbol{j}]$, $\delta z_\ell^{\mp}$ denoting the longitudinal field increments introduced above and $\delta_i z_\ell^{\pm} = (z^{\pm}(\boldsymbol{r} + \boldsymbol{\ell}) - z^{\pm}(\boldsymbol{r})) \cdot \boldsymbol{e}_i$ introducing the unit vector $\boldsymbol{e}_i$ from the orthogonal base of an arbitrary co-ordinate system. Equation (15) suggests the third-order structure function scaling $S_3^z \sim \ell$ that the IK model predicts for S_4^z.

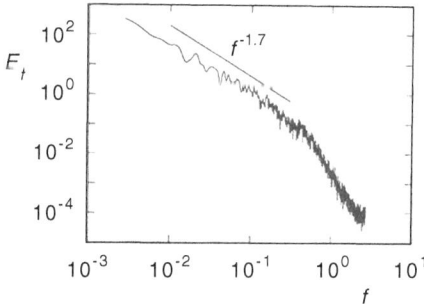

Fig. 4. Temporal energy spectrum of solar wind fluctuations measured by the WIND probe at a distance from the sun of approximately one astronomical unit [55] (from [10])

The problematic issues mentioned in this section in combination with the lack of observational evidence for 3/2-scaling (cf., e.g. [93] and Figs. 4 and 6) have led to the development of the Goldreich–Sridhar model, presented in the following.

3.3 Goldreich–Sridhar Phenomenology

In the isotropic K41 and the IK pictures turbulent fluctuations are characterized by a single length scale ℓ. However, the presence of a magnetic field renders the turbulence locally anisotropic. Alfvén wave packets of extent λ propagate along the local magnetic field $\boldsymbol{b}_0$ with the characteristic time scale $\tau_\lambda \sim \lambda/b_0$. Simultaneously, the field lines are subject to eddy-scrambling perpendicular to $\boldsymbol{b}_0$ on the turnover time-scale $\tau_l \sim l/z_l$, where l is the field-perpendicular extent and z_l the amplitude of the fluctuations. In addition, it has been found the nonlinear energy flux is much weaker along the direction of the magnetic field [32, 37, 68, 85].

Goldreich and Sridhar put forward a phenomenology which takes into account the spatial anisotropy caused by the magnetic field [35, 36] and predicts the scaling of field-perpendicular and field-parallel energy spectra. The GS-picture is based on the idea that there is an equality of the scale-dependent characteristic time scales τ_λ and τ_l. This *critical balance* expresses the fact that magnetic-field deformations associated with the field-perpendicular turnover time τ_l propagate with Alfvén speed, b_0, over the parallel distance $\lambda = b_0 \tau_A$ in the same time. Consequently, the field-perpendicular cascade rate is determined by the turnover timescale τ_l leading to a Kolmogorov-like field-perpendicular energy spectrum

$$E_{k_\perp} = \frac{1}{2} \int dk_1 \int dk_2 \left(|v_{\boldsymbol{k}}|^2 + |b_{\boldsymbol{k}}|^2 \right) \sim k_\perp^{-5/3} . \tag{16}$$

where $k_\perp \sim l^{-1}$ while $k_1 = k_\parallel \sim \lambda^{-1}$ and $k_2 \perp k_\parallel, \; k_\perp$.

An interesting and robust consequence of the balance assumption is a relation between the perpendicular spatial scales l and the corresponding parallel scales λ of the turbulent eddies [22, 34, 89]. By $\lambda/b_0 \sim l/z_l$ in combination with (16) one obtains

$$\lambda \sim l^{2/3} \,, \tag{17}$$

which implies that eddies become elongated along the local field direction with decreasing spatial scale.

Another consequence of (17) is an energy cascade parallel to $\boldsymbol{b}_0$ since perpendicular and parallel directions are tied to each other by the balance condition $\tau_\lambda \sim \tau_l$ with the associated one-dimensional energy spectrum

$$E_{k_\parallel} \sim k_\parallel^{-2} \,.$$

Recently, it was shown by direct numerical simulations [61, 68, 69] that MHD turbulence permeated by a strong mean magnetic field exhibits a field-perpendicular energy spectrum $\sim k_\perp^{-3/2}$. A possible explanation is an extension of the Goldreich–Sridhar picture due to Boldyrev [15]. There it is suggested that an increasingly parallel polarization of Alfvénic fluctuations results in a weakening of nonlinear turbulent interaction and, consequently, leads to the observed IK-like scaling of the field-perpendicular energy spectrum.

3.4 Numerical Results

Experimental observations of scaling in MHD turbulence are rare due to the great difficulty of obtaining sufficiently high Reynolds numbers in terrestrial experiments and because of the scarcity and yet insufficient precision of astronomical observations. Only high-resolution direct numerical simulations yield an inertial range broad enough to differentiate the different numerical values of the observed scaling exponents with sufficient accuracy. Yet, the numerically achievable Reynolds numbers in direct numerical simulations, $\mathcal{O}(10^3)$, are at present far below realistic values, $\mathcal{O}(10^8)$–$\mathcal{O}(10^{22})$.

Macroscopically Isotropic MHD Turbulence

Figure 6 shows an energy spectrum from a pseudospectral simulation of incompressible and isotropic decaying MHD turbulence with vanishing magnetic and cross helicity (cf. Fig. 5).

The system is represented by a finite set of Fourier modes associated with a flow in a periodic box of edge length 2π. The initial ratio $E^{\mathrm{K}}/E^{\mathrm{M}}$ is unity. Leapfrog scheme is applied to evolve (1) and (2) in time. In this setup the dimensionless dissipation parameters which are set to $\hat{\mu} = \hat{\eta} = 10^{-4}$ correspond to Reynolds numbers $\mathsf{Re} = \mathsf{Rm} \simeq 5800$.

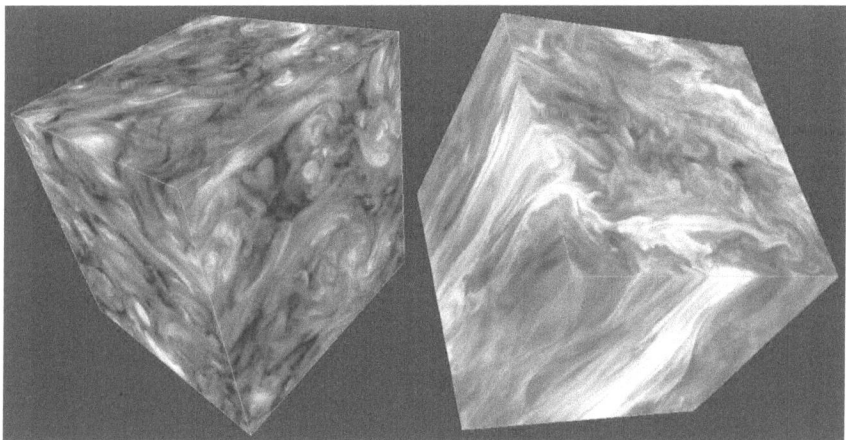

Fig. 5. Magnetic fluctuations in three-dimensional pseudospectral DNS of isotropic (*left*) and anisotropic (*right*) MHD turbulence (bright colours indicate high values). In the anisotropic case a strong and constant mean magnetic field normal to the top side of the cube is applied

The run was performed over nine eddy turnover times defined as the time required to reach the maximum of dissipation when starting from smooth initial fields. The spectrum is normalized in wavenumber using the Kolmogorov dissipation length (14) and in amplitude assuming a Kolmogorov spectrum (13). It was time-averaged over the period of self-similar decay, $t = 6 - 8.9$. The simulation involves 1024^3 collocation points and is one of the largest runs carried out so far.

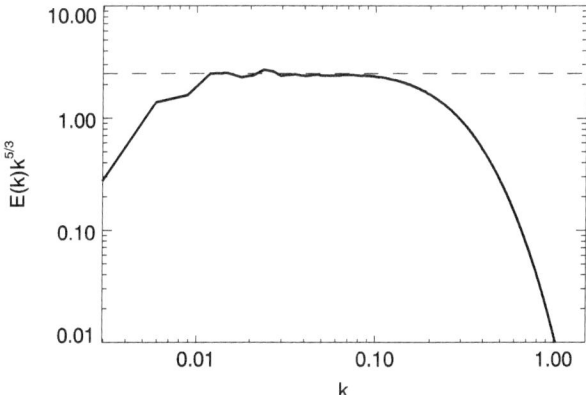

Fig. 6. Compensated total energy spectrum in DNS of decaying isotropic MHD turbulence which shows a Kolmogorov-like $\sim k^{-5/3}$ inertial range [69]. Amplitude and abscissa are normalized to prevent secular changes caused by the attenuation of the turbulence

The spectrum shows a well-developed inertial range with an associated scaling exponent $\alpha = 5/3$ in agreement with solar wind measurements, see for example [55]. The same observation has been made in related work [42, 67]. This invalidates the Iroshnikov–Kraichnan model for isotropic turbulence without mean magnetic field although Alfvén waves are present in the system (see below).

In forced three-dimensional compressible supersonic and super-Alfvénic MHD turbulence the picture is not so clear with the kinetic energy spectrum observed to be steeper, $E^{K}(k) \sim k^{-1.74}$ [16] for $M \sim 10$ and $M_{A} \sim 3$, where $M = v/a$ is the sonic Mach number defined with the sound speed a and $M_{A} = v/b$ is the Alfvénic Mach number. The scaling exponent also shows a dependence on the sonic Mach number [72].

Macroscopically Anisotropic MHD Turbulence

The numerical data on isotropic turbulence is in agreement with Goldreich and Sridhar's phenomenology. To scrutinize this model the system is made globally anisotropic by imposing a mean magnetic field $\boldsymbol{b}_0$ (for an illustration, see Fig. 7).

In this case, a pseudospectral $1024^2 \times 256$ forced turbulence simulation is carried out (see Fig. 5). Due to the stiffness of the magnetic field, turbulent fluctuations are depleted in the field-parallel direction (see below), allowing a reduced numerical resolution along the corresponding axis. The forcing is realized by freezing all modes with $k \leq k_f = 2$. Its purpose is to keep the value of fluctuating field to mean field approximately constant. The simulation with $\hat{\mu} = \hat{\eta} = 3 \times 10^{-4}$, i.e. Re = Rm $\simeq 3300$ and $|\boldsymbol{b}_0| = 5$ covers about 25 eddy turnover times. Kinetic and magnetic energy as well as E^{K}/E^{M} are of order

(a)   (b)

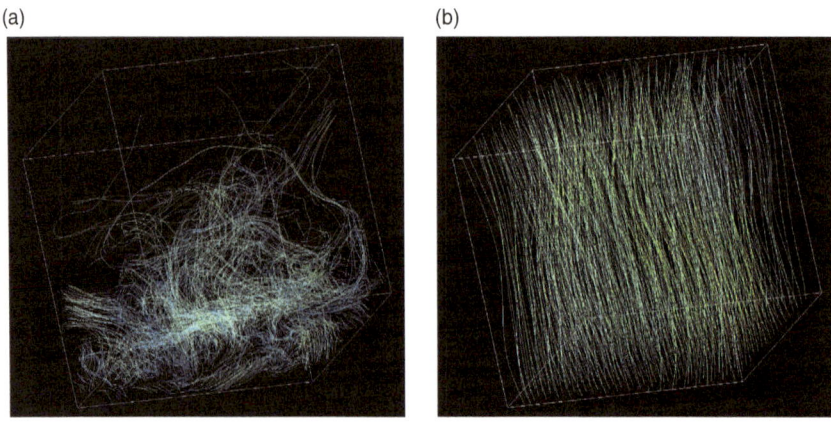

Fig. 7. Magnetic field lines in globally isotropic turbulence as in the simulation underlying Fig. 6 (*left*) and in a system with strong mean field component, $b_0 = 5$, (*right*), cf. Fig. 8

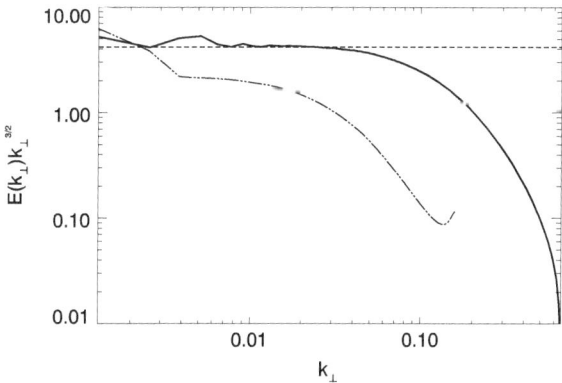

Fig. 8. Normalized and time-averaged one-dimensional parallel (*dot-dashed*) and perpendicular (*solid*) energy spectra in DNS of three-dimensional MHD turbulence with strong mean magnetic field [69]. The dashed line indicates IK-like scaling $\sim k^{-3/2}$ (in this compensation horizontal)

one. The system has relaxed to a state with cross helicity around 15% and with the magnetic helicity around 0.2 of the theoretical maximum $\sim E^{\mathrm{M}}/k_f$.

Figure 8 depicts the normalized parallel and perpendicular energy spectra averaged over $t = 20 - 25$ when the system is quasi-stationary. The parallel spectrum shows a significant reduction of turbulence compared to the perpendicular spectrum since the nonlinear energy transfer is depleted in the parallel direction as theoretically expected (cf. Sect. 3.3). The drop in amplitude of the field-parallel fluctuations compared to the field-perpendicular ones has also been observed in shell-model calculations of MHD turbulence [20]. An inertial range is not clearly discernible.

The perpendicular energy spectrum exhibits IK-like scaling, $E(k_\perp) \sim k_\perp^{-3/2}$. Alfvénic fluctuations evidently dominate the energy cascade in planes perpendicular to the mean field as is also seen in simulations of two-dimensional MHD turbulence [14]. Their existence can be inferred from the approximate equipartition of kinetic and magnetic energy on all field-perpendicular scales of motion (cf. Fig. 9).

The main effect of the mean field, b_0, is to restrict turbulent fluctuations to field-perpendicular planes.

In [61], $\alpha = 3/2$ is also observed for a much stronger mean field (the ratio of fluctuations to mean component is about 3×10^{-3}). However, there it is speculated that the scaling is due to bottleneck or intermittency effects [28, 59] caused by the use of higher-order dissipation terms, e.g. $\mu_\nu (-1)^{\nu-1} \Delta^\nu \boldsymbol{\omega}$ for dissipativities $\nu > 1$. Hyperviscosities of this kind are used to enlarge the inertial scaling range but result in a nonphysical steepening of the spectrum close to the dissipative fall-off (see, e.g. [12]). The simulations presented here use normal viscosities, i.e. $\nu = 1$, and do not exhibit a significant bottleneck effect.

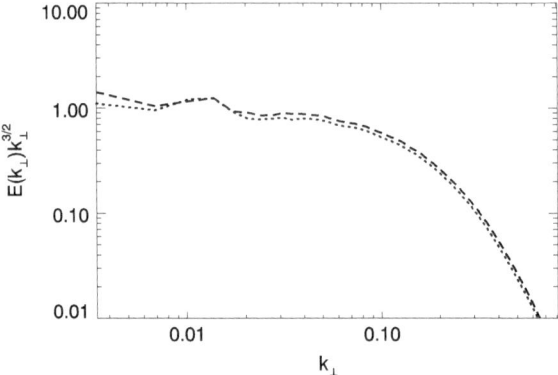

Fig. 9. Magnetic (*dashed*) and kinetic (*dotted*) perpendicular energy spectra (normalized and averaged) for the same anisotropic run as depicted in Fig. 8

The numerical results are at variance with recent simulations which claim to support the Goldreich–Sridhar model by observing $\alpha = 5/3$ in the perpendicular energy spectrum [21, 22]. However, the mean magnetic field in these simulations, carried out with lower numerical resolution (256^3) and involving fourth- and eighth-order hyperviscosities, has approximately the same amplitude as the turbulent fluctuations, $v \sim b \sim b_0$.

The simulations presented here, with $b_0 \sim 5b$, put the Goldreich–Sridhar picture for configurations with a strong mean magnetic field in question and can at present only be explained with Boldyrev's approach (cf. Sect. 3.3).

3.5 Dynamical Equilibrium of Kinetic and Magnetic Energy

The residual energy spectrum, $E_k^R(k) = \left|E_k^M(k) - E_k^K(k)\right|$, is of interest because it sheds some light on the spectral interplay of kinetic and magnetic energy and exhibits self-similar scaling. For isotropic decaying turbulence and anisotropic-forced turbulence with a mean magnetic field it displays fundamentally different behaviour which becomes evident when comparing kinetic and magnetic energy spectra for the two cases (Figs. 9 and 10). In both the systems the Alfvén effect is present. However, while it dominates the anisotropic system and leads to approximate energetic equipartition at all scales of the flow, in the isotropic simulation this is only true for the dissipation range.

The excess of magnetic energy with increasing spatial scale, visible in Fig. 10, is due to the turbulent small-scale dynamo. This mechanism amplifies the magnetic field locally through the stretching of field lines by turbulent fluid motions. A generalization of previous theoretical work [38] allows to correctly predict the resulting scaling exponent of the residual energy spectrum in both

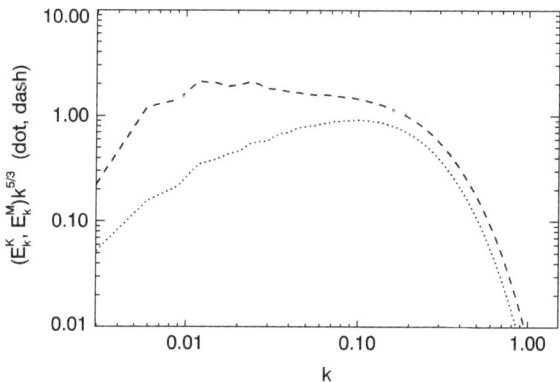

Fig. 10. Kinetic (*dotted*) and magnetic (*dashed*) energy spectra (normalized and averaged) for the isotropic run as in Fig. 6

the cases and gives some insight into the turbulent interplay between kinetic and magnetic energy.

Closure Theory

Among the different approaches towards a statistical theory of turbulence (see, e.g. [26, 30, 43, 51, 53, 54, 70]), the eddy-damped quasi-normal Markovian approximation (EDQNM), originally derived for incompressible hydrodynamic turbulence [70], has proven to be a useful compromise between mathematical rigor and phenomenological flexibility. In its magnetohydrodynamic form [80], neglecting helicity effects, the equations governing the spectral dynamics of kinetic and magnetic energy read

$$\left(\partial_t + 2\hat{\mu}k^2\right) E_k^{\mathrm{K}} = \int_{\triangle} \mathrm{d}p\mathrm{d}q \Theta_{kpq} \left(T_{\mathrm{kin}}^{\mathrm{K}} + T_{\mathrm{mag}}^{\mathrm{K}} + T_{\mathrm{crs}}^{\mathrm{K}}\right) , \tag{18}$$

$$\left(\partial_t + 2\hat{\eta}k^2\right) E_k^{\mathrm{M}} = \int_{\triangle} \mathrm{d}p\mathrm{d}q \Theta_{kpq} \left(T_{\mathrm{mag}}^{\mathrm{M}} + T_{\mathrm{crs}}^{\mathrm{M}}\right) , \tag{19}$$

with the flux-density contributions

$$T_{\mathrm{kin}}^{\mathrm{K}} = b_{kpq}\frac{k}{pq} \left(k^2 E_p^{\mathrm{K}} E_q^{\mathrm{K}} - p^2 E_q^{\mathrm{K}} E_k^{\mathrm{K}}\right), T_{\mathrm{mag}}^{\mathrm{K}} = c_{kpq}\frac{k^3}{pq} E_p^{\mathrm{M}} E_q^{\mathrm{M}} ,$$

$$T_{\mathrm{crs}}^{\mathrm{K}} = c_{kpq}\frac{kp}{q} E_q^{\mathrm{M}} E_k^{\mathrm{K}} , T_{\mathrm{mag}}^{\mathrm{M}} = -c_{kpq}\frac{k^3}{pq} E_q^{\mathrm{M}} E_k^{\mathrm{M}} ,$$

$$T_{\mathrm{crs}}^{\mathrm{M}} = h_{kpq}\frac{k}{pq} \left(k^2 E_p^{\mathrm{M}} E_q^{\mathrm{K}} - p^2 E_q^{\mathrm{K}} E_k^{\mathrm{M}}\right) + c_{kpq}\frac{k^5}{p^3q} E_p^{\mathrm{K}} E_q^{\mathrm{M}} .$$

The geometric coefficients $b_{kpq}, c_{kpq}, h_{kpq}$ defined, e.g. in [80] enforce solenoidality of the turbulent fields. The triangle symbol, '$\triangle$', denotes integration over mode numbers which fulfill $k = p + q$.

The time Θ_{kpq} is characteristic of the relaxation of the nonlinear energy flux involving the modes k, p and q and can be approximated by $\Theta_{kpq} = t/(1 + \mu_{kpq}t)$. Here, μ_{kpq} is a phenomenological expression for the damping rate of the flux by higher order moments with $\mu_{kpq} = \mu_k + \mu_p + \mu_q$ ensuring energy conservation.

A straightforward choice for the damping rates is $\mu_k = \tau_{NL}^{-1} + \tau_A^{-1} + \tau_D^{-1}$ which combines the three physical processes that underlie turbulent energy dynamics in MHD: field-line deformation by turbulent motions on the time-scale $\tau_{NL} \sim \ell/\sqrt{v_\ell^2 + b_\ell^2} \sim (k^3 E_k)^{-1/2}$, energy equipartition in interacting shear Alfvén waves characterized by $\tau_A \sim \ell/b_0 \sim (kb_0)^{-1}$ and molecular dissipation, $\tau_D \sim (\mu + \eta)^{-1}k^{-2}$.

Under realistic conditions, diffusion is associated with the longest time-scale of the turbulent system. Thus, for $t \gg \tau_D$ one has $\Theta_{kpq} \simeq \mu_{kpq}^{-1} \simeq \min(\tau_{NL}, \tau_A)$.

The evolution equation for the residual energy spectrum, E_k^R, can be derived similarly as for (18) and (19) [39] and reads in the case of negligible v–b alignment:

$$\left(\partial_t + (\hat{\mu} + \hat{\eta}) k^2\right) E_k^R = \int_\triangle dpdq\Theta_{kpq} \left(T_{res}^R + T_{crs}^R + T_{loc}^R\right) \tag{20}$$

with

$$T_{res}^R = -m_{kpq}\frac{k^2}{p}E_p^R E_q^R + r_{kpq}\frac{p^2}{q}E_q^R E_k^R ,$$

$$T_{crs}^R = m_{kpq}pE_q E_k^R + t_{kpq}pE_q^R E_k ,$$

$$T_{loc}^R = -\frac{s_{kpq}}{k} \left(k^2 E_p E_q - p^2 E_q E_k\right) .$$

The geometric coefficients m_{kpq}, r_{kpq}, s_{kpq}, t_{kpq} are defined in [39].

Spectral interactions as expressed by the right-hand side of (20) are non-local if the modulus of one wavenumber, say k, in the interacting triad differs significantly from the wavenumbers of the other two, $p \sim q$. Nonlocal interactions are associated with mutual Alfvén-wave scattering and for this special case a simplified version of (20) can be derived:

$$\partial_t E_k^R = -\Gamma_k k \left(E_k^M - E_k^K\right), \tag{21}$$

where $\Gamma_k = \frac{4}{3}k \int_0^{ak} dq\Theta_{kpq}E_q^M$ [80].

A Phenomenology for the Residual Energy

It is assumed that the right-hand side of (20) can be written as $T_{nonloc}^R + T_{loc}^R$ [38]. This states that E_k^R is a result of a dynamic equilibrium between the (spectrally local) dynamo effect which amplifies the magnetic field and the (spectrally nonlocal) Alfvén effect tending towards energetic equipartition. For stationary conditions in the inertial range ($k \ll k_{\mathrm{D}}$) dimensional analysis yields (note that E^{M} is the total magnetic energy introduced in Sect. 2)

$$\underbrace{k^3 E_k^2}_{(20)} \sim \underbrace{k^2 E^{\mathrm{M}} E_k^{\mathrm{R}}}_{(21)} . \tag{22}$$

With the definitions of τ_{NL} and τ_{A} given above and assuming that the large-scale magnetic field sets the time scale for Alfvénic interactions, $k^2 E^{\mathrm{M}} \simeq (kb_0)^2$, this expression can be re-written as [69]

$$E_k^{\mathrm{R}} \sim \left(\frac{\tau_{\mathrm{A}}}{\tau_{\mathrm{NL}}} \right)^2 E_k . \tag{23}$$

For $E_k \sim k^{-3/2}$ as seen in the simulation with mean magnetic field (see Fig. 8), the known result $E_k^{\mathrm{R}} \sim k^{-2}$ [38] is obtained. This is in good agreement with the field-perpendicular residual energy spectrum of the same run shown in Fig. 11 and with two-dimensional simulations of MHD turbulence (cf. Fig. 12).

The simulation with vanishing mean magnetic field displays Kolmogorov-like inertial range scaling (cf. Fig. 6) for which (23) predicts $E_k^{\mathrm{R}} \sim k^{-7/3}$. As in this simulation the mean magnetic field vanishes, the b_0 term in the expression above denotes the mean magnetic field carried by large-scale fluctuations. Figure 13, depicting E_k^{R} for this case, confirms that the theoretical prediction is well fulfilled.

Apart from its meaning for the fundamental mechanism converting kinetic into magnetic energy (and vice versa), (23) also serves a more practical task. It enlarges the inertial range scaling exponent of the total energy spectrum by a factor of two via the residual energy exponent. Thus, the difference between

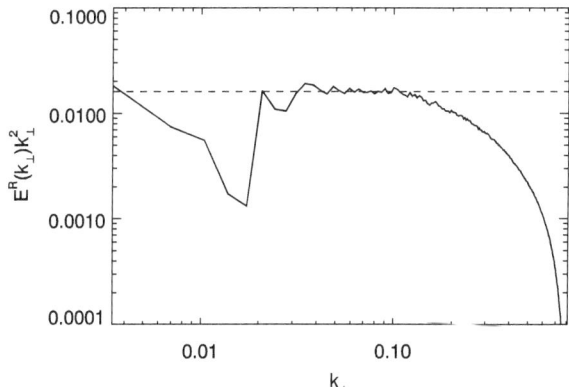

Fig. 11. Averaged and normalized perpendicular residual energy spectrum, $E_{k_\perp}^{\mathrm{R}} = \left| E_{k_\perp}^{\mathrm{M}} - E_{k_\perp}^{\mathrm{K}} \right|$, for the simulation with mean magnetic field (cf. Fig. 8)

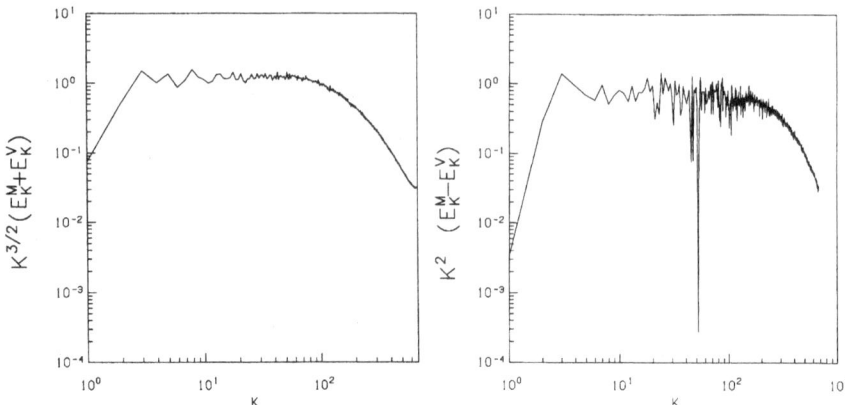

Fig. 12. Spectra of total (*left*) and residual (*right*) energy in a two-dimensional MHD turbulence simulation [8] agreeing with (22) and (23)

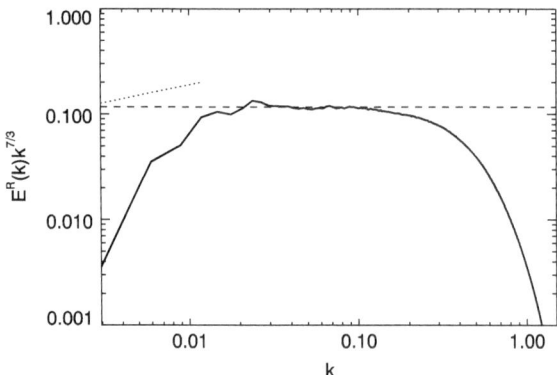

Fig. 13. Averaged and normalized residual energy spectrum, $E_k^{\mathrm{R}} = \left| E_k^{\mathrm{M}} - E_k^{\mathrm{K}} \right|$, for the isotropic simulation (cf. Fig. 6). The dotted line adumbrates IK-like scaling $\sim k^{-2}$

IK- and K41-like scaling which amounts to only 1/6 is enlarged to a much more obvious difference of 1/3 in the residual energy spectrum (see dotted line in Fig. 13). Consequently, the finding of $E_{k_\perp}^{\mathrm{R}} \sim k_\perp^{-2}$ can be regarded as an independent indication for $E_{k_\perp} \sim k_\perp^{-3/2}$ as in Fig. 8.

4 Spatial Structure

The phenomenologies presented in the previous section assume that turbulence is spatially self-similar, i.e. that it has no characteristic internal length scale. This implies that the spatial distribution of turbulent structures is space-filling and statistically uniform. Experimental data, however, shows significant deviations from this behaviour in hydrodynamic turbulence (cf. [88] for a review), in the turbulent solar wind [18, 45] as well as in DNS of incompressible [12, 21, 67, 77] and compressible [72] MHD turbulence

The departure from self-similarity can be related to the spatial distribution of dissipative turbulent structures by Kolmogorov's refined similarity hypothesis [48]. In contrast to (11), it postulates that $S_p^v \sim \langle \varepsilon_\ell^{p/3} \rangle \ell^{p/3}$ with ε_ℓ denoting the local energy dissipation in a sphere of radius ℓ. This approach is motivated by the observation that turbulent energy dissipation is not homogeneously distributed in space. Instead, small regions of intense dissipation are embedded into a weakly dissipative environment making the associated spatial distribution intermittent.

The Kolmogorov and Iroshnikov–Kraichnan models, which do not take intermittency into account, predict the isotropic structure–function exponents as $\zeta_p^{K41} = p/3$ and $\zeta_p^{IK} = p/4$ respectively. Intermittency corrections to the linear predictions are found by examination of structure–function scaling exponents ζ_p of various orders p and become more and more pronounced as p increases.

The finite domains in space and time available to experiment and numerical simulation limit the statistical convergence of the associated averages necessary for determining structure functions. This problem is particularly pronounced for functions of higher order since they react more sensitively to extreme fluctuations of the turbulent fields. The statistical noise can be reduced by exploiting the fact that structure functions of different order deviate qualitatively in the same way from their 'ideal' shape, a property termed extended self-similarity (ESS) [4].

Hence, the scaling range of a structure function, S_p, can substantially be enlarged by regarding S_p in dependence of a reference structure function S_r with a scaling exponent ζ_r that is known with sufficient precision,

$$S_p(S_r(\ell)) \sim \left(\ell^{\zeta_r} \right)^{\zeta_p} \sim \ell^{\xi_{p,r}} .$$

The absolute scaling exponents can then be found via $\zeta_p = \xi_{p,r}/\zeta_r$.

In a pseudospectral simulation with 512^3 collocation points the use of ESS allows to compute scaling exponents up to order eight with sufficient precision. The second-order value ζ_2 is related to the inertial-range behaviour of the energy spectrum, $E(k) \sim k^{-(1+\zeta_2)}$. The whole family of exponents gives more general information about the small-scale structure of the corresponding turbulent fields and represents a framework for the verification of intermittency phenomenologies.

4.1 Intermittency Modelling

There are a number of different phenomenologies [29] which predict the characteristic change of two-point scaling exponents with increasing order.

The Log-Poisson model, which was first proposed for hydrodynamic turbulence by She and Lévêque [83], takes a unique position among them as it achieves very good agreement with experiments and simulations and only contains parameters which can be estimated by physical reasoning.

The Log-Poisson phenomenology assumes a hierarchical relation between the functions $\varepsilon^{(p)} = \langle \varepsilon^{p+1} \rangle / \langle \varepsilon^p \rangle$ which reads

$$\frac{\varepsilon^{(p+1)}}{\varepsilon^{(\infty)}} \sim \left[\frac{\varepsilon^{(p)}}{\varepsilon^{(\infty)}} \right]^{\beta}, \qquad \beta \in [0,1]. \tag{24}$$

The quantity $\varepsilon^{(\infty)}$ stands for the dissipation due to the topologically most singular structures while β parameterizes the degree of intermittency: $\beta \to 1$ corresponds to spatially homogeneous dissipation, $\beta \to 0$ stands for the most intermittent configuration where ε is concentrated in one singular structure. Relation (24) expresses a generalized scale-covariance of dissipation [25, 84] which is equivalent to a logarithmic Poisson distribution of the ε_ℓ [25],

In the inertial range, where $\langle \delta v_\ell^p \rangle \sim \ell^{\zeta_p}$ and $\langle \varepsilon_\ell^p \rangle \sim \ell^{\tau_p}$, the refined similarity hypothesis leads to

$$\langle \delta v_\ell^p \rangle \sim \left\langle \varepsilon^{p/g} \right\rangle \ell^{p/g} \sim \ell^{\tau_{p/g}} \ell^{p/g} \sim \ell^{\zeta_p} \quad \Rightarrow \zeta_p = p/g + \tau_{p/g}. \tag{25}$$

Equation (25) links the scaling exponents of turbulent fields, ζ_p and dissipation, τ_p.

Assuming that the energy to be dissipated in the most singular structures, E^∞, is scale-independent,

$$\varepsilon^{(\infty)} \sim E^\infty / t_\ell^\infty \sim \ell^{-x} \quad \text{with} \quad t_\ell^\infty \sim \ell^x,$$

gives

$$\lim_{p \to \infty} (\tau_{p+1} - \tau_p) = -x \quad \Rightarrow \tau_p = -xp + C_0 + f(p). \tag{26}$$

With (24) and $\tau_0 = 0$ one obtains $f(p) = -C_0 \beta^p$ yielding

$$\tau_p = -xp + C_0(1 - \beta^p).$$

Furthermore, $\tau_1 = 0$ results in $\beta = 1 - x/C_0$ and consequently

$$\tau_p = -xp + C_0 \left(1 - (1 - x/C_0)^p \right)$$

$$\overset{(25)}{\Rightarrow} \zeta_p = (1 - x)p/g + C_0 \left(1 - (1 - x/C_0)^{p/g} \right) \tag{27}$$

This is the general Log-Poisson model (see, e.g. [74]). It depends on the parameters x, g and C_0 which have to be determined on physical grounds. The nonintermittent scaling, $\langle \delta v_\ell \rangle \sim \ell^{1/g}$, fixes g (Kolmogorov $g = 3$, Iroshnikov-Kraichnan $g = 4$).

Equation (26) is analogous to a Legendre transformation. Therefore, C_0 can be interpreted as the co-dimension of a set of singularities of strength ℓ^{τ_∞} which is equivalent to the most singular dissipative structures. The parameter x is related to the dissipation rate in these structures $t_\ell^\infty \sim \ell^x$.

Hydrodynamics

The hydrodynamic She–Lévêque model [83] is obtained for Kolmogorov scaling, $g = 3$ and quasi one-dimensional dissipative structures, i.e. vorticity filaments with $C_0 = 2$ (cf. Fig. 14). The most singular dissipation rate is assumed to be set by the flux rate of the energy cascade,

$$E^{(\infty)}/t_\ell^{(\infty)} \sim \langle \varepsilon \rangle \sim \langle v_\ell^2 \rangle / t_\ell \quad \Rightarrow t_\ell^{(\infty)} \sim \ell^{2/g} \ (x = 2/g)$$

$$\Rightarrow \zeta_p = p/9 + 2 \left(1 - (2/3)^{p/3} \right) .$$

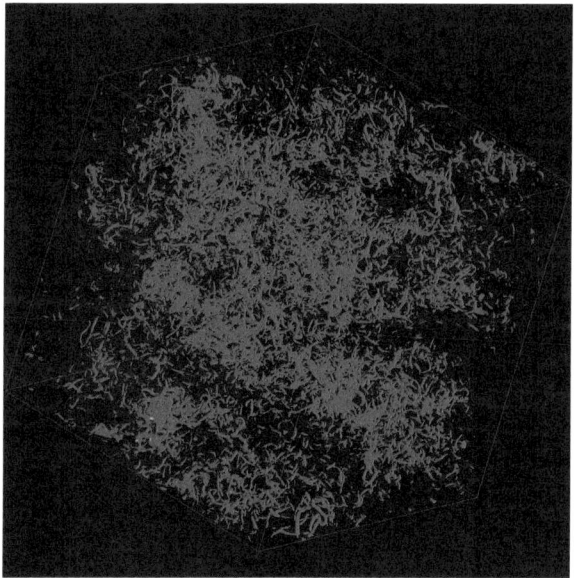

Fig. 14. Vorticity isocontours near the maximum value of $|\boldsymbol{\omega}|$ in a 512^3-resolution DNS of decaying hydrodynamic turbulence [12]

Isotropic MHD

In the case of isotropic MHD Kolmogorov-like scaling also applies ($g = 3$). Maintaining the dissipation rate assumption ($x = 2/g$) and observing that the most singular dissipative structures are current and vorticity *sheets* ($C = 1$, cf. Fig. 16) one arrives at the isotropic MHD model,

$$\zeta_p = p/9 + 1 - (1/3)^{p/3} . \tag{28}$$

The equation predicts structure–function scaling in isotropic three-dimensional incompressible MHD turbulence with good precision and is moreover consistent with the finding of a K41-like energy spectrum [12, 67]. It also agrees well with solar wind data [5, 45] and simulations of compressible [16] MHD turbulence. In contrast, the relation based on Iroshnikov–Kraichnan scaling, $g = 4$ [40, 74], is not in accordance with experimental and numerical data.

The MHD model is indicated in Fig. 15 by the solid line. The dotted line shows the nonintermittent Kolmogorov prediction, $\zeta_p = p/3$, for comparison.

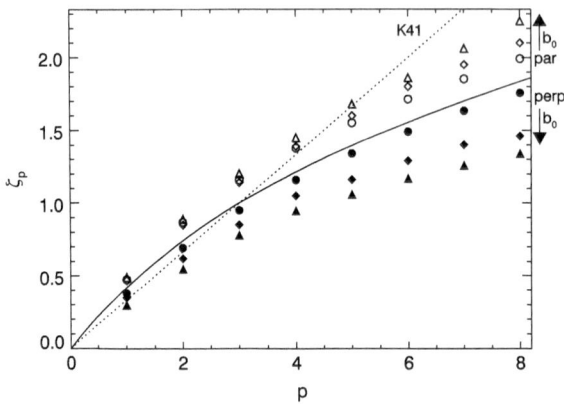

Fig. 15. Scaling exponents ζ_p of the $|z^+|$-structure functions against the corresponding order. The MHD Log-Poisson model (28) is depicted by the *solid line* and agrees very well with isotropic MHD values [67] (not shown). Nonintermittent K41 scaling is represented by the *dotted line*. Anisotropic exponents are shown for $b_0 = 0, 5, 10$ (circles, squares, triangles). Structure functions parallel to the mean magnetic field are given by open symbols, field-perpendicular data by filled symbols

Anisotropic MHD

Spatial anisotropy is displayed by higher-order statistics in MHD turbulence with and without a mean magnetic field as shown in Fig. 15 [68] (see also [21, 22, 63, 85] and [62] for the compressible case). The data stems from a globally isotropic 512^3-simulation ($b_0 = 0$, circles) of decaying turbulence and from forced anisotropic $512^2 \times 256$-simulations with mean magnetic field ($b_0 = 5, 10$, squares/triangles). For the anisotropic runs, the resolution along b_0 can be reduced to 256 collocation points as the turbulence is depleted in this direction (cf. Fig. 8) due to the stiffness of the magnetic field lines.

Structure functions are calculated with the space increment, ℓ, taken either parallel or perpendicular to the magnetic field. The direction of the local field is found by applying a top-hat filter of width ℓ to the magnetic fluctuations (for details, cf. [68]).

Fig. 16. Dissipative micro current sheets in isotropic (*left*) and anisotropic (*right*) turbulence. The anisotropic system is permeated by a mean magnetic field of strength $b_0 = 5$ pointing upwards (cf. Fig. 7). Each current sheet is surrounded by vorticity sheets in quadrupolar configuration

In the field-parallel direction the structure function exponents show an asymptotic approach towards a straight line with increasing b_0. This indicates a decrease of intermittency in this direction, i.e. more homogeneous dissipation. The field-perpendicular exponents display a gradual transition towards values known from two-dimensional turbulence simulations; in fact the perpendicular exponents for $b_0 = 10$ are identical within the error margin with two-dimensional results [13, 78].

Figure 16 suggests a plausible explanation: while the dissipative current sheets (and the associated vorticity sheets) show no preferred orientation in the isotropic case, they tend to align with an applied mean magnetic field. The aligned configuration results in increased homogeneity of dissipation in the b_0-direction and makes the system look two-dimensional in field-perpendicular planes. A generalized version of the isotropic MHD intermittency model (28), $\zeta_p = p/g^2 + 1 - (1/g)^{p/g}$, where g is a free parameter setting the energy cascade rate in the respective parallel/perpendicular direction, is able to reproduce the observed scaling exponents [68].

5 Summary

The understanding of the nonlinear energy cascade and the spatial small-scale structure of incompressible MHD turbulence is largely based on phenomenology and direct numerical simulation. This chapter briefly summarizes the current level of knowledge and highlights recent developments in this field of research. Particular stress is put on the difference between isotropic turbulence and configurations permeated by a mean magnetic field. Properly taking into account spatial anisotropy of the turbulence induced by the magnetic field is an important challenge to MHD turbulence theory at present.

The physical picture of the turbulent energy cascade is still under discussion. While there is ample evidence of Kolmogorov scaling in numerical simulations of globally isotropic MHD turbulence, recent high-resolution simulations where a strong mean magnetic field is imposed on the turbulent flow show Iroshnikov–Kraichnan-like scaling for field-perpendicular fluctuations (cf. Sect. 3.4). This rules out the Goldreich–Sridhar model which is the first anisotropic phenomenology of MHD turbulence and predicts field-perpendicular Kolmogorov spectra. These findings have been corroborated by results of EDQNM closure theory calculations which give a simple relation between the residual and the total energy spectrum verified by numerical simulation. A recent enhancement of the GS-model by Boldyrev might however explain the observed behaviour.

The intermittent small-scale structure, which is probed by higher-order two-point statistics, is visible in the structure–function scaling exponents. In isotropic MHD turbulence, their characteristic behaviour can be matched well by a Log-Poisson model which takes into account that the energy cascade is Kolmogorov-like and that the most singular dissipative structures are quasi-two-dimensional current and vorticity sheets. The model can be generalized to reproduce the structure–function scalings parallel and perpendicular to an applied mean magnetic field. Numerical simulations with increasing field strength show, furthermore, that the system becomes gradually two-dimensional in the field-perpendicular direction while the dissipative structures turn out to be more homogeneous and less intermittent along the mean field. This behaviour is the consequence of the alignment of dissipative current and vorticity sheets with the mean magnetic field.

Acknowledgements

The author would like to thank Dieter Biskamp, Roland Grappin, and Jacques Léorat for stimulating collaboration and gratefully acknowledges illuminating discussions with Patrick Diamond and Stanislav Boldyrev.

References

1. Batchelor, G.K.: Computation of the energy spectrum in homogeneous two-dimensional turbulence. Phys. Fluids Suppl. II **12**(3), 233–239 (1969)
2. Batchelor, G.K.: The Theory of Homogeneous Turbulence. Cambridge University Press, Cambridge (1993)
3. Baumjohann, W., Treumann, R.A.: Basic Space Plasma Physics. Imperial College Press, London (2004)
4. Benzi, R., Ciliberto, S., Tripiccione, R., Baudet, C., Massaioli, F., Succi, S.: Extended self-similarity in turbulent flows. Phys. Rev. E **48**(1), R29–R32 (1993)

5. Bershadskii A.: Three-dimensional isotropic magnetohydrodynamic turbulence and thermal velocity of the solar wind ions. Phys. Plasmas. **10**(12), 4613–4615 (2003)

6. Birkhoff G.D.: Proof of the ergodic theorem. Proc. Natl. Acad. Sci. USA **17**, 656–660, (1931)

7. Biskamp D.: Nonlinear Magnetohydrodynamics. Cambridge University Press, Cambridge (1993)

8. Biskamp D.: Scaling properties in MHD turbulence. Chaos, Solitons & Fractals, **5**(10), 1779–1793 (1995)

9. Biskamp D.: Magnetic Reconnection in Plasmas. Cambridge University Press, Cambridge (2000)

10. Biskamp D.: Magnetohydrodynamic Turbulence. Cambridge University Press, Cambridge (2003)

11. Biskamp, D., Müller W.-C.: Decay laws for three-dimensional magnetohydrodynamic turbulence. Phys. Rev. Lett. **83**(11), 2195–2198 (1999)

12. Biskamp, D., Müller W.-C.: Scaling properties of three-dimensional isotropic magnetohydrodynamic turbulence. Phys. Plasmas **7**(12), 4889–4900 (2000)

13. Biskamp, D., Schwarz E.: On two-dimensional magnetohydrodynamic turbulence. Phys. Plasmas **8**(7), 3282–3292 (2001)

14. Biskamp, D., Welter H.: Dynamics of decaying two-dimensional magnetohydrodynamic turbulence. Phys. Fluids B **1**(10), 1964–1979 (1989)

15. Boldyrev S.: On the spectrum of magnetohydrodynamic turbulence. Astrophys. J. **626**, L37–L40 (2005)

16. Boldyrev, S., Nordlund, Å., Padoan P.: Scaling relations of supersonic turbulence in star-forming molecular clouds. Astrophys. J. **573**, 678–684 (2002)

17. Brandenburg, A., Subramanian K.: Astrophysical magnetic fields and nonlinear dynamo theory. Phys. Rep. **417**, 1–209 (2005)

18. Burlaga L.F.: Intermittent turbulence in the solar wind. J. Geophy. Res. **96**(A4), 5847–5851 (1991)

19. Busse F.H.: Homogeneous dynamos in planetary cores and in the laboratory. Annu Rev. Fluid Mech. **32**, 383–408 (2000)

20. Carbone, V., Veltri P.: A shell model for anisotropic magnetohydrodynamic turbulence. Geophy. Astrophys. Fluid Dyn. **52**, 153–181 (1990)

21. Cho, J., Lazarian, A., Vishniac E.T.: Simulations of magnetohydrodynamic turbulence in a strongly magnetized medium. Astrophys. J. **564**, 291–301 (2002)

22. Cho, J., Vishniac E.T.: The anisotropy of magnetohydrodynamic Alfvénic turbulence. Astrophys. J. **539**, 273–282 (2000)

23. Comte-Bellot, G., Corrsin S.: The use of a contraction to improve the isotropy of grid-generated turbulence. J. Fluid Mech. **25**(4), 657–682 (1966)

24. Dobrowolny, M., Mangeney, A., Veltri P.: Fully developed anisotropic hydromagnetic turbulence in interplanetary space. Phys. Rev. Lett. **45**(2), 144–147 (1980)

25. Dubrulle B.: Intermittency in fully developed turbulence: Log-Poisson statistics and generalized scale covariance. Phys. Rev. Lett. **73**(7), 959–962 (1994)

26. Edwards S.F.: The statistical dynamics of homogeneous turbulence. J. Fluid Mech. **18**, 239–273 (1964)

27. Elsässer W.M.: The hydromagnetic equations. Phys. Rev. **79**, 183 (1950)

28. Falkovich G.: Bottleneck phenomenon in developed turbulence. Phys. Fluids **6**(4), 1411–1414 (1994)

29. Frisch U.: Turbulence. Cambridge University Press, Cambridge (1996)
30. Frisch, U., Lesieur, M., Brissaud A.: A Markovian random coupling model for turbulence. J. Fluid Mech. **65**(1), 145–152 (1974)
31. Frisch U., Pouquet, A., Léorat, J., Mazure A.: Possibility of an inverse cascade of magnetic helicity in magnetohydrodynamic turbulence. J. Fluid Mech. **68**(4), 789–778 (1975)
32. Galtier, S., Nazarenko, S.V., Newell, A.C., Pouquet A.: A weak turbulence theory for incompressible magnetohydrodynamics. J. Plasma Phys. **63**(5), 447–488 (2000)
33. Galtier, S., Politano, H., Pouquet A.: Self-similar energy decay in magnetohydrodynamic turbulence. Phys. Rev. Lett. **79**(15), 2807–2810 (1997)
34. Galtier, S., Pouquet, A., Mangeney A.: On spectral scaling laws for incompressible anisotropic MHD. J. Plasma Phys. **12**, 092310 (2005)
35. Goldreich, P., Sridhar S.: Toward a theory of interstellar turbulence. II. Strong Alfvénic turbulence. Astrophys. J. **438**, 763–775 (1995)
36. Goldreich, P., Sridhar S.: Magnetohydrodynamic turbulence revisited. Astrophys. J. **485**, 680–688 (1997)
37. Grappin R.: Onset and decay of two-dimensional magnetohydrodynamic turbulence with velocity-magnetic field correlation. Phys. Fluids **29**(8), 2433–2443 (1986)
38. Grappin, R., Pouquet, A., Léorat J.: Dependence of MHD turbulence spectra on the velocity field-magnetic field correlation. Astron. Astrophys. **126**, 51–58 (1983)
39. Grappin, R., Frisch, U., Léorat, J., Pouquet A.: Alfvénic fluctuations as asymptotic states of MHD turbulence. Astron. Astrophys. **105**, 6–14 (1982)
40. Grauer, R., Krug, J., Marliani C.: Scaling of high-order structure functions in magnetohydrodynamic turbulence. Phys. Lett. A **195**, 335–338 (1994)
41. Hatori T.: Kolmogorov-style argument for the decaying homogeneous MHD turbulence. J. Phys. Soc. Japan **53**(8), 2539–2545 (1984)
42. Haugen, N.E.L., Brandenburg, A., Dobler W.: Is nonhelical hydromagnetic turbulence peaked at small scales ? Astrophys. J. **597**, L141–L144 (2003)
43. Herring J.R.: Self-consistent-field approach to turbulence theory. Phys. Fluids **8**(12), 2219–2225 (1965)
44. Hinze J.O.: Turbulence. McGraw-Hill, New York (1987)
45. Horbury, T.S., Balogh A.: Structure function measurements of the intermittent MHD turbulent cascade. Nonlinear Processes in Geophys. **4**, 185–199 (1997)
46. Iroshnikov P.S.: Turbulence of a conducting fluid in a strong magnetic field. Soviet Astron. **7**, 566–571 (1964) [Astron. Zh., **40**, 742, 1963]
47. Kolmogorov A.N.: On the degeneration of isotropic turbulence in an incompressible viscous liquid. Doklady Akademiia Nauk SSSR **31**, 538–540 (1941)
48. Kolmogorov A.N.: A refinement of previous hypotheses concerning the local structure of turbulence in a viscous incompressible fluid at high Reynolds number. J. Fluid Mech. **13**, 82–85 (1962)
49. Kolmogorov A.N.: Dissipation of energy in the locally isotropic turbulence. Proc. Roy. Soc. A **434**, 15–17 (1991) [Dokl. Akad. Nauk SSSR, **32**(1), 1941]
50. Kolmogorov A.N.: The local structure of turbulence in incompressible viscous fluid for very large Reynolds numbers. Proc. Roy. Soc. A **434**, 9–13 (1991) [Dokl. Akad. Nauk SSSR, **30**(4), 1941]

51. Kraichnan R.H.: The structure of isotropic turbulence at very high Reynolds numbers. J. Fluid Mech. **5**, 497–543 (1959)

52. Kraichnan R.H.: Inertial-range spectrum of hydromagnetic turbulence. Phys. Fluids **8**(7), 1385–1387 (1965)

53. Kraichnan R.H.: Lagrangian-history closure approximation for turbulence. Phys. Fluids **8**(4), 575–598 (1965)

54. Kraichnan R.H.: An almost Galilean-invariant turbulence model. J. Fluid Mech. **47**(3), 513–524 (1971)

55. Leamon, R.J. Smith, C.W., Ness, N.F., Matthaeus, W.H., Wong H.K.: Observational constraints on the dynamics of the interplanetary magnetic field dissipation range. J. Geophys. Res. **103**(A3), 4775–4787 (1998)

56. Lesieur M.: Turbulence in Fluids. Kluwer Academic Publishers, Dordrecht (1997)

57. Lesieur, M., Schertzer D.: Amortissement autosimilaire d'une turbulence à grand nombre de Reynolds. Journal de Mécanique **17**(4), 609–646 (1978)

58. Leslie D.C.: Developments in the Theory of Turbulence. Clarendon Press, Oxford (1983)

59. Lohse, D., Müller-Groeling A.: Bottleneck effects in turbulence: Scaling phenomena in r versus p space. Phys. Rev. Lett. **74**(10), 1747–1750 (1995)

60. Mac Low, M.-M., Klessen, R.S., Burkert, A. Smith M.D.: Kinetic energy decay rates of supersonic and super-alfvénic turbulence in star-forming clouds. Phys. Rev. Lett. **80**(13), 2754–2757 (1998)

61. Maron, J., Goldreich P.: Simulations of incompressible magnetohydrodynamic turbulence. Astrophys. J. **554**, 1175–1196 (2001)

62. Matthaeus, W.H., Ghosh, S., Oughton, S., Roberts D.: Anisotropic three-dimensional MHD turbulence. J. Geophys. Res. **101**(A4), 7619–7629 (1996)

63. Milano, L.J., Matthaeus, W.H., Dmitruk, P., Montgomery D.C.: Local anisotropy in incompressible magnetohydrodynamic turbulence. Phys. Plasmas **8**(6), 2673–2681 (2001)

64. Moffatt H.K.: The degree of knottedness of tangled vortex lines. J. Fluid Mech. **35**(1), 117–129 (1969)

65. Monin, A.S., Yaglom A.M.: Statistical Fluid Mechanics, vol. 1. MIT Press, Cambridge, Massachusetts (1971)

66. Monin, A.S., Yaglom A.M.: Statistical Fluid Mechanics, vol. 2. MIT Press, Cambridge, Massachusetts (1981)

67. Müller, W.-C., Biskamp D.: Scaling properties of three-dimensional magnetohydrodynamic turbulence. Phys. Rev. Lett. **84**(3), 475–478 (2000)

68. Müller, W.-C., Biskamp, D., Grappin R.: Statistical anisotropy of magnetohydrodynamic turbulence. Phys. Rev. E **67**, 066302–1–066302–4 (2003)

69. Müller W.-C., Grappin R.: Energy dynamics in magnetohydrodynamic turbulence. Phys. Rev. Lett. **95**, 114502–1–114502–4 (2005)

70. Orszag S.A.: Analytical theories of turbulence. J. Fluid Mech. **41**(2), 363–386 (1970)

71. Ortolani, S., Schnack D.D.: Magnetohydrodynamics of Plasma Relaxation. World Scientific, Singapore (1993)

72. Padoan, P., Jimenez, R., Nordlund, Å., Boldyrev S.: Structure function scaling in compressible super-Alfvénic MHD turbulence. Phys. Rev. Lett. **92**(19), 191102–1–191102–4 (2004)

73. Parker E.N.: Cosmical Magnetic Fields. Clarendon Press, Oxford (1979)

74. Politano, H., Pouquet A.: Model of intermittency in magnetohydrodynamic turbulence. Phys. Rev. E **52**(1), 636–641 (1995)
75. Politano, H., Pouquet A.: Dynamical length scales for turbulent magnetized flows. Geophys. Res. Lett. **25**(3), 273–276 (1998)
76. Politano, H., Pouquet A.: Von Kármán-Howarth equation for magnetohydrodynamics and its consequences on third-order longitudinal structure and correlation functions. Phys. Rev. E **57**(1), R21–R24 (1998)
77. Politano, H., Pouquet, A., Carbone V.: Determination of anomalous exponents of structure functions in two-dimensional magnetohydrodynamic turbulence. Europhys. Lett. **43**(5), 516–521 (1998)
78. Politano, H., Pouquet, A., Sulem P.L.: Inertial ranges and resistive instabilities in two-dimensional magnetohydrodynamic turbulence. Phys. Fluids B **1**(12), 2330–2339 (1989)
79. Pope S.B.: Turbulent Flows. Cambridge University Press, Cambridge (2000)
80. Pouquet, A., Frisch, U., Léorat J.: Strong MHD helical turbulence and the nonlinear dynamo effect. J. Fluid Mech. **77**(2), 321–354 (1976)
81. Richardson L.F.: Weather Prediction by Numerical Process. Cambridge University Press, Cambridge (1922)
82. Rose, H.A., Sulem P.L.: Fully developed turbulence and statistical mechanics. Journal de Physique **39**(5), 441–483 (1978)
83. She, Z.-S., Lévêque E.: Universal scaling laws in fully developed turbulence. Phys. Rev. Lett. **72**(3), 336–339 (1994)
84. She, Z.-S., Waymire E.C.: Quantized energy cascade and log-Poisson statistics in fully developed turbulence. Phys. Rev. Lett. **74**(2), 262–265 (1995)
85. Shebalin, J.V., Matthaeus, W.H., Montgomery D.: Anisotropy in MHD turbulence due to a mean magnetic field. J. Plasma Phys. **29**(3), 525–547 (1983)
86. Smith, M.R., Donnelly, R.J., Goldenfeld, N., Vinen W.F.: Decay of vorticity in homogeneous turbulence. Phys. Rev. Lett. **71**(16), 2583–2586 (1993)
87. Soward, A.M., Jones, C.A., Hughes, D.W., Weiss, N.O. (Eds.): Fluid Dynamics and Dynamos in Astrophysics and Geophysics. CRC Press, Boca Raton, Florida (2005)
88. Sreenivasan, K.R., Antonia R.A.: The phenomenology of small-scale turbulence. Annu. Rev. Fluid Mech. **29**, 435–472 (1997)
89. Sridhar, S., Goldreich P.: Toward a theory of interstellar turbulence. I. Weak Alfvénic turbulence. Astrophys. J. **432**, 612–621 (1994)
90. Stone, J.M., Ostriker, E.C., Gammie C.F.: Dissipation in compressible magnetohydrodynamic turbulence. Astrophys. J. **508**, L99–L102 (1998)
91. Ting, A.C., Matthaeus, W.H., Montgomery D.: Turbulent relaxation processes in magnetohydrodynamics. Phys. Fluids **29**(10), 3261–3274 (1986)
92. Tritton D.J.: Physical Fluid Dynamics. Clarendon Press, Oxford (1998)
93. Tu, C.-Y., Marsch E.: MHD structures, waves and turbulence in the solar wind: observations and theories. Space Sci. Rev. **73**, 1–210 (1995)
94. Wesson J.: Tokamaks. Clarendon Press, Oxford (1997)
95. Yoshizawa, A., Itoh, S.-I., Itoh K.: Plasma and Fluid Turbulence. Institute of Physics Publishing, Bristol (2003)
96. Zeldovich, Ya.B., Ruzmaikin, A.A., Sokoloff D.D.: Magnetic Fields In Astrophysics. Gordon and Breach Science Publishers, New York (1983)

Turbulent Combustion in Thermonuclear Supernovae

F.K. Röpke and W. Schmidt

Max-Planck-Institut für Astrophysik, Karl-Schwarzschild-Str. 1, D-85741
Garching, Germany; Department of Astronomy and Astrophysics, University of
California at Santa Cruz, 1156 High Street, Santa Cruz, CA 95064, U.S.A.
fritz@mpa-garching.mpg.de
Lehrstuhl für Astronomie, Institut für Theoretische Physik und Astrophysik,
Universität Würzburg, Am Hubland, D-97074 Würzburg, Germany
schmidt@astro.uni-wuerzburg.de

1 Introduction

Supernovae have attracted attention since the dawn of astronomy. They range
among the brightest astrophysical events and the strongest explosions in the
Universe since the Big Bang. In astrophysical processes, supernovae play a
significant role. They enrich the interstellar medium with heavy elements,
drive shock waves, and influence the formation of stars and cosmic structure.
A subclass of these objects – termed type Ia supernovae (SNe Ia) – has recently been applied in cosmological distance determinations, indicating that
our Universe is currently undergoing an *accelerated* expansion [57, 71]. This
result has extensive consequences for physics since it may be interpreted as a
first glimpse on a new "dark" energy form accounting for as much as 70% of
the energy contents of the Universe (for a review see [39]).

Therefore, one of the most challenging tasks in modern astrophysics is to
explain the nature of these objects. Yet, despite decades of effort, a consistent
picture explaining all details of supernova observations is still lacking. Assuming that supernovae originate from a single stellar object, only its gravitational
binding energy [95] or its nuclear energy [29] can account for the observed explosion strength. Both mechanisms have been associated with subclasses of
supernovae – the release of nuclear energy with SNe Ia and gravitational energy release in a core collapse with all other classes. Here, we focus on the
former, since, as will be shown below, the nuclear energy release involves turbulent combustion.

A widely accepted progenitor scenario for SNe Ia is that of a white dwarf
(WD) star consisting of carbon and oxygen. White dwarfs are supported by
the electron degeneracy pressure that depends on the mass density while it is
virtually independent of the temperature. Such objects mark the final stages of

Röpke, F.K., Schmidt, W.: *Turbulent Combustion in Thermonuclear Supernovae*. Lect. Notes
Phys. **756**, 255–289 (2009)
DOI 10.1007/978-3-540-78961-1_7

the evolution of small and medium-size stars in the mass range from $0.80\,M_\odot$ to $8\,M_\odot$. Before becoming a WD, these stars are stabilized against gravitational collapse by nuclear burning – obviously a limited energy source. Once the nuclear fuel is exhausted (or cannot be burnt anymore under the prevailing conditions), a WD is formed that is stabilized by the pressure of degenerate electrons thus being a very inert object. It cools down for billions of years and finally disappears from observation. However, if dynamics is introduced into the system by a binary companion, the WD might indeed reach an explosive state due to mass accretion. Several possibilities are conceivable [27], and while none can be excluded, research in the last years focused on the so-called *single-degenerate Chandrasekhar mass model.* Here, the WD accretes hydrogen and helium from a nondegenerate main-sequence star or an asymptotic giant branch star.[1] This material is transformed into carbon in quiescent hydrostatic burning at the surface of the WD (e.g. [52]). Thus, its mass steadily increases. However, for a WD there exists a fundamental limit – the Chandrasekhar mass – beyond which the pressure of the degenerate electrons cannot stabilize it against gravitational collapse [12]. Approaching this limit, the density at the WD's centre increases dramatically so that finally nuclear reactions of carbon to heavier isotopes ignite. Unlike a normal star, the degeneracy of the WD prevents it from compensating for the energy release by expansion cooling, because an increase in temperature does not imply an increase in gas pressure and thus expansion. Since nuclear reaction rates are very sensitive to temperature, eventually a thermonuclear runaway will occur in a confined region near the centre of the star, giving rise to an outward-travelling reaction wave (for an analysis of the ignition process see [24, 28, 30, 36, 93]). This marks the ignition of a thermonuclear flame.

The reaction wave is able to propagate in two distinct modes (see Sect. 2.1) – a supersonic detonation and a subsonic deflagration. First attempts to simulate thermonuclear supernova explosions assumed a prompt detonation [3]. Since here the entire star is burnt at the original high densities, such models fail to produce sufficient amounts of intermediate mass elements (IME) like silicon, sulphur and calcium observed in the spectra of SNe Ia. If, on the other hand, the flame starts out in the subsonic deflagration mode, the material ahead of it can pre-expand before being burnt. Therefore, nuclear reactions partially take place at lower densities where IME are synthesized. A laminar deflagration flame, however, is far too slow to explain the energy release needed for an explosion of the star.

This problem can be solved noting that turbulence will be generated and affects the flame. The interaction of the flame with turbulent motions accelerates its propagation. In a parametrized way this was taken into account in the pioneering W7 model [54], which is still used as a standard explosion model for SNe Ia in many fields of astrophysics. However, being a one-dimensional

---

[1] The former are still burning hydrogen in their core while the latter have already undergone phases of helium burning in their core and off-centre in spherical shells.

simulation, it lacks a consistent description of the turbulence–flame interaction and the arbitrariness of choosing an effective flame propagation speed was exploited to fit the observations. In this way it was able to guide further developments.

With multi-dimensional models of thermonuclear supernova explosions, a self-consistent description of the flame physics has come into reach. Advances in numerical techniques and in computer power over the past decade led to a rapid development of two- and three-dimensional simulations.

The implementation of a turbulent combustion model in multi-dimensional simulations of thermonuclear supernovae is the main focus of this article and will be reviewed in the following.

2 Combustion in SNe Ia

2.1 Fundamentals of Combustion Physics

The flame propagation phenomenon is captured in the equations of reactive fluid dynamics. These combine the Navier–Stokes equations of hydrodynamics with species conversion and energy generation in reactions.

The system under consideration is therefore described by the following set of equations:

- mass conservation

$$\frac{\partial \rho}{\partial t} = -\boldsymbol{\nabla} \cdot (\rho \boldsymbol{v}), \tag{1}$$

- momentum balance

$$\frac{\partial \rho \boldsymbol{v}}{\partial t} = -\boldsymbol{\nabla} \cdot (\rho \boldsymbol{v} \boldsymbol{v}) - \boldsymbol{\nabla} \cdot \boldsymbol{\Pi} + \rho \boldsymbol{f}, \tag{2}$$

- species balance

$$\frac{\partial \rho X_i}{\partial t} = -\boldsymbol{\nabla} \cdot (\rho X_i \boldsymbol{v}) - \boldsymbol{\nabla} \cdot (\rho v_i^{\mathrm{D}} X_i) + \rho \omega_{X_i} \qquad i = 1 \ldots N, \tag{3}$$

- and energy balance

$$\frac{\partial \rho e_{\mathrm{tot}}}{\partial t} = -\boldsymbol{\nabla} \cdot (\rho e_{\mathrm{tot}} \boldsymbol{v}) - \boldsymbol{\nabla} \cdot (\boldsymbol{v} \boldsymbol{\Pi}) + \rho \boldsymbol{v} \cdot \boldsymbol{f} + \rho \sum_{i=1}^{N} X_i v_i^{\mathrm{D}} \cdot \boldsymbol{f}_i - \boldsymbol{\nabla} \cdot \boldsymbol{q} + \rho S, \tag{4}$$

where $\boldsymbol{\Pi}$, $\boldsymbol{f}$, X_i, $\boldsymbol{v}^{\mathrm{D}}$, ω, $\boldsymbol{q}$ and S denote the pressure tensor, external forces, mass fraction of species, diffusion velocity, reaction rate, heat flux and energy source terms due to reactions respectively. All other symbols bear their usual meanings (see the chapter "An Introduction to Turbulence").

Certainly, the terms included in these equations do not account for all physical effects that could be related to a combustion process. For instance, we neglect in the following the radiative heat flux for reasons given in [90]. The balance equations stated above have to be complemented by a set of auxiliary relations. The equation of state relates pressure to density, specific internal energy ($e_{\text{int}} = e_{\text{tot}} - v^2/2$) and composition:

$$p = f_{\text{EOS}}(\rho, e_{\text{int}}, X_i). \tag{5}$$

Additionally, one has to supply expressions for the diffusion velocities and the external forces. The source terms in the species and energy equation depend on density, temperature and composition as

$$\omega_{X_i} = \omega_{X_i}(\rho, T, X_i) \tag{6}$$

$$S = S(\omega_{X_i}) \tag{7}$$

Considering an ideal fluid with no internal friction, the viscosity vanishes and the pressure tensor simplifies to the scalar pressure. If we further neglect the microphysical transport terms, the equations above simplify to the well-known (reactive) Euler equations:

$$\frac{\partial \rho}{\partial t} = -\boldsymbol{\nabla} \cdot (\rho \boldsymbol{v}), \tag{8}$$

$$\frac{\partial \boldsymbol{v}}{\partial t} = -(\boldsymbol{v}\boldsymbol{\nabla}) \cdot \boldsymbol{v} - \frac{\boldsymbol{\nabla} p}{\rho} + \boldsymbol{f}, \tag{9}$$

$$\frac{\partial \rho X_i}{\partial t} = -\boldsymbol{\nabla} \cdot (\rho X_i \boldsymbol{v}) + \rho \omega_{X_i} \qquad i = 1 \ldots N, \tag{10}$$

$$\frac{\partial \rho e_{\text{tot}}}{\partial t} = -\boldsymbol{\nabla} \cdot (\rho e_{\text{tot}} \boldsymbol{v}) - \boldsymbol{\nabla}(p\boldsymbol{v}) + \rho \boldsymbol{v} \cdot \boldsymbol{f} + \rho S. \tag{11}$$

These, of course, do not account anymore for the full phenomenon of combustion. However, if the scales under consideration are much larger than the width of the reaction zone, it is justified to simplify the burning front to a moving discontinuity in the state variables, which is referred to as "flame front".

This *discontinuity approximation* allows to relate the prefront and postfront states. Under the condition of continuity in all flux densities, the Euler equations of fluid dynamics admit the formulation of certain jump conditions for the state variables (see e.g. [38]). This is achieved by integrating them over an arbitrary volume containing a part of the flame front. In this way one obtains the *Rayleigh criterion* for the square of the mass flux j_m over the front,

$$j_m^2 = (\rho_{\text{u}} v_{\text{u,n}})^2 = (\rho_{\text{b}} v_{\text{b,n}})^2 = \frac{p_{\text{u}} - p_{\text{b}}}{V_{\text{b}} - V_{\text{u}}}, \tag{12}$$

where the indices u and b indicate states in the unburnt and burnt material, respectively, and n denotes components normal to the flame front. $V := 1/\rho$ is the specific volume. As a second important relation, the *Hugoniot curve* is found:

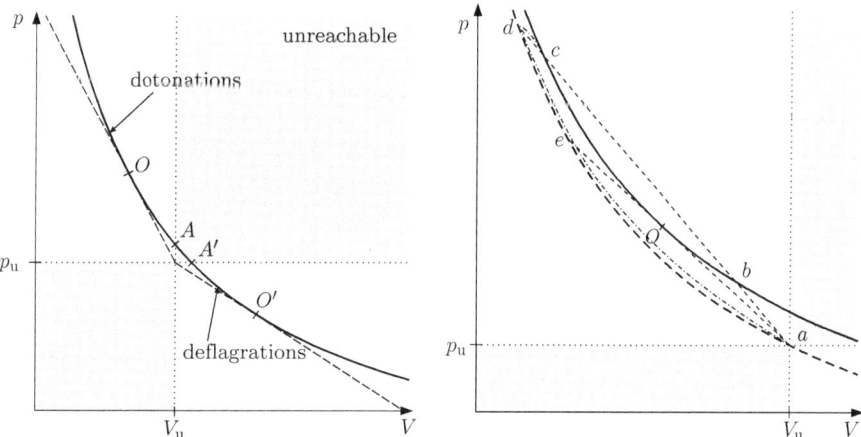

Fig. 1. Relation between unburnt (V_u, p_u) and burnt (V_b, p_b) states (*left*) and closeup of the detonation branch (*right*). The **bold solid** curve is the Hugoniot while the *dashed curve* indicates the ordinary shock adiabatic through the initial state (following [40] and [38])

$$e_{int,u} - e_{int,b} = \Delta h_0 - \frac{p_u + p_b}{2}(V_b - V_u), \qquad (13)$$

where Δh_0 denotes the difference in the formation enthalpies of burnt and unburnt material.

Using the Hugoniot curve and the Rayleigh criterion, conditions for burnt states can be derived for given unburnt states. This is illustrated in Fig. 1 assuming an ideal gas equation of state. The state of the burnt material must lie on the Hugoniot curve (sometimes also called detonation adiabatic). Because of the Rayleigh criterion (12), the unburnt and the resulting burnt states are connected by straight lines (the so-called Rayleigh lines, indicated in Fig. 1 as light dashed lines, e.g. ac), the slope of which corresponds to the negative square of the mass flux density over the front. This excludes the part AA' of the Hugoniot curve from the range of possible final states, because for states of the burnt material in this range the mass flux density would be imaginary. Separate modes of flame front propagation can be identified by constructing the tangents to the Hugoniot curve starting from (V_u, p_u). The slope of these tangents is related to the velocity of the front with respect to the unburnt material by virtue of Eq. (12). These slopes can be compared to the slopes of tangents to the ordinary shock adiabatic through the initial state (cf. right plot of Fig. 1) measuring the sound speed with respect to the unburnt material [38]. For final states on the Hugoniot above A, the flame is supersonic *with respect to the fuel* and the combustion mode is termed *detonation* and for final states below A' the so-called *deflagration* advances subsonically with respect to the fuel. However, in both modes its velocity can still be supersonic or subsonic *with respect to the ashes*. The boundaries between these possibilities are given by the points marked O and O' in Fig. 1. They are constructed by drawing the tangents to the Hugoniot starting from the initial state (V_u, p_u).

O and O' are called the upper and lower Chapman–Jouguet points respectively. The flame advances sonically with respect to the ashes in them and supersonically above O and below O' (see [38, 40]).

It turns out that the way the reaction wave is mediated in the deflagration and the detonation modes differs fundamentally. While detonations are driven by shocks, deflagrations are mediated by microphysical transport phenomena.

2.2 Thermonuclear Combustion in Degenerate C + O Matter

Obviously, reaction partners of a different species are not required in thermonuclear burning. Thus, in many respects, thermonuclear combustion in WD matter is similar to what is known as "pre-mixed combustion" in chemical processes. This is exploited below where numerical techniques developed in combustion engineering are applied to the astrophysical problem.

Yet there are in fact some differences, caused by the peculiarities of the degenerate C+O matter in which the combustion process takes place. The reactive Navier–Stokes equations given above can be reformulated in a non-dimensional way. In this case, several similarity numbers enter the equations relating specific properties of the flow. These can be used to characterize the physical processes.

The transport processes in WD matter are dominated by the electron gas. This is due to the high degeneracy of the material. The Fermi energy E_F of the electron gas is about $1\,\mathrm{MeV}$, while the thermal energy $k_B T$ is $\sim 10\,\mathrm{keV}$. Electron states below $E_F - k_B T$ are occupied and limit the final states for scattering processes to high velocities. Therefore the mean free path of the electrons is much larger than that for the baryons.

This has significant impact on the *Lewis number* comparing thermal conduction with diffusive transport and on the *Prandtl number* relating momentum transport to thermal conduction. With values given by [44, 56] and [90] these amount to $Le \sim 10^7$ and $Pr \sim 10^{-3}$, characterizing one of the main differences between thermonuclear flames in degenerate matter and chemical flames, where usually $Le \sim 1$ and $Pr \sim 1$. These differences have consequences for certain instabilities that occur in flames and on the transition between regimes of turbulent combustion [50].

The basic properties of laminar thermonuclear flame propagation can be inferred from one-dimensional calculations of planar flames with an extended reaction network. In particular, [90] fitted numerically computed values for the laminar flame speed in C+O WD matter of mass density in the range $10^7\,\mathrm{g\,cm^{-3}} \leq \rho_u \leq 10^{10}\,\mathrm{g\,cm^{-3}}$ and found the following formula:

$$s_l = 92.0 \times 10^5 \left(\frac{\rho_u}{2 \times 10^9} \right)^{0.805} \left[\frac{X(^{12}C)}{0.5} \right]^{0.889} \mathrm{cm\,s^{-1}} \qquad (14)$$

According to [90], this formula approximates the laminar flame speed to about 10% in the density range mentioned above.

2.3 Flame Instabilities

Flame propagation is subject to several instabilities. Those are of eminent importance in SN Ia models. In particular, three instabilities dominate the flame propagation on different scales in the supernova explosion, namely the Rayleigh–Taylor (RT) instability, the Kelvin–Helmholtz (KH) instability and the Landau–Darrieus (LD) instability. Other instabilities of flames play no significant role here. An example is the so-called diffusional–thermal instability which is suppressed due to the large Lewis number.

On the largest scales, the flow is dominated by the RT instability. The flame starts out near the centre of the WD and burns outwards. In the ashes behind the flame, the degeneracy is partially lifted due to the energy release in nuclear burning and therefore the density here is lower than in the cold fuel ahead of the flame. This stratification, however, is inverse to the gravitational acceleration of the star and therefore subject to a buoyancy-driven instability. The RT instability in its nonlinear stage leads to bubbles of burning material that rise into the fuel. At the interfaces of these bubbles – where the flame is located – strong shear flows are expected. These give rise to shear (KH) instabilities.

Contrary to these two general flow instabilities, the LD instability [17, 37] is specific to self-propagating fronts. In an SN Ia explosion, it applies to all scales on which the discontinuity approximation of the flame is valid. The origin of the instability is the refraction of the streamlines of the flow at the density jump over the flame. The fluid velocity component tangential to the flame front is steady and mass conservation leads to a discontinuity in the normal velocity component. This causes a broadening of the flow tubes in the vicinity of a bulge of a perturbation. Thus the local fluid velocity is lower than the fluid velocity away from the front. Therefore the burning velocity of the flame is higher than the corresponding local fluid velocity and this leads to an increment of the bulge. The opposite holds for recesses of the perturbed front. In this way the perturbation keeps growing. The LD instability, however, stabilizes in the nonlinear regime. Following the flame propagation by means of Huygens' principle, a flame front perturbed from a planar geometry will evolve into a cellular structure. Since, by a simple geometrical argument ([94]), the cusps of this structure should burn faster than the crests, this structure leads to a stabilization. [76, 77, 79] showed that this effect holds for thermonuclear flames in SNe Ia.

Such stabilizing effect in the nonlinear regime exists neither for the RT nor for the KH instability. Therefore, the flame propagation is determined by these effects. On small scales, however, both are suppressed. For the RT instability this results from the competition of the growth time scale of perturbations with the burning time scale. This constitutes a minimum length scale capable of deforming the flame front (sometimes called "fire polishing scale") [90]:

$$\lambda_{\min} = \frac{s_l^2}{2\pi} \left(g \frac{\rho_u - \rho_b}{\rho_u + \rho_b} \right)^{-1}. \tag{15}$$

On larger scales scales, the flow field will very soon be dominated by nonlinear effects and a typical structure found here are mushroom-shaped bubbles. The preference of large structures as a result of the buoyancy instability originates from the tendency of initially smaller bubbles to merge into larger ones while rising. For the situation in SN Ia explosions this has been simulated by [21, 65–67, 69, 74].

The general picture of a flow field dominated by the KH instability is that in the vicinity of the tangential discontinuity vortices will develop in the nonlinear regime. However, for the KH instability to occur, a tangential discontinuity is necessary. This requires the mass flux across the surface of the discontinuity to vanish, which is, of course, not the case for burning fronts. Here, the finite mass flux across the discontinuity stabilizes the flame against the KH instability. Nevertheless the situation changes if the flow field around a burning RT-bubble is dominated by buoyant acceleration. Then, the mass flux can become negligibly small compared to the tangential velocity components [45]. Such a situation can be regarded as being similar to a tangential shear flow in a viscous fluid. Here, the mass flux across the surface does not vanish either, because of microscopic transport, and this leads to a shear layer of finite thickness. The question of the stability of this modified configuration was addressed in numerical simulations by [45] and [49]. The authors concluded that the flames become unstable when the shear velocities reach the laminar burning velocities of the flame fronts.

2.4 Turbulent Combustion

Owing to the relatively small viscosity in combination with the large scales of stellar objects, very high Reynolds numbers are common in astrophysics; the burning process of SNe Ia, however, is extreme. Turbulence is generated in a generic way. An estimate for the Reynolds number around a typical burning Rayleigh–Taylor bubble is derived from its typical size ($L \sim 10^7$ cm), a shear flow velocity of $v_{\text{shear}} \sim 10^7$ cm s^{-1}, a typical density of 10^9 g cm^{-3} and a shear viscosity $\mu \sim 10^9$ cm^2 s^{-1}:

$$\text{Re} = \frac{\rho L v_{\text{shear}}}{\mu} \sim 10^{14}.$$

Given the huge value of the Reynolds number, it is evident that shear instabilities will lead to the generation of turbulence in addition to the buoyancy effect and a turbulent energy cascade is expected to establish (see the chapter "An Introduction to Turbulence"). The large Reynolds number implies that a huge range of scales is dominated by turbulence effects – a severe complication in numerical approaches to the problem as will be discussed below. As has been pointed out in the literature many times and recently has been confirmed by large scale supernova simulations ([67]), turbulent combustion is the key to deflagration models of SN Ia.

The theory of turbulent combustion, however, is still under continuous development and only a fraction of the phenomena can be regarded as being well understood. Certainly, one of the most important notions in this subject is that turbulent combustion takes place in distinct regimes, which are accessible by different methods of theoretical modelling. A suitable way of classification is to relate the turbulent velocity fluctuations (normalized to the laminar burning velocity) to the corresponding length scale (normalized to the flame width). A diagram of this classification was introduced by [7] and [58] and is shown in Fig. 2. Here, the transition regions between the regimes are determined by comparing the velocity of turbulent eddies at certain scales characteristic for the flame to the laminar burning velocity. As long as the turbulent eddies of a given size are much slower than the laminar burning velocity, the flame will burn through the eddies before they can affect its structure. The corresponding regime is called the *wrinkled flamelet regime*. The turbulent flow bends the flame slightly on large scales compared to the flame width and thereby increases the flame surface. However, it does not affect the inner structure of the flame (and thus microscopically the flame velocity remains s_l). In the *corrugated flamelet regime* turbulent eddies noticeably alter the shape of the flame, but still leave its inner structure unaffected. The transition scale l_{Gibs} between the regimes is defined by comparing the eddy turnover time to the flame crossing time at that length:

$$\tau_{\text{eddy}}(l_{\text{Gibs}}) = \tau_{\text{flame}}(l_{\text{Gibs}}) \quad \Rightarrow \quad v(l_{\text{Gibs}}) = s_l. \tag{16}$$

This scale l_{Gibs} is identified as the *Gibson scale* [58]. If it becomes smaller than the thermal width of the flame, then turbulent motion starts to modify the structure of the preheat zone and the flame enters the *thin reaction zone* regime. With smaller Gibson scales eddies become able to distribute material

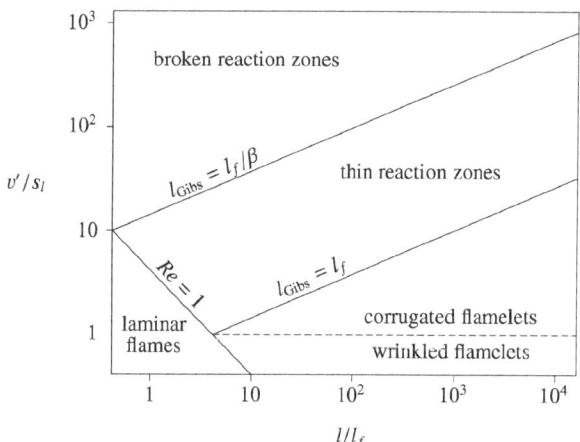

Fig. 2. Regimes of premixed turbulent combustion (following [60])

out of the flame structure and to disrupt the closed flame front. Therefore, the region $l_{\mathrm{Gibs}} < l_f$ is also termed the *distributed burning regime*. For a Gibson scale smaller than the reaction zone ($l_{\mathrm{Gibs}} \lesssim 0.1 l_f$), turbulent eddies finally start to dominate this part of the flame as well. The corresponding regime is that of *broken reaction zones* which ultimately turns into the so-called well-stirred reactor. Except for the flamelet regimes all other turbulent burning modes modify the inner structure of the flame, whose (microscopic) propagation velocity then deviates from s_l accordingly.

We would like to emphasize at this point, that the above classification and in particular, the transition regions strictly apply to chemical combustion processes only. Here, the Prandtl number is $Pr \sim 1$. This is not true for the situation in SN Ia explosions (cf. Sect. 2.2). The possible changes have been investigated by [50]. The authors point out that the transition between the flamelet and the thin reaction zones regimes may shift towards stronger turbulent fluctuations for lower Prandtl numbers. This transition, however, is expected to take place at fuel densities of $\lesssim 10^7\,\mathrm{g\,cm^{-3}}$, which are reached only in the very late stages of the explosion. A preliminary approach to take this stage into account was described by [75], but here we focus on the burning in the flamelet regime, which accounts for the major part of energy generation and nucleosynthesis. This results from the fact that the Gibson scale in SN Ia explosions is much larger than the flame width which is only of the order of millimetres at fuel densities above $10^7\,\mathrm{g\,cm^{-3}}$ [90]. Typically, the laminar burning velocities are around $50\,\mathrm{km\,s^{-1}}$ [90]. Assuming the energy input into a turbulent cascade following Kolmogorov scaling to take place at the length scales of Rayleigh–Taylor bubbles (10^7 cm) with shear velocities of the order of $10^7\,\mathrm{cm\,s^{-1}}$, the Gibson scale is of the order of $10\,\mathrm{km}$.

In the flamelet regime, the corrugation of the flame by turbulent eddies enlarges the flame surface and consequently increases the net burning rate accelerating the propagation of the mean flame front. In large simulations, the corrugated flame surface cannot be fully resolved. This is usually compensated by assigning an effective *turbulent flame speed* s_t to the artificially smoothed front. One goal in modelling flamelet combustion is to relate this quantity to the turbulent velocity fluctuations v'. From the experimental study of Bunsen flames, Damköhler already suggested in the 1940s that $s_t \sim v'$ [16]. We adhere to the following relation found from a theoretical analysis [63]:

$$s_t = s_{\mathrm{lam}} \sqrt{1 + C_t \left(\frac{q_{\mathrm{sgs}}}{s_{\mathrm{lam}}} \right)^2},\tag{17}$$

An exact definition of the characteristic velocity of sub-grid scale turbulence, $q_{\mathrm{sgs}} = \sqrt{3} v'$, will be given in Sect. 3.3. We set $C_t = 4/3$. Then $s_t \simeq 2v'$ in the asymptotic regime of turbulent burning, which is consistent with the result found in [59].

3 Modelling Approach

Modeling SN Ia explosions numerically is a very challenging task. This is mainly due to the many different scales involved in the problem. Since the phenomenon of turbulent combustion is one of the key features, the wide range of scales covered by the turbulent cascade sets the range of relevant spatial scales – starting at the size of the WD star (radius $\sim 2,000$ km) down to the Kolmogorov length (less than a millimetre). The internal structure of the flame itself is thinner than a centimetre and thus not resolvable in current three-dimensional simulations. Most relevant nuclear reactions take place on an extremely short timescale compared to the explosion process, which itself is extremely short compared to the progenitor evolution which determines the initial conditions for the explosion. The pre-ignition evolution is an intricate issue by itself and not well understood so far. It requires numerical techniques different from the one usually applied in the explosion models, and will therefore be elided in the following discussion. All multi-dimensional explosion simulations so far ignited the flame in some arbitrarily chosen configuration.

Hydrodynamics, flame propagation and turbulence clearly require modelling when implemented in numerical schemes and a number of approximations have to be made. These have led to a wide variety of implementations in numerical simulations. For the hydrodynamics several standard methods exist. Eulerian finite volume techniques (e.g. [10, 21, 67]) as well as Lagrangian smooth particle hydrodynamics codes [23] have been applied. Implicit schemes [41] and special low-Mach number implementations [2] provide different approaches.

Two distinct strategies of flame modelling (apart from resolved flames in small-scale simulations) have been suggested. One scheme exploits the fact that from the scale of the WD the flame appears as a sharp discontinuity [68] and will be described in more detail in Sect. 3.2. The other approach artificially broadens the flame structure such that it becomes resolvable on the computational grid [32]. In the following sections we discuss a particular implementation approach.

3.1 Hydrodynamics

Because of the wide range of scales the only way to model full-star SN Ia explosions in multiple dimensions is to treat the flame propagation and hydrodynamics of the flow separately. The former is usually modelled in a parametrized way. In the spirit of this operator splitting approach, the remaining parts of the reactive Navier–Stokes equations are the simpler reactive Euler equations.

Implementation of Fluid Dynamics

The reactive Euler equations are solved applying a finite volume technique. Based on the integral formulation of the balance equations, this approach provides conservativity. The state variables are discretized in "control volumes".

Thus, each value represents an average over some interval ("grid cell"). Our discretization is strictly spoken neither Eulerian nor Lagrangian, but follows the more general moving grid approach, of which both are limiting cases. This technique is particularly useful in connection with thermonuclear supernova explosion simulations, since in this way the expansion of the WD star can be followed for arbitrary times without compromising initial resolution [72]. In an improved version, the flame and the WD are followed with nested grids, enabling a fine resolution of the flame in a uniform grid part (which is necessary for the implementation of the sub-grid-scale turbulence model, see Sect. 3.3) and a coarse outer grid with exponentially growing grid size covering the star [78].

A cell-average of a density a of an extensive quantity can be followed in the moving grid discretization by a combination of the balance equation in Eulerian form

$$\frac{\partial}{\partial t} a = -\boldsymbol{\nabla} \cdot (a\boldsymbol{v}) + s(a),$$

with $s(a)$ denoting the source term, and the *moving grid transport theorem* (see e.g. [43, 72])

$$\frac{\mathrm{d}}{\mathrm{d}t} \int_{V_{\mathrm{grid}}} (a \, \mathrm{d}V_{\mathrm{grid}}) = \int_{V_{\mathrm{grid}}} \left\{ \frac{\partial a}{\partial t} + \boldsymbol{\nabla} \cdot (a\boldsymbol{v}_{\mathrm{grid}}) \right\} \mathrm{d}V_{\mathrm{grid}}, \qquad (18)$$

where $\mathrm{d}/\mathrm{d}t$ denotes the derivative with respect to the moving grid and moving grid quantities are marked with the index "grid". From this, an integral formulation of the balance equation

$$\int_{V_{\mathrm{grid}}} s(a) \, \mathrm{d}V_{\mathrm{grid}}$$

$$= \frac{\mathrm{d}}{\mathrm{d}t} \int_{V_{\mathrm{grid}}} a \, \mathrm{d}V_{\mathrm{grid}} + \int_{V_{\mathrm{grid}}} \boldsymbol{\nabla} \cdot [a(\boldsymbol{v} - \boldsymbol{v}_{\mathrm{grid}})] \, \mathrm{d}V_{\mathrm{grid}}$$

$$= \frac{\mathrm{d}}{\mathrm{d}t} \int_{V_{\mathrm{grid}}} a \, \mathrm{d}V_{\mathrm{grid}} + \iint_{\partial V_{\mathrm{grid}}} a\boldsymbol{v}_{\mathrm{rel}} \cdot \mathrm{d}\boldsymbol{S},$$

arises, which, applied to the appropriate hydrodynamical quantities, defines the set of equations that is solved numerically. This is achieved by updating the cell-averages according to the source terms and the fluxes over the grid cell interfaces. These fluxes are determined by solving the Riemann problems at the cell interfaces. To this end, the average values at the cell centres are extrapolated to the boundaries. Following the Piecewise Parabolic Method (PPM) as introduced by [13], this is done in a parabolic reconstruction corresponding to a higher order Godunov scheme. The Riemann problems are solved utilizing the iterative scheme suggested by [14].

To avoid multi-dimensional Riemann problems, directional splitting is applied in the multi-dimensional simulations. Time steps for the integration are determined according to the CFL criterion with a time step reduction by a factor of $C_{\mathrm{CFL}} = 0.8$. The numerical implementation is based on the PROMETHEUS code [19].

We note, however, that turbulent combustion in SNe Ia is highly subsonic. Therefore, PPM is certainly not the most efficient scheme available. A future implementation of alternative hydrodynamics solvers that are tailored to the specific situation, such as incompressible low-Mach number approximations, would be desirable (for a recent approach see [2]).

The Equation of State for White Dwarf Matter

White dwarfs are the relics of stars that have ceased nuclear burning and are therefore not supported against gravitational collapse by thermal pressure anymore. By contracting from their former size they reach extremely high densities and are thus part of the class of compact astrophysical objects [11]. The interior of a WD resembles a very dense solid with an ion lattice surrounded by degenerate electrons. These provide the cold degenerate fermion pressure that supports the WD against further contraction up to the limiting Chandrasekhar mass (about $\sim 1.4 M_\odot$) [12].

The degenerate electron gas governs the equation of state. The equations for energy and pressure are given by the usual expressions of quantum statistics. Taking into account electron–positron pair creation at high densities, the Fermi-integrals are modified according to [15]. Since the calculation of the Fermi-integrals is numerically expensive, the corresponding values are read off a precalculated table via bilinear interpolation.

Further contributions come from the photon gas following a black-body spectrum, and the completely ionized nuclei which are assumed to be in thermal equilibrium with the electron gas and described by the Maxwell–Boltzmann statistics.

Simulating a stage after convective carbon burning, solidification effects are neglected, allowing the treatment of the nuclei as an ideal gas. Corrections due to Coulomb interaction between electrons and baryons are only marginal and thus they are also neglected. Neutronization occurs in the reaction products at very high densities (e.g. [8, 53]), but this effect is not yet taken into account in the implementation presented here [66, 73].

External Forces: Gravity

Obviously, the self-gravity of the WD star introduces an external force to the hydrodynamical equations (cf. Sect. 2.1):

$$f = -\nabla \Phi, \tag{19}$$

where Φ is the gravitational potential. Since for Chandrasekhar-mass WDs $\mathcal{G}M/Rc^2 \sim 10^{-3}$, general relativistic effects are irrelevant. It is thus well-justified to apply the Newtonian limit, where Φ is given by Poisson's equation $\Delta\Phi = 4\pi\mathcal{G}\rho$. The numerical solution of this elliptic partial differential equation introduces some complications especially in parallel computational code designs. Fortunately, in SN Ia explosion matters are simplified by the low density contrast over the flame. Moreover, an overall isotropic expansion in the explosion is a reasonable assumption. Therefore, one can neglect angular variations and approximate the gravitational potential of the WD by its monopole moment

$$\Phi_0(r) = -\frac{\mathcal{G}}{r} \int_0^r 4\pi r'^2 \bar{\rho}(r') \mathrm{d}r' - \mathcal{G} \int_r^\infty 4\pi r' \bar{\rho}(r') \mathrm{d}r',$$

with $\bar{\rho}(r)$ denoting the angular average of the density, which can be numerically implemented in a straightforward way.

Source Terms: Thermonuclear Reactions

The initial WD material consists of a mixture of mainly ^{12}C and ^{16}O. These nuclei are burnt in thermonuclear reactions forming heavier species. The composition of the ashes depends on the density and on the temperature during burning and in principle many isotopes play a role here. A correct treatment of the nuclear reactions would require a reaction network including hundreds of isotopes and reaction rates between them. Such an approach, however, cannot be realized with current computational resources. Therefore, simplified networks are usually applied to describe the reactions. Since the focus of the implementation described here is to follow the explosion hydrodynamics, a particularly simple treatment of the reactions is chosen. Apart from the initial carbon and oxygen, only three other species are taken into account.

At high fuel densities, burning proceeds to what is known as nuclear statistical equilibrium (NSE). In our approximation this is described as a mixture of iron group elements (represented by ^{56}Ni) and α-particles. Once the fuel density drops below $5.25 \times 10^7 \, \mathrm{g\,cm}^{-3}$, burning is assumed to be incomplete and to terminate in intermediate mass elements (IME, represented by ^{24}Mg). At very low fuel densities, reaction rates will eventually become too slow to be relevant in the explosion process (currently, a threshold of $10^7 \, \mathrm{g\,cm}^{-3}$ is set in most models but burning may continue to densities of a few times $10^6 \, \mathrm{g\,cm}^{-3}$).

In order to handle the energy release due to the reactions, the differences in binding energies between the exemplary species are released in material that has been crossed by the flame. The change in composition of the NSE material depending on temperature is taken into account by adjusting the $\alpha/^{56}$Ni proportion according to tabulated values.

We note, however, that a far more detailed description of the nuclear reactions and yields is still possible in the described approach. Concurrently with the explosion hydrodynamics, a number of tracer particles is advected, which record the temporal evolution of temperature and density in co-moving fluid elements. This data can then be used to reconstruct the nucleosynthesis in a postprocessing step [73, 92] involving an extended nuclear reaction network.

3.2 Flame Model

The Level-Set Technique

The method that is applied in our simulations to parameterize the flame propagation is based on the so-called level-set technique that was introduced by [55]. The central idea of the level-set method is to associate the flame with a moving hypersurface $\Gamma(t)$, which is the zero-level set of a function $G(\boldsymbol{x}, t)$:

$$\Gamma(t) := \{\boldsymbol{x} \mid G(\boldsymbol{x}, t) = 0\}. \tag{20}$$

In principle, there is no constraint on G away from the front and it thus could be chosen arbitrarily. However, it is convenient to prescribe G to be a signed distance function

$$|\boldsymbol{\nabla} G| \equiv 1 \tag{21}$$

with respect to the flame front and with $G < 0$ in the unburnt material and $G > 0$ in the ashes.

The G-Equation

Follow the path $\boldsymbol{x}_P$ of a point P attached to the propagating front in Eulerian frame of reference. Its motion will obviously be determined by the advection due to fluid motion and propagation due to burning of the front:

$$\dot{\boldsymbol{x}}_P = \boldsymbol{v}_\mathrm{u} + s_\mathrm{u} \boldsymbol{n}, \tag{22}$$

where $\boldsymbol{v}_\mathrm{u}$, s_u and $\boldsymbol{n}$ denote the fluid velocity in the fuel region, the flame propagation speed with respect to the unburnt material, and the normal vector to the front respectively. The latter will be defined to point towards the unburnt material by

$$\boldsymbol{n} = -\frac{\boldsymbol{\nabla} G}{|\boldsymbol{\nabla} G|}, \tag{23}$$

which becomes possible by fixing G as in Eq. (21).

The value of G on the trajectory of such a point is zero by definition. Hence the total time derivative of G on the trajectory vanishes:

$$\frac{\mathrm{d}G(\boldsymbol{x}_P)}{\mathrm{d}t} = \frac{\partial G(\boldsymbol{x}_P)}{\partial t} + \boldsymbol{\nabla} G(\boldsymbol{x}_P) \cdot \dot{\boldsymbol{x}}_P := 0. \tag{24}$$

This condition, together with Eqs. (22) and (23), yields the temporal evolution of G,

$$\frac{\partial G}{\partial t} = -\boldsymbol{\nabla} G(\boldsymbol{x}_P) \cdot \dot{\boldsymbol{x}}_P = -(\boldsymbol{v}_\mathrm{u} + s_\mathrm{u} \boldsymbol{n})(-\boldsymbol{n}\,|\boldsymbol{\nabla} G|) = (\boldsymbol{v}_\mathrm{u} \boldsymbol{n} + s_\mathrm{u})|\boldsymbol{\nabla} G| \tag{25}$$

for points located on the front. This equation is often termed the "G-equation" in literature. It bears physical meaning only for points at the front, since s_u is undefined elsewhere.

To apply this formula to the region of fuel and ashes, the velocities have to be spread out from the front. Since the G-values are relevant only in the proximity of the flame front, the evaluation can be restricted to a narrow band around the zero-level set [86].

Re-initialization

Due to the advection of the G-field with the hydrodynamical flow, the signed distance constraint (21) is usually not conserved. This can cause problems in simulations, since parts of the G-function may bulge up and artificially appear as new flames. Thus, additional measures have to be taken in order to preserve the G-function's property of being a distance function. One way, according to [89], is to employ a pseudo time iteration of

$$\frac{\partial G}{\partial \tau} = \frac{G}{|G| + \epsilon} \left(1 - |\boldsymbol{\nabla} G|\right), \tag{26}$$

until convergence to $|\boldsymbol{\nabla} G| = 1$ is reached. Values used in our numerical implementation are $\Delta \tau \approx 0.1 \, \Delta x$ and $\epsilon \approx \Delta x$. A particularly efficient alternative method has been suggested by [1], but this is difficult to parallelize.

The iterative re-initialization algorithm needs to leave the zero-level set of G unaffected, since otherwise an unphysical shift of the flame front would occur. Also topological changes of the flame require a correction of the G field. For a detailed discussion of suitable methods we refer to [68, 70].

Flame/Flow Coupling

In the context of the finite-volume method we apply to discretize the hydrodynamics, the cells cut by the flame front ("mixed cells" in the following) contain a mixture of burnt and unburnt states. Therefore the quantity $\boldsymbol{v}_{\mathrm{u}}$ needed in Eq. (25) is not readily available. One strategy to circumvent this problem is the so-called "passive implementation" of the level-set method [68]. There it is assumed that the velocity jump is small compared to the laminar burning velocity and $\boldsymbol{v}_{\mathrm{u}}$ is approximated by the average flow velocity. An operator splitting approach for the time evolution of G (25) yields the advection term due to the fluid velocity in conservative form which is identical to the advection equation of a passive scalar. This part can be treated by the PROMETHEUS implementation of the PPM method. Front propagation, energy release and species conversion due to burning are performed in an additional step.

An alternative strategy is the "complete implementation" (see [68]). It was developed by [88] and facilitates the reconstruction of the exact burnt and unburnt states in mixed cells. On the basis of the jump conditions over the flame front (see Sect. 2.1) together with geometrical information derived from the flame intersections with the cell boundaries, a set of equations can be formulated and solved for the burnt and unburnt states. This allows for an accurate treatment of (25). The main advantage is that it now becomes possible to treat flows of burnt and unburnt material over cell boundaries separately in the implementation of the fluid dynamics. This prevents the flame front from smearing out over several cells as it does for the passive implementation. In this way, the flame front is resolved as a sharp discontinuity without any mesh refinement (which would lead to very small CFL timesteps). Unfortunately, it turned out that a three-dimensional implementation of the reconstruction scheme for burning in degenerate matter is very complicated and unavoidable discretization errors in the geometrical information lead to unphysical states in the solution of the system of equations.

Therefore, the passive implementation is applied in large-scale SN Ia simulations, which provides a reasonable approximation given the small density contrast over the flame in the degenerate WD material.

3.3 Sub-grid-Scale Model

The Hydrodynamical Equations in the Germano Consistent Decomposition

The hydrodynamical equations (8, 9, 10 and 11) follow from first principles (see the chapter "An Introduction to Turbulence"). However, so far we have ignored the fact that it is infeasible to solve these equations including all dynamical scales. For this reason, the equations are decomposed into a large-scale and a fluctuating part. Henceforth, we put an ∞ on top of symbols corresponding to the exact solution that includes the fluctuations. The numerically computed solution, on the other hand, is interpreted in terms of filtered quantities, for which fluctuations are smoothed out. In generic form, we write

$$q(\boldsymbol{x}, t) = \left\langle \overset{\infty}{q} \right\rangle_G \equiv \int \mathrm{d}^3 x' \, G(\boldsymbol{x} - \boldsymbol{x}', t) \overset{\infty}{q}(\boldsymbol{x}', t), \qquad (27)$$

where $\overset{\infty}{q}(\boldsymbol{x}, t)$ is an *ideal* quantity defined with infinite spatiotemporal resolution and $G(\boldsymbol{x} - \boldsymbol{x}', t)$ is the kernel of a low-pass filter that smooths out the fluctuations of $\overset{\infty}{q}(\boldsymbol{x}, t)$ at length scales smaller than a prescribed characteristic length. For brevity, we use the notation $q = \langle \overset{\infty}{q} \rangle_{\mathrm{eff}}$. It is conjectured that the numerical discretization of the hydrodynamical equations corresponds to an *implicit filter* of characteristic length $\Delta_{\mathrm{eff}} \sim \Delta$. This is the basic idea of a large-eddy simulation (LES).

Let us now consider the filtered mass density $\rho = \langle \overset{\infty}{\rho} \rangle_{\mathrm{eff}}$. Assuming that the implicit filter is homogeneous and independent of time, i.e. the kernel is a function of $|\boldsymbol{x} - \boldsymbol{x}'|$ only, the operation of filtering commutes with time derivatives and spatial gradients. Then the smoothed mass density ρ obeys a conservation law of exactly the same form as the continuity equation for $\overset{\infty}{\rho}$:

$$\frac{\partial}{\partial t}\rho + \boldsymbol{\nabla} \cdot \rho \boldsymbol{v} = 0. \qquad (28)$$

The conservation law for the momentum density $\overset{\infty\infty}{\rho\,v}$ reads

$$\frac{\partial}{\partial t}\overset{\infty\infty}{\rho\,v} + \boldsymbol{\nabla}\cdot\overset{\infty\infty}{\rho\,v}\otimes\overset{\infty}{v} = \overset{\infty}{\boldsymbol{F}}, \tag{29}$$

where the effective force density $\overset{\infty}{\boldsymbol{F}}$ is defined by

$$\overset{\infty}{\boldsymbol{F}} = -\boldsymbol{\nabla}\overset{\infty}{P} + \boldsymbol{\nabla}\cdot\overset{\infty}{\sigma} + \overset{\infty\infty}{\rho\,\boldsymbol{g}}. \tag{30}$$

The first term on the right-hand side is the pressure gradient, the second and the third terms are, respectively, the viscous and gravitational force per unit volume. The viscous dissipation tensor $\overset{\infty}{\sigma}$ is proportional to the trace-free part of the rate of strain:

$$\overset{\infty}{\sigma}_{ij} = 2\nu\overset{\infty}{\rho}\overset{\infty}{S}^{*}_{ij} = 2\nu\overset{\infty}{\rho}\left(\overset{\infty}{S}_{ij} - \frac{1}{3}\overset{\infty}{d}\delta_{ij}\right), \tag{31}$$

where ν is the microscopic viscosity of the fluid,

$$\overset{\infty}{S}_{ij} = \frac{1}{2}\left(\frac{\partial\overset{\infty}{v}_i}{\partial x_j} + \frac{\partial\overset{\infty}{v}_j}{\partial x_i}\right), \tag{32}$$

and $\overset{\infty}{d} = \overset{\infty}{S}_{ii}$ is the divergence of the velocity $\overset{\infty}{v}$.

For compressible flows, it turns out that the momentum equation can be written in terms of filtered quantities if a *Favre filtered* velocity field is introduced:

$$v = \frac{\left\langle\overset{\infty\infty}{\rho\,v}\right\rangle_{\text{eff}}}{\left\langle\overset{\infty}{\rho}\right\rangle_{\text{eff}}}. \tag{33}$$

Favre filtering the conservation law (29), one obtains

$$\frac{\partial}{\partial t}\left\langle\overset{\infty\infty}{\rho\,v}\right\rangle_{\text{eff}} + \boldsymbol{\nabla}\cdot\left\langle\overset{\infty\infty}{\rho\,v}\otimes\overset{\infty}{v}\right\rangle_{\text{eff}} = F, \tag{34}$$

where $\boldsymbol{F} = \left\langle\overset{\infty}{\boldsymbol{F}}\right\rangle_{\text{eff}}$. Using the identity $v\rho = \left\langle\overset{\infty\infty}{\rho\,v}\right\rangle_{\text{eff}}$, which follows immediately form definition (33), and substituting

$$\left\langle\overset{\infty\infty}{\rho\,v}\otimes\overset{\infty}{v}\right\rangle_{\text{eff}} = \rho v\otimes v - \tau\left(\overset{\infty\infty}{\rho\,v},\overset{\infty}{v}\right),$$

equation (34) becomes

$$\frac{\partial}{\partial t}\rho v + \boldsymbol{\nabla}\cdot\rho v\otimes v = \boldsymbol{F} + \boldsymbol{\nabla}\cdot\tau\left(\overset{\infty\infty}{\rho\,v},\overset{\infty}{v}\right). \tag{35}$$

Defining the filtered pressure $P = \left\langle\overset{\infty}{P}\right\rangle_{\text{eff}}$ and the Favre filtered gravity $\boldsymbol{g} = \left\langle\overset{\infty\infty}{\rho\,\boldsymbol{g}}\right\rangle_{\text{eff}}/\rho$, we have

$$\frac{\partial}{\partial t}\rho\boldsymbol{v} + \boldsymbol{\nabla}\cdot\rho\boldsymbol{v}\otimes\boldsymbol{v} = -\boldsymbol{\nabla}P + \rho\boldsymbol{g} + \boldsymbol{\nabla}\cdot\left(\left\langle\overset{\infty}{\sigma}\right\rangle_{\mathrm{eff}} + \tau\left(\overset{\infty\infty}{\rho\boldsymbol{v}},\overset{\infty}{\boldsymbol{v}}\right)\right). \tag{36}$$

Note that all terms in this equation are expressed in terms of computable filtered quantities except for the divergence of the filtered viscous dissipation tensor and the *generalized turbulence stress tensor*

$$\tau\left(\overset{\infty\infty}{\rho\boldsymbol{v}},\overset{\infty}{\boldsymbol{v}}\right) = -\left\langle\overset{\infty\infty}{\rho\boldsymbol{v}}\otimes\overset{\infty}{\boldsymbol{v}}\right\rangle_{\mathrm{eff}} + \rho\boldsymbol{v}\otimes\boldsymbol{v}. \tag{37}$$

This tensor is a generalization of the second moment of the velocity fluctuations in Reynolds-stress models of turbulence. The notion of a generalized turbulence stress was introduced by Germano [25] for incompressible fluids. If the Mach numbers are smaller than unity, it is still reasonable to use the order of magnitude estimate $\tau(\overset{\infty\infty}{\rho\boldsymbol{v}},\overset{\infty}{\boldsymbol{v}}) \sim \rho v'^2$, where $\boldsymbol{v}' = \overset{\infty}{\boldsymbol{v}} - \boldsymbol{v}$. In this case, the velocity fluctuations obey the 2/3-law, i.e. $v'^2 \sim \epsilon^{2/3}\Delta^{2/3}$, where ϵ is the rate of viscous energy dissipation [18]. On the other hand, with the definition of the Kolmogorov scale $\eta = (\nu^3/\epsilon)^{1/4}$, it follows that

$$\left\langle\overset{\infty}{\sigma}\right\rangle_{\mathrm{eff}} = 2\nu\left\langle\rho S^*\right\rangle_{\mathrm{eff}} \sim \left(\eta^4\epsilon\right)^{1/3}\rho(\epsilon\Delta)^{1/3}/\Delta \sim \rho\epsilon^{2/3}\eta^{4/3}\Delta^{-2/3}.$$

Thus, the magnitude of the viscous dissipation tensor scales with $(\eta/\Delta)^{4/3}$ relative to the turbulence stress tensor. For this reason, viscous dissipation will become negligible at the level of the filtered momentum equation for a sufficiently large numerical cutoff scale in comparison to the length scale of viscous dissipation. Since this criterion applies in virtually all LES, we can express the filtered momentum equation as

$$\frac{\partial}{\partial t}\rho\boldsymbol{v} + \boldsymbol{\nabla}\cdot\rho\boldsymbol{v}\otimes\boldsymbol{v} = -\boldsymbol{\nabla}P + \rho\boldsymbol{g} + \boldsymbol{\nabla}\cdot\tau\left(\overset{\infty\infty}{\rho\boldsymbol{v}},\overset{\infty}{\boldsymbol{v}}\right). \tag{38}$$

Scalar multiplication of the momentum equation with the velocity $\boldsymbol{v}$ yields the dynamical equation for the resolved kinetic energy density, $K_{\mathrm{res}} = \frac{1}{2}\rho|\boldsymbol{v}|^2$:

$$\frac{\partial}{\partial t}K_{\mathrm{res}} + \boldsymbol{\nabla}\cdot\rho\boldsymbol{v}K_{\mathrm{res}} = \boldsymbol{v}\cdot\left[\boldsymbol{F} + \boldsymbol{\nabla}\cdot\tau\left(\overset{\infty\infty}{\rho\boldsymbol{v}},\overset{\infty}{\boldsymbol{v}}\right)\right]. \tag{39}$$

In addition to the above equation, a conservation law for the total energy density $E_{\mathrm{tot}} = E_{\mathrm{int}} + K_{\mathrm{res}}$ associated with the resolved scales can be obtained [84].

The difference between the filtered kinetic energy, $K = \frac{1}{2}\langle\overset{\infty}{\rho}|\overset{\infty}{\boldsymbol{v}}|^2\rangle_{\mathrm{eff}}$ and K_{res} is called the sub-grid scale turbulence energy K_{sgs}. It is related to the trace of the turbulence stress tensor:

$$K_{\mathrm{sgs}} = -\frac{1}{2}\mathrm{Tr}\,\tau\left(\overset{\infty\infty}{\rho\boldsymbol{v}},\overset{\infty}{\boldsymbol{v}}\right) = \frac{1}{2}\left(\left\langle\overset{\infty}{\rho}|\overset{\infty}{\boldsymbol{v}}|^2\right\rangle_{\mathrm{eff}} - \rho|\boldsymbol{v}|^2\right). \tag{40}$$

The corresponding velocity scale is $q_{\mathrm{sgs}} = \sqrt{2K_{\mathrm{sgs}}/\rho}$. The derivation of the conservation law for K_{sgs} is somewhat more involved [84]. Here we only state the result:

$$\frac{\partial}{\partial t} K_{\mathrm{sgs}} + \nabla \cdot \rho \boldsymbol{v} K_{\mathrm{sgs}} - \mathfrak{D} = \Gamma + \Sigma - \rho(\lambda + \epsilon), \tag{41}$$

where the source contributions on the right-hand side are

$$\Gamma = \left\langle \overset{\infty}{\rho}, \overset{\infty}{v}_i \overset{\infty}{g}_i \right\rangle_{\mathrm{eff}} - \rho v_i g_i, \tag{42}$$

$$\Sigma = \tau_{ij} S_{ij}, \tag{43}$$

$$\rho \lambda = - \left\langle \overset{\infty\infty}{d P} \right\rangle_{\mathrm{eff}} + dP, \tag{44}$$

$$\rho \epsilon = \left\langle \overset{\infty\infty}{\nu} \overset{\infty}{\rho} |\overset{\infty}{S}{}^*|^2 \right\rangle_{\mathrm{eff}}. \tag{45}$$

The nonlocal transport term $\mathfrak{D}$ is given by

$$\mathfrak{D} = \frac{\partial}{\partial x_i} \left[\frac{1}{2} \tau_{ijj} + \mu_i \right], \tag{46}$$

where the generalized moment

$$\boldsymbol{\mu} = - \left\langle \overset{\infty}{v} \overset{\infty}{P} \right\rangle_G + \boldsymbol{v} P. \tag{47}$$

accounts for the transport of turbulence energy due to pressure fluctuations.

The dynamical equations (39) and (41) constitute the *Germano consistent decomposition* of the kinetic energy budget. In order to close the equations, several modelling assumptions for the terms depending on nonfiltered quantities have to be made. In the following, we will discuss only a subset of these closures. For a complete account of the SGS model, see [84] and [85].

Closures for the Production of Turbulence Energy

There are two different sources of SGS turbulence production. First, the SGS buoyancy term Γ_{sgs} (42) and second, the rate of energy transfer $\Sigma_{\mathrm{sgs}} = \tau_{ij} S_{ij}$ across the length scale Δ_{eff} due to nonlinear turbulent interactions. With regard to simulations of SNe Ia, there has been a lively controversy whether buoyancy effects or the energy transfer dominates the production of turbulence energy on unresolved scales.

A simple scaling argument can be invoked in favour of the dominance of turbulent energy transfer [48]. According to the Kolmogorov theory, the root mean square turbulent velocity fluctuations obey the scaling law $v'(l) \propto l^{1/3}$ [18]. On the other hand, the Sharp–Wheeler relation for the velocity scale associated with the RT instability implies $v_{\mathrm{RT}}(l) \propto l^{1/2}$ [87]. As a consequence, we have $v_{\mathrm{RT}}(l)/v'(l) \propto l^{1/6} \to 0$ towards decreasing length scales.

Of course, both scaling laws are based on assumptions that do not necessarily hold in the course of a supernova explosion. For this reason, [85] attempted to include both effects in the SGS model.

The Localized Eddy-Viscosity Closure

The most commonly used closure for the the trace-free part of the SGS turbulence stress tensor,

$$\tau_{ij}^* = \tau_{ij} - \frac{1}{3}\tau_{ii}\delta_{ij} = \tau_{ij} + \frac{2}{3}K_{\text{sgs}}\delta_{ij}, \tag{48}$$

is formulated analogous to the viscous stress tensor in a Newtonian fluid [64]:

$$\tau_{ij}^* \overset{\circ}{=} 2\rho\nu_{\text{sgs}}S_{ij}^* = 2\rho\nu_{\text{sgs}}\left(S_{ij} - \frac{1}{3}d\delta_{ij}\right). \tag{49}$$

Hence, the microscopic viscosity ν in the fully resolved equation (29) is effectively replaced by an *eddy-viscosity* ν_{sgs} in the filtered equation (38).

A dimensional ansatz for the eddy-viscosity is to set ν_{sgs} proportional to the product of the effective cutoff length Δ_{eff} and the characteristic velocity of SGS turbulence, i.e. $k_{\text{sgs}}^{1/2}$ [64, 80]:

$$\nu_{\text{sgs}} \overset{\circ}{=} C_\nu \Delta_{\text{eff}} k_{\text{sgs}}^{1/2}. \tag{50}$$

The coefficient C_ν is the unknown closure parameter of the model. Although the mean value of C_ν can be estimated from approximate theories or numerical simulations of isotropic turbulence, substantial local deviations from the mean are expected to occur in nonstationary and inhomogeneous flows.

This problem was tackled with the advent of dynamical procedures [26]. The basic idea of a dynamical procedure is the local determination of C_ν from structural properties of the numerically computed flow on the smallest resolved length scales. To that end a *test filter* of characteristic length $\Delta_{\text{T}} = \gamma_{\text{T}}\Delta_{\text{eff}}$ is introduced. The test filter probes velocity fluctuations within the small range of length scales between Δ_{eff} and $\gamma_{\text{T}}\Delta_{\text{eff}}$, where the ratio of filter lengths, γ_{T}, is typically chosen in the range from about 1.5–4. For PPM, $\gamma_{\text{T}} \approx 3.75$ has turned out to be optimal [81].

The turbulence stress of the velocity fluctuations smoothed out by the test filter is given by

$$\tau_{\text{T}}(\rho\boldsymbol{v}, \boldsymbol{v}) = -\langle\rho\boldsymbol{v}\otimes\boldsymbol{v}\rangle_{\text{T}} + \rho^{(\text{T})}\boldsymbol{v}^{(\text{T})}\otimes\boldsymbol{v}^{(\text{T})}, \tag{51}$$

where $\rho^{(\text{T})} = \langle\rho\rangle_{\text{T}}$ and $\boldsymbol{v}^{(\text{T})} = \langle\rho\boldsymbol{v}\rangle_{\text{T}}/\langle\rho\rangle_{\text{T}}$. Note that the right-hand side of equation (51) can be evaluated completely from the numerically computed variables ρ and $\boldsymbol{v}$. Following [34], we apply the eddy-viscosity closure to $\tau_{\text{T}}(\rho\boldsymbol{v}, \boldsymbol{v})$, i.e.

$$\tau_{\text{T}}(\rho v_i, v_k) = \rho_{\text{T}}C_\nu\Delta_{\text{T}}k_{\text{T}}^{1/2}S_{ik}^{*\,(\text{T})}. \tag{52}$$

Here k_{T} is given by $\rho_{\text{T}}k_{\text{T}} = -\frac{1}{2}\tau_{\text{T}}(\rho v_i, v_i)$ and $S_{ik}^{(\text{T})}$ is the symmetrized Jacobian matrix of $\boldsymbol{v}^{(\text{T})}$.

Contracting the closure (52) with $S_{ik}^{(\text{T})}$, a scalar equation that determines C_ν is obtained:

$$C_\nu = \frac{\tau_T^*(v_i, v_k) S_{ik}^{(T)}}{\rho_T \Delta_T k_T^{1/2} |S^{*\,(T)}|^2}. \tag{53}$$

Since turbulence tends to become asymptotically self-similar towards smaller scales, one can substitute the local value of C_ν calculated from Eq. (53) into the eddy-viscosity closure (48) for the SGS turbulence stress. Therefore, the outcome of the dynamical procedure in the form proposed by Kim et al. [34] is

$$\frac{1}{\rho}\Sigma_{\mathrm{sgs}} + \frac{2}{3}k_{\mathrm{sgs}}d \doteq \frac{\tau_T^*(v_i, v_k) S_{ik}^{(T)}}{\gamma_T \rho_T} \frac{|S^*|^2}{|S^{*\,(T)}|^2} \sqrt{\frac{k_{\mathrm{sgs}}}{k_T}}. \tag{54}$$

Note that $k_{\mathrm{sgs}} = K_{\mathrm{sgs}}/\rho$ is the SGS turbulence energy per unit mass.

Using flow realizations from direct numerical simulations of isotropic turbulence, we performed a priori tests of the localized eddy-viscosity closure [84]. The results indicate a more accurate prediction of the rate of energy transfer if the dynamical procedure for calculating C_ν rather than a statistical value is applied. The performance of the localized SGS model in LES of turbulent burning in a periodic box appeared promising as well [81]. In those simulations, two simplifications were made: first, the turbulence stress term in the momentum equation (38) was neglected due to the numerical dissipation of PPM (also see [82]). The SGS model was still coupled to the resolved flow via the energy budget and, particularly, the turbulent flame speed relation. Second, the inverse energy transfer for $C_\nu < 0$ was suppressed, because, in combination with PPM, this so-called backscattering would produce spurious energy dissipation [81].

Archimedian Production

Given the definition (42) of SGS buoyancy, it is rather difficult to come up with a physically sensible closure. In general, correlations between fluctuations of the velocity and the gravitational potential have to be taken into account. In thermonuclear supernovae, self-gravity is not significant on small scales. However, the density contrast between burnt material and nuclear fuel, respectively, induces convective motions due to the RT instability in the bulk gravitational field of the exploding star. The lower threshold for the growth of RT instabilities is given by the fire polishing length λ_{fp} defined by equation (15), which can be identified with the Gibson length l_G in the case of developed turbulence. In LES of thermonuclear supernovae, both l_G and λ_{fp} are small compared to the numerical resolution Δ_{eff}, except in the late stage of the burning process.

The Archimedian force per unit mass generated by the RT instability can be written as

$$g_{\mathrm{eff}} = \mathrm{At}\, g, \tag{55}$$

where the density contrast between fuel and burnt material is specified by the Atwood number

$$At = \frac{\rho_f - \rho_b}{\rho_f + \rho_b}. \tag{56}$$

Combining the acceleration g_{eff} with the characteristic magnitude of unresolved velocity fluctuations, yields a tentative closure for SGS buoyancy:

$$\Gamma_{\text{sgs}} = \frac{1}{\sqrt{2}} C_A \rho g_{\text{eff}} q_{\text{sgs}}. \tag{57}$$

In order to constrain the contributions from SGS buoyancy to the vicinity of flame fronts, we use the following definition:

$$g_{\text{eff}} = \chi_{\pm\delta}(G = 0)\theta(\Delta - \lambda_{\text{fp}})At\, g, \tag{58}$$

where $\chi_{\pm n\Delta}(G = 0)$ is the characteristic function of all cells for which the distance from the flame front (represented by $G(\boldsymbol{x}, t) = 0$) is less than δ, and θ is the Heaviside step function, i.e. $\theta(\Delta - \lambda_{\text{fp}}) = 1$ for $\Delta > \lambda_{\text{fp}}$ and zero otherwise. The Atwood number is calculated by fitting tabulated numerical data [84, 91].

An indication in support of closure (57) is the asymptotics obtained in the limiting case of a stationary and homogeneous distribution of SGS turbulence energy. Neglecting nonlocal transport, turbulent energy transfer and compression effects, the dynamical equation (41) becomes

$$\frac{d}{dt} q_{\text{sgs}} \simeq \frac{1}{\sqrt{2}} C_A g_{\text{eff}} - \frac{q_{\text{sgs}}^2}{\ell_\epsilon}. \tag{59}$$

for a fluid parcel in the vicinity of the flame front. Assuming equilibrium, this equation has the fixed point solution

$$q_{\text{sgs}} \simeq \sqrt{\frac{2 C_A \Delta_{\text{eff}} g_{\text{eff}}}{C_\epsilon}} = v_{\text{RT}}(\Delta_{\text{eff}}). \tag{60}$$

Setting $C_A = C_\epsilon/8$, this is just the Sharp–Wheeler relation $v_{\text{RT}}(l) = 0.5\sqrt{lg_{\text{eff}}}$ for the characteristic velocity of the RT instability on length scales $\ell \sim \Delta_{\text{eff}}$. Since $C_\epsilon \approx 0.5\ldots 1.0$ for developed turbulence [81], we conclude that $C_A \approx 0.1$.

4 Numerical Simulations

4.1 Full-Star Supernova Explosion Simulations

Snapshots from a full-star supernova simulation performed in the modelling approach described above are shown in Fig. 3, where the logarithm of the density is volume rendered and the zero-level set of the G-field (associated with the flame front) is indicated by the blue isosurface. In this simulation, the WD star was constructed by choosing a central density of $2.9 \times 10^9 \, \text{g cm}^{-3}$

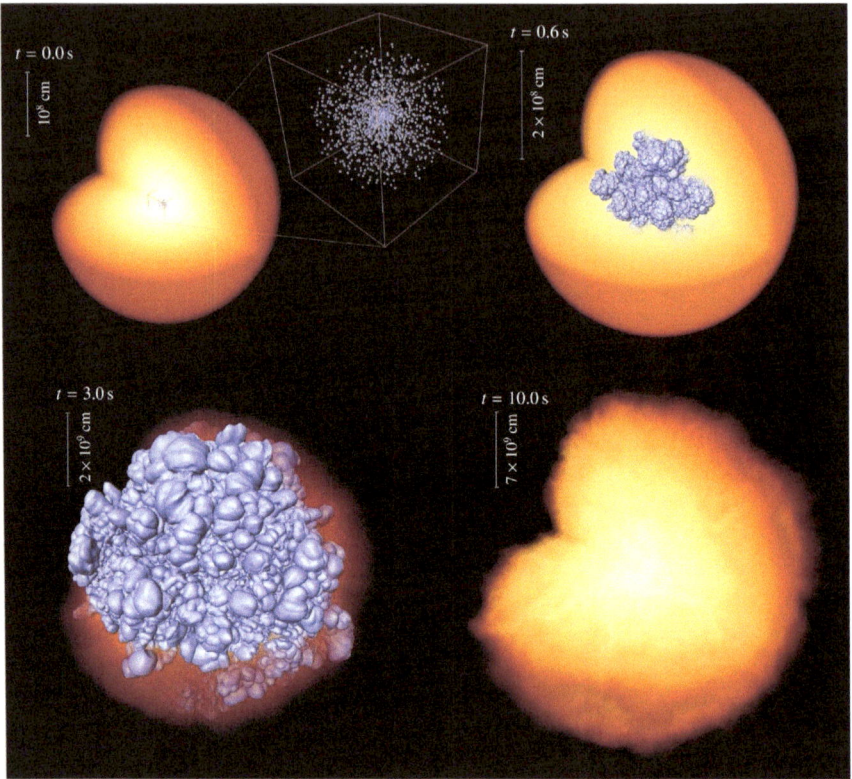

Fig. 3. Thermonuclear supernova explosion simulation

and a composition of equal parts of carbon and oxygen. Since the modelled description of flame propagation does not depend on the temperature, and, due to the high degeneracy of the material, its effect on the WD's structure is marginal, a cold WD ($T = 5 \times 10^5$ K) was set up by integrating the equations of hydrostatic equilibrium in one dimension and mapping the obtained density profile to the grid. Since the equation of state applied in the WD setup is the same as used in the explosion simulation, this procedure provides sufficient stability of the WD over the time simulated ([70]).

The flame was started in a large number of spherical ignition spots within the first 180 km from the centre of the WD star (cf. upper left snapshot of Fig. 3 and close-up of the ignition region).

Shortly after ignition the bubbles assume a "mushroom-cap"-like structure due to the effect of the buoyancy instability. The complicated and disordered flow field establishing in the burning region quickly leads to merging of the small bubbles so that a connected flame surface emerges. This flame is wrinkled on large scales by the RT instability and interacts with the turbulence generated (cf. upper right snapshot of Fig. 3). This interaction is modelled by the sub-grid-scale model and will be discussed in detail in Sect. 4.2. The energy release in nuclear burning expands the star as indicated by the scales given in Fig. 3.

The lower left snapshot of Fig. 3 was taken 3.0 s after ignition and corresponds to an evolution stage where burning ceases because of the expansional decrease of density. Consequently, the zero-level set of G has no physical meaning anymore, but still indicates an approximate interface between left-over fuel and ashes. As visible in the snapshot, burnt material reaches out to the surface of the WD.

Due to the energy release in the burning process, the remnant, whose density is shown in the lower right corner of Fig. 3 is no longer gravitationally bound and continues expansion. About 10 s after ignition, this expansion proceeds in a homologous way to a good approximation (see [72]), i.e. the velocity of a fluid element is proportional to its radius.

Photons leave the remnant only at much later times when the medium becomes less dense and the outer layers get optically thin. Therefore, the homologous expansion reached in the late stages of the explosion simulation is scaled to these times and (together with the results of the nucleosynthesis postprocessing) serves as input for the derivation of observable quantities such as spectra and light curves. These can then be compared to observations and ultimately provide a way to validate the explosion model (see Sect. 5).

4.2 Sub-grid-Scale Dynamics

For the numerical computation of the SGS turbulent velocity q_{sgs}, the conservation law (41) for $K_{sgs} = \frac{1}{2}\rho q_{sgs}^2$ is divided by ρq_{sgs}. Upon substituting the closures described in Sect. 3.3 and in [84], respectively, the following dynamical equation is obtained:

$$\left(\frac{\partial}{\partial t} + \boldsymbol{v} \cdot \nabla\right) q_{sgs} - \frac{1}{\rho} \nabla \cdot (\rho \ell_\kappa q_{sgs} \nabla q_{sgs}) - \ell_\kappa |\nabla q_{sgs}|^2$$
$$= \frac{1}{\sqrt{2}} C_A g_{eff} + \ell_\nu |S^*|^2 - \frac{7}{30} q_{sgs} d - \frac{q_{sgs}^2}{\ell_\epsilon}. \tag{61}$$

The characteristic length scales ℓ_κ, ℓ_ν and ℓ_ϵ are related to SGS turbulent transport, the rate of energy transfer from resolved towards sub-grid scales and the rate of viscous dissipation. Each characteristic length can be expressed in terms of the effective cutoff length Δ_{eff} and a similarity parameter:

$$\ell_\nu = \frac{C_\nu \Delta_{eff}}{\sqrt{2}}, \qquad \ell_\epsilon = \frac{2\sqrt{2}\Delta_{eff}}{C_\epsilon}, \qquad \ell_\kappa = \frac{C_\kappa \Delta_{eff}}{\sqrt{2}}. \tag{62}$$

Due to the dissipative effects of this numerical scheme on the smallest resolved length scales, we set $\Delta_{eff} \approx 1.6\Delta$ [82]. The dynamical calculation of the closure parameters C_ν is outlined in Sect. 3.3. For the determination of C_ϵ and C_κ, see [84].

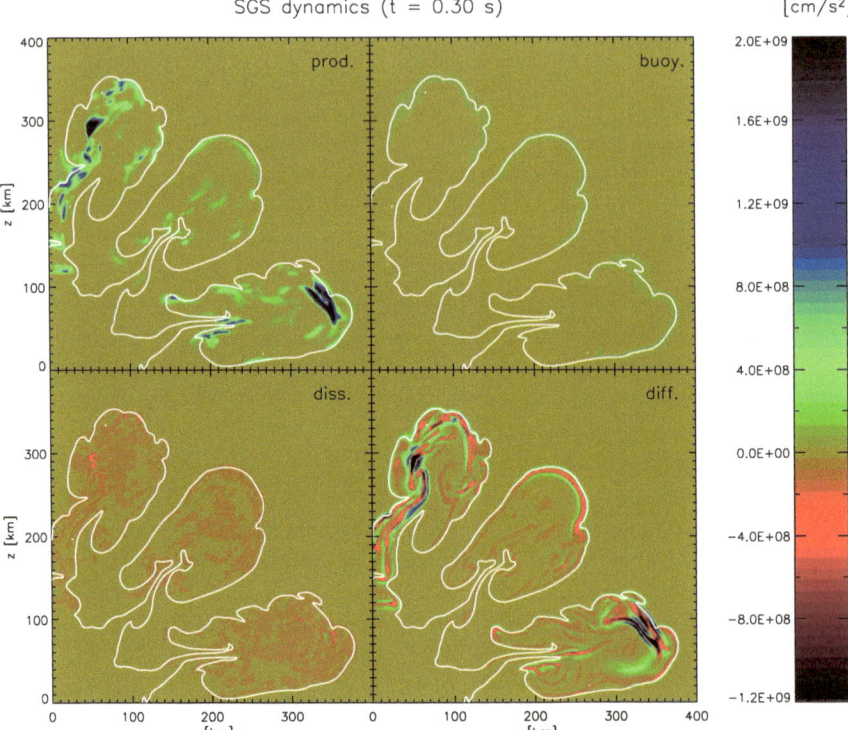

Fig. 4. Contour sections showing the contributions to the evolution of the SGS turbulent velocity q_{sgs} given by Eq. (61) at $t = 0.3\,\mathrm{s}$. Only the inner region of the numerical grid with $N = 384^3$ cells is shown. The white contours represent the sections through the flame surface

Two-dimensional spatial sections at subsequent instants of time in a simulation with $N = 384^3$ grid cells are shown in Figs. 4 and 5. In each figure, the following dynamical terms of equation (61) are plotted:

1. Rate of production caused by strain, $\ell_\nu |S^*|^2$ (left top panel).
2. Specific Archimedian force $0.1 g_{\mathrm{eff}}$ (right top panel).
3. Rate of dissipation $-\frac{7}{30} q_{sgs} d - q_{sgs}^2/\ell_\epsilon$ (left bottom panel).
4. Rate of diffusion $\frac{1}{\rho} \boldsymbol{\nabla} \cdot (\rho \ell_\kappa q_{sgs} \boldsymbol{\nabla} q_{sgs}) - \ell_\kappa |\boldsymbol{\nabla} q_{sgs}|^2$ (right bottom panel).

Note that these quantities have the dimension of acceleration. The flame surface as given by the zero-level set is indicated by the contours in white. Figure 4 shows the typical Rayleigh–Taylor mushroom shapes which have formed out of initially axisymmetric sinusoidal perturbations at time $t = 0.3\,\mathrm{s}$. Significant energy transfer is concentrated in small regions and there is little dissipation yet. At $t = 0.45\,\mathrm{s}$, the rate of energy transfer has reached its maximum and is spread all over the interior of the flames (see Fig. 5). The SGS buoyancy is typically by an order of a magnitude smaller. Both dissipation and transport

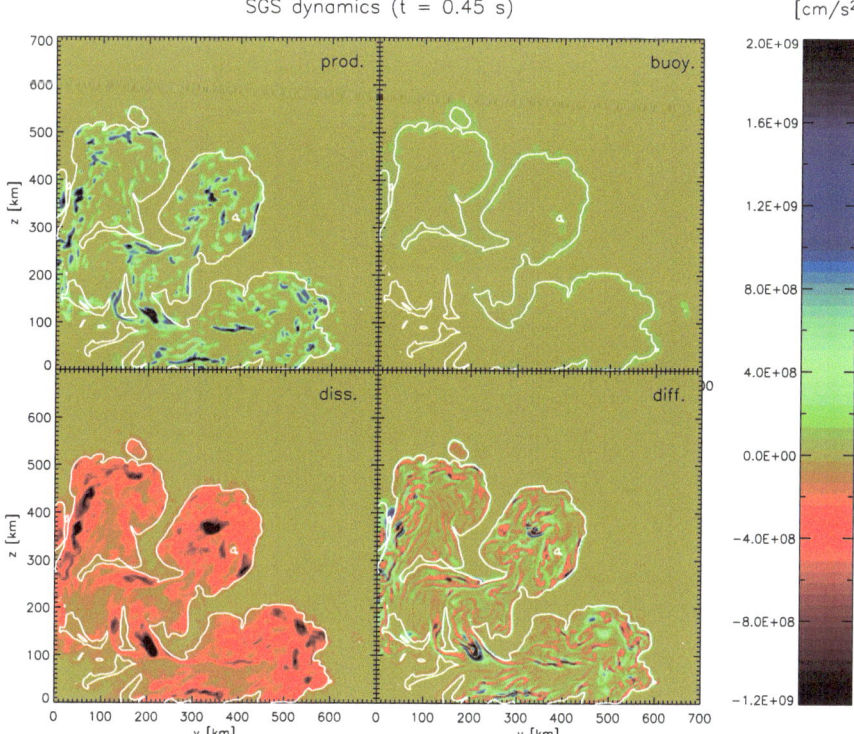

Fig. 5. The same plot as in Fig. 4 at $t = 0.45$ s

due to SGS turbulent diffusion are comparable to the rate of energy transfer at this time. In the unburnt material outside, on the other hand, there is still virtually no SGS turbulence. Thus, it appears that the flow is anisotropic in the vicinity of the flames. This highlights the necessity of a localized SGS model. We have utilized this behaviour by restricting SGS turbulence production to the equidistant part of the grid. A uniform cell size allows for an efficient parallelization.

In comparison to the SGS model that was used in earlier simulations of thermonuclear supernovae [48], the localized model increases the yield of nuclear energy significantly. This is demonstrated by the graphs of the integrated total energy in the left top panel of Fig. 6. The final kinetic energy of $0.472 \cdot 10^{51}$ erg $= 4.72 \cdot 10^{43}$ J in the simulation with the localized SGS model is about 25% greater than in the reference simulation with the old model (WPF). The total amount of burning products increases noticeably as well. We also plotted the outcome of suppressing Archimedian production. Although slightly less energy is produced, the burning process is still significantly enhanced due to the localized energy transfer. In the simulations with the localized SGS model, there is initially only little SGS turbulence

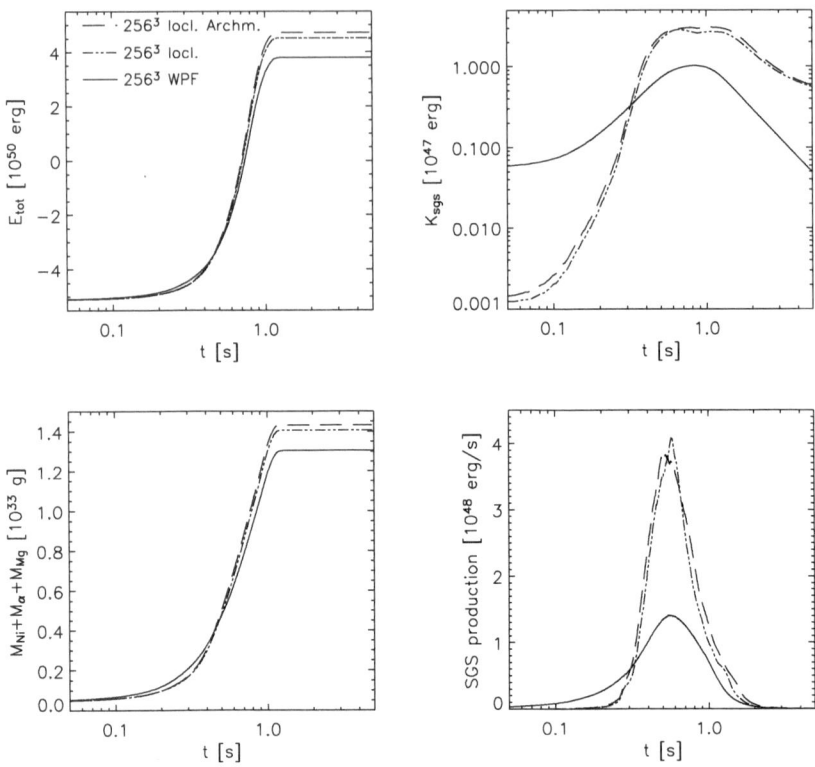

Fig. 6. Time evolution of integrated quantities for three simulations with identical initial conditions and resolution 256^3. In one case, Clement's SGS model with wall proximity functions (WPF) was used. For the other two simulations we applied the localized SGS model with Archimedian production, respectively, switched off and on

energy followed by a steep rise. The rapid growth of turbulence energy can be attributed to the substantially stronger production within the time interval from 0.3 to 1.0 s (see right bottom panel in Fig. 6). In the second half of the explosive burning, the total SGS turbulence energy is almost one order of magnitude larger, which enhances the flame propagation speed accordingly and results in a stronger explosion.

4.3 Complementary Small-Scale Simulations

In addition to simulations on the scales of the WD, complementary studies are necessary to validate the modelling assumptions entering here.

As mentioned in Sect. 2.2, one-dimensional simulations of the fully resolved laminar thermonuclear flame [90] provided the laminar flame speed as a function of fuel composition and density, which sets the lower limit in the supernova simulations.

Several studies addressed the multi-dimensional flame structure on small scales. In particular, below the Gibson scale flame propagation is not dominated by turbulence but by the LD instability (see Sect. 2.3). From first simulations [47] and semi-analytical models [6] it appeared unclear if the non-linear stabilization of the flame in a cellular structure would hold also at low fuel densities, and "self-turbulization" (also named "active turbulent combustion") has been suggested [51]. With the level-set based flame model (here in the full implementation to model the flame/flow interaction correctly), the stabilization could be shown to be effective under the conditions of SN Ia explosions [76, 77, 79]. Similar results were obtained with resolved flame simulations applying a low-Mach number hydrodynamics solver [4].

At the current stage, these complementary simulations provide confidence, that the flame propagation on scales not resolved in global SN Ia supernova simulations is adequately modelled and flame propagation proceeds in a stable way.

5 Conclusion

Numerical simulations of SN Ia explosions can be regarded as a challenging test case for advanced models of turbulent fluid dynamics and combustion physics. In the model we have presented, there are two essential ingredients: first, the level-set prescription for the propagation of flame fronts and second, a sub-grid scale model for the computation of the turbulent flame propagation speed. The level-set method combines the advantages of its ability to handle complex topologies and avoiding artificial diffusion. The sub-grid scale model is particularly suitable for a highly inhomogeneous and transient process such as a supernova explosion, because the localized closure for the production of turbulence energy does not rely on a priori assumptions about flow properties. In addition, we proposed to include unresolved buoyancy effects in a tentative way. Further parameters of the sub-grid scale model are treated in a semi-statistical fashion or were calibrated with data from separate numerical studies [81].

The described modelling approach leads to a self-consistent picture of turbulent deflagrations in SNe Ia. Therefore, the deflagration stage can be regarded as well represented in the simulations and the results of such simulations can be directly confronted to observations. In this way it is possible to address the question of whether the modelling concept reflects reality and of whether the astrophysical scenario is complete.

To begin with there is no doubt that pure turbulent deflagration scenarios are capable of producing a viable explosion of the WD star [21, 69, 74]. However, the outcome is sensitive to the ignition configuration and a failure to gravitationally unbind the star may arise in some setups [42, 62].

A first check of the models is provided by the global quantities. The energy released in the (exploding) models – typically up to 8×10^{50} erg – falls into the range of observational expectations for "normal" SNe Ia [9], although on the side of the weaker objects. The brightness of a SN Ia is determined by the amount of radioactive ^{56}Ni, one of the main products of the explosive nuclear burning. This unstable isotope decays to ^{56}Co and ultimately to ^{56}Fe and the released gamma photons are downscattered in the explosion ejecta finally giving rise to the observable event. Typically, the deflagration models produce around $0.4\,M_\odot$ of ^{56}Ni, but this may vary depending on the ignition conditions of the flame ([22, 78, 83]).

Synthetic lightcurves derived from explosion models ([5]) show reasonable agreement with the observations (cf. Fig. 7) This leads to the conclusion, that at least weaker SNe Ia can be well reproduced by the turbulent deflagration model. However, a potential problem arises in spectral features. Late time spectra show indication of unburnt material in the central parts of the ejecta, which disagrees with observations [35]. Also the very strong supernova events (producing more than one $M_\odot$ of iron group elements) remain a puzzle. It has been suggested, that a transition of the flame propagation mode from deflagration to detonation in later stages of the explosion could solve some of these issues [20, 31, 33]. A mechanism providing such a transition in thermonuclear supernovae, however, could not be identified yet [46] and therefore, this scenario remains hypothetical.

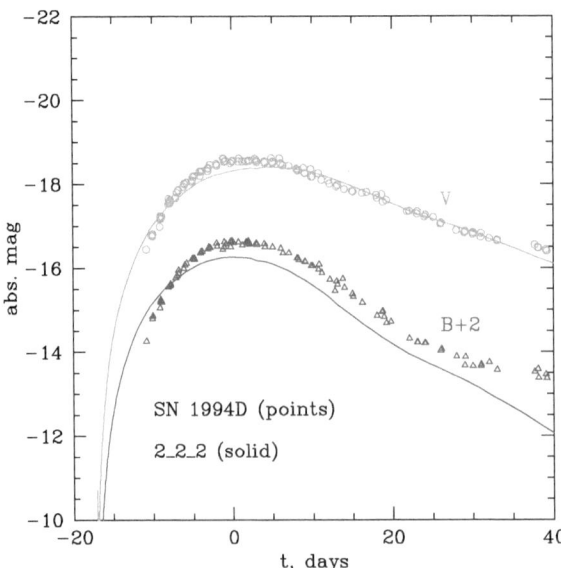

Fig. 7. Synthetic light curves in the B and V band derived from an explosion model in comparison with observations of SN1994D [5]. Roughly speaking, B and V are logarithmic measures of colours. A more negative value of B, for instance, implies a brighter appearance of the event in blue light

Nonetheless, the success of the pure turbulent deflagration scenario reproducing at least gross features of some SNe Ia indicates that it captures main aspects of the explosion mechanism. Further steps involve the exploration of different initial conditions of the WD star, possibly explaining the diversity observed in SNe Ia [73]. This is of fundamental interest for the cosmological application of SNe Ia, since their use as distance indicators relies on methods of calibrating these diversities [61].

In any case, we have been confronted with the most challenging aspects of turbulence modelling in numerical simulations of SNe Ia: reactive fluid dynamics that is neither stationary nor homogeneous, a nonideal equation of state and turbulence at extremely high Reynolds numbers. The insights that have resulted so far are of considerable worth in their own right. An example is the prospect of adopting the localized sub-grid scale model to adaptive methods. This might very well trigger new exciting developments in the field of astrophysical fluid dynamics.

References

1. Adalsteinsson D., Sethian J.A.: The fast construction of extension velocities in level set methods. J. Comp. Phys. **148**, 2–22 (1999)
2. Almgren A.S., Bell J.B., Rendleman C.A., Zingale M.: Low mach number modeling of Type Ia supernovae. I. Hydrodynamics. ApJ **637**, 922–936 (February 2006)
3. Arnett W.D.: A possible model of supernovae: detonation of 12C. Ap&SS **5**, 180–212 (March 1969)
4. Bell J.B., Day M.S., Rendleman C.A., Woosley S.E., Zingale M.: Direct numerical simulations of Type Ia supernovae flames. I. The Landau-Darrieus instability. ApJ **606**, 1029–1038 (May 2004)
5. Blinnikov S.I., Röpke F.K., Sorokina E.I., Gieseler M., Reinecke M., Travaglio C., Hillebrandt W., Stritzinger M.: Theoretical light curves for deflagration models of Type Ia supernova. A&A **453**, 229–240 (July 2006)
6. Blinnikov S.Iv., Sasorov P.V.: Landau-Darrieus instability and the fractal dimension of flame fronts. Phys. Rev. E. **53**(5), 4827–4841 (1996)
7. Borghi R.W.: On the structure and morphology of turbulent premixed flames. In C. Casci (ed.) Recent Advances in the Aerospace Sciences, Chap. 7, pp. 117–138. Plenum Publishing Corporation, New York (1985)
8. Brachwitz F., Dean D. J., Hix W. R., Iwamoto K., Langanke K., Martínez-Pinedo G., Nomoto K., Strayer M.R., Thielemann F., Umeda H.: The role of electron captures in Chandrasekhar-mass models for Type Ia supernovae. ApJ **536**, 934–947 (June 2000)
9. Branch D., Fisher A., Nugent P.: On the relative frequencies of spectroscopically normal and peculiar Type Ia supernovae. AJ **106**, 2383–2391 (December 1993)
10. Calder A.C., Plewa T., Vladimirova N., Lamb D.Q., Truran J.W.: Type Ia supernovae: An asymmetric deflagration model. astro-ph/0405126 (2004)
11. Carroll B.W., Ostlie D.A.: An Introduction to Modern Astrophysics. Addison-Wesley, New York (1996)

12. Chandrasekhar S.: The maximum mass of ideal white dwarfs. Astrophys. J. **74**, 81 (1931)
13. Colella P., Woodward P.R.: The piecewise parabolic method (PPM) for gas-dynamical simulations. J. Comp. Phys. **54**, 174–201 (1984)
14. Colella P., Glaz H.M.: Efficient solution algorithms for the Riemann problem for real gases. J. Comp. Phys. **59**, 264–289 (1985)
15. Cox J.P., Giuli R.T.: Principles of Stellar Structure, Vol. 2. Gordon and Breach, New York (1968)
16. Damköhler G.: Der Einfluß der Turbulenz auf die Flammengeschwindigkeit in Gasgemischen. Z. f. Elektroch. **46**(11), 601–652 (1940)
17. Darrieus G.: Propagation d'un front de flame. Communication presented at La Technique Moderne, Unpublished (1938)
18. Frisch U.: Turbulence. Cambridge University Press, Cambridge (1995)
19. Fryxell B.A., Müller E. Arnett W.D.: Hydrodynamics and nuclear burning. MPA Green Report 449. Max-Planck-Institut für Astrophysik, Garching (1989)
20. Gamezo V.N., Khokhlov A.M., Oran E.S.: Deflagrations and detonations in thermonuclear supernovae. Phys. Rev. Lett. **92**(21), 211102 (May 2004)
21. Gamezo V.N., Khokhlov A.M., Oran E.S., Chtchelkanova A.Y., Rosenberg R.O.: Thermonuclear supernovae: Simulations of the deflagration stage and their implications. Science **299**, 77–81 (2003)
22. García-Senz D., Bravo E.: Type Ia supernova models arising from different distributions of igniting points. A&A **430**, 585–602 (February 2005)
23. Garcia-Senz D., Bravo E., Serichol N.: A Particle code for deflagrations in white dwarfs. I. Numerical techniques. ApJS, **115**, 119–139 (March 1998)
24. Garcia-Senz D., Woosley S.E.: Type Ia supernovae: The flame is born. ApJ **454**, 895–900 (December 1995)
25. Germano M.: Turbulence: The filtering approach. J. Fluid Mech. **238**, 325–336 (1992)
26. Germano M., Piomelli U., Moin P., Cabot W.H.: A dynamic subgrid-scale eddy viscosity model. Phys. Fluids **3**, 1760–1765 (July 1991)
27. Hillebrandt W., Niemeyer J.C.: Type Ia supernova explosion models. ARA&A 38, 191–230 (2000)
28. Höflich P., Stein J.: On the thermonuclear runaway in Type Ia supernovae: How to run away? ApJ **568**, 779–790 (April 2002)
29. Hoyle F., Fowler W.A.: Nucleosynthesis in supernovae. ApJ **132**, 565–590 (November 1960)
30. Iapichino L., Brüggen M., Hillebrandt W., Niemeyer J.C.: The ignition of thermonuclear flames in Type Ia supernovae. A&A **450**, 655–666 (May 2006)
31. Ivanova L.N., Imshennik V.S., Chechetkin V.M.: Pulsation regime of the thermonuclear explosion of a star's dense carbon core. Ap&SS **31**, 497–514 (December 1974)
32. Khokhlov A.: Flame modeling in supernovae. ApJ **419**, L77–L80 (December 1993)
33. Khokhlov A.M.: Delayed detonation model for Type IA supernovae. A&A **245**, 114–128 (May 1991)
34. Kim W., Menon S., Mongia H.C.: Large-eddy simulation of a gas turbine combustor flow. Combust. Sci. Tech. **143**, 25–62 (1999)
35. Kozma C., Fransson C., Hillebrandt W., Travaglio C., Sollerman J., Reinecke M., Röpke F.K., Spyromilio J.: Three-dimensional modeling of Type Ia supernovae – the power of late time spectra. A&A **437**, 983–995 (July 2005)

36. Kuhlen M., Woosley S.E., Glatzmaier G.A.: Carbon ignition in Type Ia supernovae. II. A three-dimensional numerical model. ApJ **640**, 407–416 (March 2006)

37. Landau L.D.: On the theory of slow combustion. Acta Physicochim. URSS **19**, 77–85 (1944)

38. Landau L.D., Lifshitz E.M.: Fluid Mechanics, Vol. 6 of Course of Theoretical Physics. Pergamon Press, Oxford (1959)

39. Leibundgut B.: Cosmological implications from observations of Type Ia supernovae. ARA&A **39**, 67–98 (2001)

40. Liñan A., Williams F.A.: Fundamental aspects of combustion. Oxford University Press, Oxford, New York (1993)

41. Livne E.: Numerical simulations of the convective flame in white dwarfs. ApJ **406**, L17–L20 (March 1993)

42. Livne E., Asida S.M., Höflich P.: On the sensitivity of deflagrations in a Chandrasekhar mass white dwarf to initial conditions. ApJ **632**, 443–449 (October 2005)

43. Müller E.: Fundamentals of gasdynamical simulations. In Contopoulos G., Spyrou N.K., Vlahos L. (eds.) *Galactic Dynamics and N-Body Simulations*, Lect. Notes Phys. **433**, 313–363. Springer-Verlag, Berlin Heidelberg (1994)

44. Nandkumar R., Pethick C.J.: Transport coefficients of dense matter in the liquid metal regime. MNRAS **209**, 511–524 (August 1984)

45. Niemeyer J.C.: On the propagation of thermonuclear flames in Type Ia supernovae. PhD thesis, Technical University of Munich (1995). also available as MPA Green Report 911

46. Niemeyer J.C.: Can deflagration-detonation transitions occur in Type Ia supernovae? ApJ **523**, L57–L60 (September 1999)

47. Niemeyer J.C., Hillebrandt W.: Microscopic instabilities of nuclear flames in Type Ia supernovae. ApJ **452**, 779–784 (October 1995)

48. Niemeyer J.C., Hillebrandt W.: Turbulent nuclear flames in Type Ia supernovae. Astrophys. J. **452**, 769 (October 1995)

49. Niemeyer J.C., Hillebrandt W.: Microscopic and macroscopic modeling of thermonuclear burning fronts. In Ruiz-Lapuente P., Canal R., Isern J. (eds.) Thermonuclear Supernovae, pp. 441–456. Kluwer Academic Publishers, Dordrecht (1997)

50. Niemeyer J.C., Kerstein A.R.: Burning regimes of nuclear flames in SN Ia explosions. New Astron. **2**, 239–244 (August 1997)

51. Niemeyer J.C., Woosley S.E.: The thermonuclear explosion of Chandrasekhar mass white dwarfs. ApJ **475**, 740–753 (February 1997)

52. Nomoto K.: Accreting white dwarf models for Type I supernovae. I – presupernova evolution and triggering mechanisms. ApJ **253**, 798–810 (February 1982)

53. Nomoto K., Kondo Y.: Conditions for accretion-induced collapse of white dwarfs. Astrophys. J. Lett. **367**, L19–L22 (January 1991)

54. Nomoto K., Thielemann F.-K., Yokoi K.: Accreting white dwarf models of Type I supernovae. III—Carbon deflagration supernovae. ApJ **286**, 644–658 (November 1984)

55. Osher S., Sethian J.A.: Fronts propagating with curvature-dependent speed: Algorithms based on Hamilton-Jacobi formulations. J. Comp. Phys. **79**, 12–49 (November 1988)

56. Paquette C., Pelletier C., Fontaine G., Michaud G.: Diffusion coefficients for stellar plasmas. ApJS **61**, 177–195 (May 1986)

57. Perlmutter S., Aldering G., Goldhaber G., Knop R.A., Nugent P., Castro P.G., Deustua S., Fabbro S., Goobar A., Groom D.E., Hook I.M., Kim A.G., Kim M.Y., Lee J.C., Nunes N.J., Pain R., Pennypacker C.R., Quimby R., Lidman C., Ellis R.S., Irwin M., McMahon R.G., Ruiz-Lapuente P., Walton N., Schaefer B., Boyle B.J., Filippenko A.V., Matheson T., Fruchter A.S., Panagia N., Newberg H.J.M., Couch W.J.: The Supernova Cosmology Project.: Measurements of Omega and Lambda from 42 high-redshift supernovae. ApJ, **517**, 565–586 (June 1999)
58. Peters N.: Laminar flamelet concepts in turbulent combustion. In Twenty-First Symposium (International) on Combustion, pp. 1231–1250. The Combustion Institute, Pittsburgh (1986)
59. Peters N.: The turbulent burning velocity for large-scale and small-scale turbulence. J. Fluid Mech. **384**, 107–132 (April 1999)
60. Peters N.: The turbulent burning velocity for large-scale and small-scale turbulence. J. Fluid Mech. **384**, 107–132 (1999)
61. Phillips M.M.: The absolute magnitudes of Type Ia supernovae. ApJ **413**:L105–L108 (August 1993)
62. Plewa T., Calder A.C., Lamb D.Q.: Type Ia supernova explosion: Gravitationally confined detonation. ApJ **612**, L37–L40 (September 2004)
63. Pocheau A.: Scale invariance in turbulent front propagation. Phys. Rev. E., **49**, 1109–1122 (February 1994)
64. Pope S.B.: Tur Bulent Flows. Cambridge University Press, Cambridge (2000)
65. Reinecke M., Hillebrandt W., Niemeyer J.C.: Thermonuclear explosions of Chandrasekhar-mass C+O white dwarfs. A&A **347**, 739–747 (July 1999)
66. Reinecke, M., Hillebrandt, W., Niemeyer, J.C.: Refined numerical models for multidimensional Type Ia supernova simulations. A&A **386,** 936–943 (May 2002)
67. Reinecke, M., Hillebrandt, W., Niemeyer, J.C.: Three-dimensional simulations of Type Ia supernovae. A&A **391,** 1167–1172 (September 2002)
68. Reinecke, M., Hillebrandt, W., Niemeyer, J.C., Klein, R., Gröbl, A.: A new model for deflagration fronts in reactive fluids. A&A, **347,** 724–733 (July 1999)
69. Reinecke, M., Niemeyer, J.C., Hillebrandt, W.: On the explosion mechanism of SNe Type Ia. New Astron. Rev. **46**, 481–486 (July 2002)
70. Reinecke, M.A.: Modeling and simulation of turbulent combustion in Type Ia supernovae. PhD thesis, Technical University of Munich (2001). available at http://tumb1.biblio.tu-muenchen.de/publ/diss/allgemein.html
71. Riess, A.G., Filippenko, A.V., Challis, P., Clocchiatti, A., Diercks, A., Garnavich, P.M., Gilliland, R.L., Hogan, C.J., Jha, S., Kirshner, R.P., Leibundgut, B., Phillips, M.M., Reiss, D., Schmidt, B.P., Schommer, R.A., Smith, R.C., Spyromilio, J., Stubbs, C., Suntzeff, N.B., Tonry, J.: Observational evidence from supernovae for an accelerating universe and a cosmological constant. AJ. **116**, 1009–1038 (September 1998)
72. Röpke, F.K.: Following multi-dimensional type Ia supernova explosion models to homologous expansion. A&A **432**, 969–983 (March 2005)
73. Röpke, F.K., Gieseler, M., Reinecke, M., Travaglio, C., Hillebrandt, W.: Type Ia supernova diversity in three-dimensional models. A&A **453**, 203–217 (July 2006)
74. Röpke, F.K., Hillebrandt, W.: Full-star type Ia supernova explosion models. A&A **431**, 635–645 (February 2005)

75. Röpke, F.K., Hillebrandt, W.: The distributed burning regime in type Ia super-nova models. A&A **429**, L29–L32 (January 2005)
76. Röpke, F.K., Hillebrandt, W., Niemeyer, J.C.: The cellular burning regime in type Ia supernova explosions. I. Flame propagation into quiescent fuel A&A **420**, 411–422 (June 2004)
77. Röpke, F.K., Hillebrandt, W., Niemeyer, J.C.: The cellular burning regime in type Ia supernova explosions. II. Flame propagation into vortical fuel. A&A **421**, 783–795 (July 2004)
78. Röpke, F.K., Hillebrandt, W., Niemeyer, J.C., Woosley, S.E.: Multi-spot ignition in type Ia supernova models. A&A **448**, 1–14 (March 2006)
79. Röpke, F.K., Niemeyer, J.C., Hillebrandt, W.: On the small-scale stability of thermonuclear flames in type Ia supernovae. ApJ **588**, 952–961 (May 2003)
80. Sagaut, P.: Large Eddy Simulation for Incompressible Flows. Springer, (2001)
81. Schmidt, W., Hillebrandt, W., Niemeyer, J.C.: Level set simulations of turbulent thermonuclear combustion in degenerate carbon and oxygen. Combust. Theory Modell. 9(4), 693–720 (2005)
82. Schmidt, W., Hillebrandt, W., Niemeyer, J.C.: Numerical dissipation and the bottleneck effect in simulations of compressible isotropic turbulence. Comp. Fluids. 35, 353–371 (2006)
83. Schmidt, W., Niemeyer, J.C.: Thermonuclear supernova simulations with stochastic ignition. A&A **446,** 627–633 (February 2006)
84. Schmidt, W., Niemeyer, J.C., Hillebrandt, W.: A localised subgrid scale model for fluid dynamical simulations in astrophysics. I. Theory and numerical tests. A&A **450**, 265–281 (April 2006)
85. Schmidt, W., Niemeyer, J.C., Hillebrandt, W., Röpke, F.K.: A localised subgrid scale model for fluid dynamical simulations in astrophysics. II. Application to type Ia supernovae. A&A **450**, 283–294 (April 2006)
86. Sethian, J.A.: Level Set Methods and Fast Marching Methods. University Press, Cambride (1999)
87. Sharp, D.H.: An overview of Rayleigh-Taylor instability. Physica D Nonlinear Phenomena **12**, 3–3 (July 1984)
88. Smiljanovski, V., Moser, V., Klein, R.: A capturing-tracking hybrid scheme for deflagration discontinuities. Combustion Theory and Model. **1**,183–215 (1997)
89. Sussman, M., Smereka, P., Osher, S.: A level set approach for computing solutions to incompressible two-phase flow. J. Comp. Phys. **114**(1), 146–159 (1994)
90. Timmes, F.X., Woosley, S.E.: The conductive propagation of nuclear flames. I. degenerate C+O and O+Ne+Mg white dwarfs. ApJ **396**, 649–667 (March 1992)
91. Timmes, F.X., Woosley, S.E.: The conductive propagation of nuclar flames. I. Degenerate C+O and O+Ne+Mg white dwarfs. Astrophys. J. **396**, 649–667 (1992)
92. Travaglio, C., Hillebrandt, W., Reinecke, M., Thielemann, F.-K.: Nucleosynthesis in multi-dimensional SN Ia explosions. A&A **425**, 1029–1040 (October 2004)
93. Woosley, S.E., Wunsch, S., Kuhlen, M.: Carbon Ignition in Type Ia Supernovae: An Analytic Model. ApJ **607**, 921–930 (June 2004)
94. Zel'dovich, Ya.B.: An effect which stabilizes the curved front of a laminar flame. J. Appl. Mech. Tech. Phys. **1**, 68–69 (1966). English translation
95. Zwicky, F.: On collapsed neutron stars. ApJ **88**, 522–525 (November 1938)

One-Dimensional Turbulence Stochastic Simulation of Multi-Scale Dynamics

A.R. Kerstein

Combustion Research Facility, Sandia National Laboratories, Livermore, California
94551-0969 USA
arkerst@sandia.gov

1 A New Modelling Paradigm

1.1 Motivation

The historical roots of the current paradigm for numerical simulation of turbulent flows can be traced to early attempts at weather prediction. The mesh that is used is typically far too coarse to resolve all relevant processes, so sub-grid parameterizations are introduced to represent unresolved processes and their coupling to the resolved flow.

This approach has been successful in many contexts, enabling useful predictions of the unresolved processes as well as the resolved flow. However, as computing power increases and expectations of model performance increase commensurately, it is not self-evident that this paradigm will continue to be the optimal choice for all cases of interest. A particular challenge that is emphasized here is turbulent flow coupled to multiple physical and chemical processes at small scales.

Several recent developments in numerical flow simulation suggest the emergence of an alternate paradigm, here termed 'autonomous microstructure evolution' (AME). In Sect. 1.2, this paradigm is introduced by describing several methods of this type. The focus of this chapter is the proposal of a new AME-type method for simulation of turbulent flows that is based on a stochastic model, 'one-dimensional turbulence' (ODT).

After introducing the AME paradigm, its desirable attributes from the perspective of turbulence modelling are outlined in Sect. 2. The remainder of the chapter describes the proposed simulation method.

1.2 Autonomous Microstructure Evolution

At the molecular level, viscous fluid flow is strictly a local process. Molecular collisions are the elementary mechanism of momentum and heat transfer,

Kerstein, A.R.: *One-Dimensional Turbulence Stochastic Simulation of Multi-Scale Dynamics.*
Lect. Notes Phys. **756**, 291–333 (2009)
DOI 10.1007/978-3-540-78961-1_8   © Springer-Verlag Berlin Heidelberg 2009

and they likewise control mass transfer, flow energetics and chemical change in reacting flows. This is recognized in derivations of the continuum equations of fluid flow (e.g. the Navier–Stokes equation) from kinetic theory using elementary statistical hypotheses (e.g. the Boltzmann chaos assumption).

Accordingly, the continuum equations governing compressible fluid flow are local in nature. However, in low-Mach-number flows (Ma $\ll$ 1), the sound speed becomes irrelevant to the dominant flow processes and it is physically more appropriate, and computationally more efficient, to adopt an incompressible formulation. This formulation treats the sound speed as effectively infinite and thereby allows flow evolution to be represented as an elliptic problem, in which all fluid elements and constraints (e.g. boundary conditions) are coupled instantaneously.

At this level of description, it is entirely appropriate, and generally quite advantageous, to dispense with the local character of the physical processes that govern low-Ma flow evolution. However, there is an alternative, local formulation that is sometimes advantageous at low Ma, called the pseudo-compressible formulation [1]. In this formulation, an artificially low sound speed is introduced in order to reduce the time-scale disparity between acoustic and solenoidal flow processes, thus mitigating the severe time-step constraints for compressible-flow time advancement at low Ma.

To summarize, the continuum-level governing equations need not obey locality in order to capture the governing physics at low Ma, although a local formulation may be a viable option. These considerations provide a useful context for defining and illustrating the AME paradigm.

The most direct way to simulate fluid flow is to remain as faithful as possible to its occurrence in nature, i.e. by simulating the underlying molecular motions and interactions. The most common and successful method of this type is 'molecular dynamics' (MD) [2, 3], whose virtue in this regard is that it captures noncontinuum effects when they are important, as well as flow evolution describable by continuum methods.

MD in this context is a direct molecular simulation, subject to idealization of molecular collision processes, e.g. through the adoption of a molecular pair potential. However, the MD concept has been generalized through the development of models in which computational molecules are pseudo-particles. Their properties and interactions are defined so that macroscopic flows can be simulated using much fewer than Avogadro's number of particles. Examples of this approach are 'smoothed particle hydrodynamics' (SPH) [3] and 'dissipative particle dynamics' (DPD) [4]. Another method of this type is 'lattice-gas hydrodynamics' (LGH) [5], whose distinguishing feature is that it is designed to yield controlled (i.e. arbitrarily accurate) approximations to the continuum governing equations although the particles do not represent physical molecules.

The common feature of the methods described thus far, and the method proposed here, is that a representation, exact or idealized, of small scale processes is adopted that yields, through process evolution, collective behaviours that correspond to continuum flow, with varying degrees of accuracy. This is the defining feature of the AME paradigm in the context of flow simulation.

Another AME-type method, the 'lattice Boltzmann model' (LBM) [5], illustrates that these methods are not exclusively particle-based. LBM evolves probability density functions (PDFs) of particle properties rather than particles per se. It thus retains a link to particle properties though particles are not explicit within the method. In this regard it may be viewed as intermediate between particle and continuum methods. Another notable feature of LBM is that turbulence modelling has been incorporated into the LBM framework [5].

The link between LBM evolution and the implied particle evolution can be formalized by noting that evolution of a PDF represents the ensemble evolution of a collection of particles governed by coupled stochastic differential equations (SDEs). In this context, the latter is a more detailed level of description from which the former can be deduced. However, in a class of turbulent flow models, the relationship is reversed. An unclosed hierarchy of evolution equations for the PDF of flow properties in turbulence can be closed by modelling to obtain a single-point evolution equation for the joint PDF of velocity and scalar fields [6]. Though elliptic in character in its usual low-Ma formulation, it can be solved using an algorithm of AME type. Namely, particle SDEs are formulated whose details, apart from the conformance of their ensemble properties to the PDF evolution equation, need not be physically realistic. These SDEs are solely numerical devices for efficient solution of the PDF evolution equation. In this sense, they are analogous to LGH, in which particle evolution is strictly a device for solving continuum equations, albeit the exact equations in that case.

The foregoing AME-type methods, whether used as complete (particle through continuum regime) flow simulations (e.g. MD), as numerical devices for solving exact or modelled continuum equations, or as models in their own right, are all explicitly or implicitly particle based. The AME paradigm also accommodates processes rather than particles as its primitive elements. A notable example is vortex dynamics (VD) [7], which in its two-dimensional (2D) implementation evolves discrete point vortices or vortex blobs. (In 3D, vortex filaments, arrows and particles have been used [7–9].) The Biot–Savart equation that couples the discrete elements is nonlocal, illustrating that the AME paradigm is not limited to local interactions. Vortex dynamics is generally applied to low-Ma flow, and captures the nonlocality of that flow regime in a natural way. In 2D flow, the large-scale organization of vorticity is elegantly reproduced by discrete-vortex simulations. To represent unresolved motions in VD simulations of turbulence, vortex blobs can execute random walks and/or undergo evolution of their internal structure (vorticity profile).

The formulation proposed here is akin to VD in that its primitive elements are processes rather than particles, and as in VD, a process represented in this manner introduces some form of nonlocality. As in VD, the primitive elements are associated with vortical motion, but unlike the discrete vortices of VD, they represent the outcome of vortical motion rather than vortices per se.

Closely related to this distinction is the key attribute of the primitive elements introduced here: they are processes implemented on a 1D spatial domain.

2 Implications for Turbulent Flow Modelling

2.1 Large-Eddy Simulation: Capabilities and Limitations

The motivation for adopting the AME paradigm for turbulent flow modelling, and implications concerning the structure of such a model, are now considered. These questions are addressed by first examining the conventional paradigm outlined in Sect. 1.1.

Specifically, consider 'large-eddy simulation' (LES) of constant-property flow. The LES strategy is to resolve scales far enough below the flow-dependent energy-containing scales so that the unresolved motions are within the inertial subrange, whose properties are presumed to be universal [10]. Moreover, the main role of the unresolved motions is presumed to be cascading of mesh-resolved kinetic energy to smaller, unresolved scales. This is represented within LES by dissipation of mesh-resolved kinetic energy, at a rate commensurate with the cascading mechanism. The dissipation is typically incorporated using eddy viscosity, or a generalization thereof (tensor viscosity, spectral viscosity, etc.) [10].

Though this strategy has proven to be quite successful thus far and holds great promise for the future, it is subject to two types of limitations that motivate consideration of an alternative approach. One type of limitation is generic to all applications of this strategy, while the other type is flow specific.

The generic limitations are associated with the LES representation of cascade physics. Intermittency of the turbulent cascade [11] has several consequences whose representation within LES is not yet fully satisfactory. One is backscatter of kinetic energy from unresolved to resolved scales. Modelling of backscatter within LES is an active research topic, and there has been useful progress in this regard [10]. Another is a spectrally nonlocal contribution to downscale (forward cascade) energy transfer. Spectral viscosity methods can account for this, but are not necessarily advantageous or practical from other viewpoints. Intermittency effects depend on the turbulence Reynolds number Re in a manner that has not yet been convincingly captured as LES is applied to flows at successively higher Re [12].

Apart from these physics concerns, there is the practical concern of devising an LES closure that is numerically robust within the time advancement of a nonsmooth discretized velocity field as well as physically sound. There is room for improvement in this regard as well.

These limitations arise in the context of applications that satisfy the basic axiom of the LES strategy: mesh refinement sufficient to resolve flow-specific phenomena. There are (at least) two classes of applications that challenge this axiom: wall-bounded flows and flows coupled to dissipation-scale processes.

In wall-bounded flows, the scale of near-wall flow-specific phenomena is proportional to distance from the wall, hence decreases as the wall is approached, until the viscous sublayer is reached. This requires refinement to full flow resolution, in effect, direct numerical simulation (DNS), near walls in order to maintain fidelity consistent with the LES strategy. Though costly, this near-wall refinement may be feasible for some applications. In general, however, the cost of this approach is prohibitive, so instead, near-wall parameterizations are introduced. The consequences of introducing parameterizations are considered shortly, after other examples of parameterization are noted.

Those examples are parameterizations that represent the coupling of unresolved flow scales to dissipation-scale (or in general, sub-grid-scale) processes, such as thermodynamic fluctuations (in compressible flow), mixing of dynamically active scalars (e.g. density in buoyant variable-density flow), chemical reactions (including heat-release effects on density and hence on the flow field), and multiphase couplings. Multiphase couplings include diverse phenomena such as momentum, heat, and mass transfer between dispersed and continuum phases, and surface tension at interfaces between immiscible liquids.

The limitations of parameterizations of these coupled, highly nonlinear, multivariate, spatially distributed processes are well known and are not elaborated here. Certainly, they are at least as challenging as the parameterization of near-wall constant-property flow, so the latter is examined to illustrate the difficulties that can arise.

In near-wall flow, an obvious modelling concern is prediction of separation and reattachment. One mitigating factor in an LES formulation is that adaptive meshing can resolve the vicinities of separation and reattachment loci along a wall at much less cost than resolving the entire near-wall flow. An application that is less amenable to adaptive meshing is near-wall flow subject to transient bulk forcing. An example is near-surface flow in the atmospheric boundary layer (ABL) subject to shifts of wind speed and direction. The time-lagged response of the near-wall flow to this transient forcing can, for example, result in nonmonotonic wall-normal profiles of ensemble-averaged velocity components, and related flow-specific features, that defy representation by a parameterization. (Another canonical example of near-wall nonmonotonicity is buoyancy-driven flow near a heated vertical wall [13].)

This considerations point inexorably to the conclusion that parameterization of unresolved flow-specific phenomena, in contradiction of the axiom underlying the LES modelling strategy, imposes inherent limitations on the breadth and accuracy of predictive capability that this strategy can ultimately achieve. Within the scope of these limitations, much can and will be accomplished. Nevertheless, it is apparent that there is a compelling fundamental as well as practical imperative to pursue alternate strategies that might not be subject to these limitations.

2.2 An Alternative to Parameterization

It is useful to define what is meant by parameterization in order to delineate a possible alternative. Here, a broad definition is adopted. Namely, a parameterization is any mathematical construct associated with a mesh control volume (or more generally, a localized stencil of control volumes) that exchanges information during the simulation only with values of mesh-resolved variables in that control volume (or stencil). This definition includes some formulations that are termed 'dynamically active' LES sub-grid models [10]. Here, the distinguishing feature is taken to be the nature of the communication among modules (i.e. the flow solver and the parameterization) rather than the internal content of the parameterization.

This definition is adopted because it addresses the strategy of devising better parameterizations. One can in principle improve the parameterization to the point of performing DNS within each control volume. Nevertheless, this is not equivalent to DNS of the whole flow if the information exchange between control volumes is based solely on mesh-resolved variables. This restricted information exchange introduces an inherent information loss that does not occur in whole-flow DNS.

In this regard, it is useful to compare constant-property LES of unbounded flow to the more challenging cases discussed in Sect. 2.1. The downscale information transfer in LES is straightforward in principle because it involves the loss of information (about flow structure whose scale is compressed below the resolution scale) that is no longer needed in the simulation (if only the forward cascade is considered). The upscale information transfer is a more delicate issue because it requires retention of sub-grid-scale information (e.g. by using a parameterization) that is sufficient to characterize backscatter through transfer of this information to the resolved variables. On this basis, upscale transfer is generally viewed as a more challenging modelling problem than downscale transfer.

Compare this to the modelling requirements for more complicated flows (variable property, chemically reacting, etc.), supposing in these cases that an elaborate, accurate parameterization is available. For quantities such as chemical species in a reacting flow, the upscale information transfer may consist of a straightforward averaging or spatial filtering procedure. However, the downscale information transfer may require, for example, adjustment of small-scale species concentrations resolved by the parameterization, where the adjustment is based on spatially filtered information at the mesh scale. This adjustment can be problematic with regard to either realizability (e.g. causing mass fractions to be negative or exceed unity) or chemical consistency (e.g. creating spurious nonequilibrium mixtures).

Thus, as sub-grid parameterizations become more elaborate in order to address increasingly complex problems, the fidelity of the overall formulation may be constrained, to an increasing degree, by the sparse information content (relative to the sub-grid formulation) at mesh-resolved scales. It can be anticipated that this problematic downscale information transfer will prove to be the most enduring constraint on the ultimate utility of parameterization, as broadly defined here.

Accordingly, the basic axiom that guides the present pursuit of a better alternative is that parameterization requiring downscale information transfer should play a minimal role, if any. AME, as defined and exemplified in Sect. 1.2, is precisely the paradigm that adheres to this principle.

Adopting the AME paradigm on this basis, a formulation of this type is desired that preserves the essential characteristics of AME as it is generalized to multi-physics problems. As noted in Sect. 1.2, LBM and VD have been applied to turbulence by appending treatments that are parameterizations (as defined here) underneath the model. These extended formulations are thus hybrids that are subject to the same limitations as other parameterizations.

This is a generic outcome of efforts to isolate the model representation of individual subprocesses within limited scale ranges. As this inference suggests, a robust remedy would be to implement all subprocesses at all scales. Superficially this defines DNS, which is fully accurate but unaffordable for most problems. However, there is an alternate, more affordable realization of this strategy. Namely, implement all subprocesses at all scales in a lower-dimensional space. 2D examples of dimension reduction include VD and 2D Eulerian solution of the exact evolution equations. However, 2D turbulence has qualitatively different characteristics from 3D turbulence, in addition to the obvious limitation that general 3D initial and boundary conditions cannot be represented in 2D.

Nevertheless, there is a form of dimensional reduction that can both preserve the physics of 3D turbulence and accommodate general 3D flow configurations, while providing an all-scale representation of all subprocesses. Description and assessment of this formulation is the focus of the remainder of this chapter.

2.3 Superparameterization and its AME Reformulation

Simulation of global atmospheric circulation is a salient application that confronts the challenges outlined in Sects. 2.1 and 2.2. An emerging strategy for addressing this problem exemplifies the all-scale AME paradigm. An important caveat in this context is that geophysical-scale flows cannot be affordably simulated with resolution to viscous scales, even in a formulation with reduced dimensionality. However, the modelling concept can be recast so as to obtain a fully resolved formulation applicable to engineering-scale flows.

The AME-type atmospheric simulation strategy is a variant of 'superparameterization' (SP) [14]. As noted in Sect. 2.2, the ultimate parameterization is a DNS associated with each mesh control volume. A step back from this would be a fully resolved 2D simulation associated with each mesh control volume. The qualitative as well as quantitative limitations of 2D simulations (Sect. 2.2) would counteract the benefits of this degree of detail for

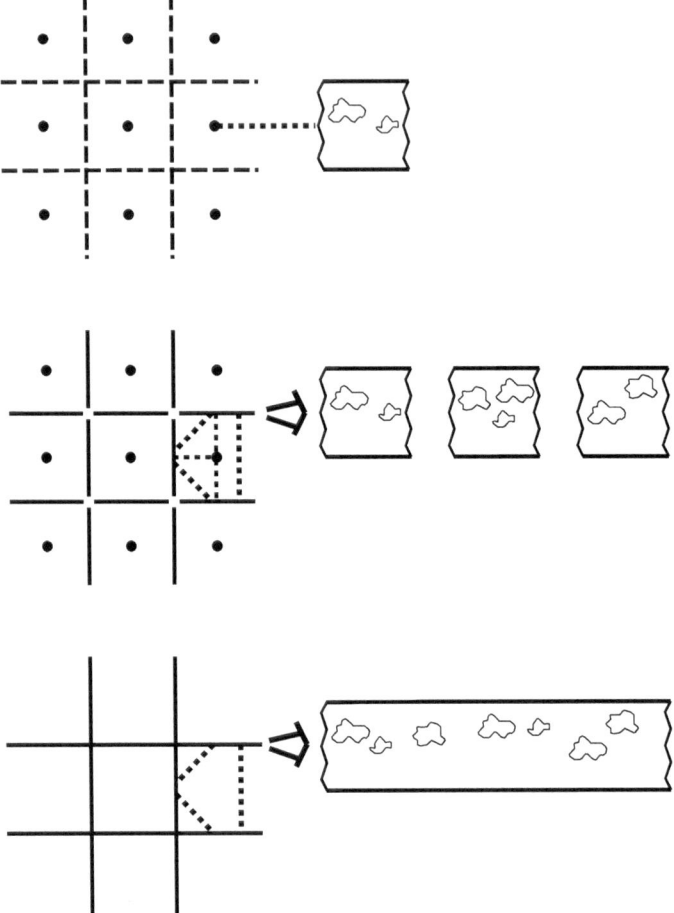

Fig. 1. Alternate implementations of a superparameterization strategy for atmospheric simulation. *Left*, top views; *right*, side views. In the top views, *dots* are nodes of the large-scale simulation (general circulation model, GCM), *long-dash lines* demarcate simulation control volumes, and solid lines show the placements of the cloud-system resolving model (CSRM), which live on vertically oriented domains. Short-dash lines indicate couplings. *Top case*: Each CSRM domain (side view on *right*) is coupled to a GCM node and simulates cloud evolution within the associated control volume (here ignoring the multiple vertical layers of GCM control volumes). Lateral boundary conditions are periodic (*jagged lines*). *Middle case*: A distinct CSRM domain, configured as in the top case, is associated with each GCM control–volume lateral interface. CSRM domains are coupled to each other and to GCM nodes, as indicated. Couplings between CSRM domains that do not border a common control volume are also allowed, but omitted for clarity. *Bottom case*: The atmosphere is simulated using a coupled ensemble of CSRM domains, each of which encircles the earth (so these domains have no lateral boundaries). Another possible variant is the CSRM domain structure of the lower case combined with GCM evolution as in the middle case

many applications. However, for typical convective flow regimes in the ABL, 'cloud-system resolving models' (CSRMs) implemented on planar vertically oriented domains have proven to be cost-effective alternatives to 3D simulation [15–17]. In effect, 2D is found to be the dimensionality that optimizes the cost-performance tradeoff for this class of flows. Accordingly, SP implements a 2D CSRM-type simulation associated with each control volume of a general circulation model (GCM) of the Earth's atmosphere.

SP as such is subject to the inherent limitations of parameterization, but a variant of this formulation that adheres to the AME paradigm is under development [18]. To visualize this variant, imagine that the CSRM simulations tile the vertical faces of the GCM control volumes in a hypothetical Cartesian geometry involving one planar layer of rectangular GCM control volumes. Thus, the control volume height is the vertical extent of the simulations. Instead of implementing an independent CSRM on each vertical control-volume face, suppose that a CSRM is implemented on each 2D domain corresponding to a vertical sidewall of each row or column (the two horizontal coordinates) of the array of GCM control volumes. Then the height of each CSRM domain is the vertical extent of the simulated atmosphere, and its horizontal extent spans one of the two horizontal directions (e.g. the Earth's circumference if the given direction is the Cartesian analogue of a great circle). With a suitable coupling among these 2D domains, it is possible (and desirable, for the reasons explained in Sect. 2.2), to dispense with the GCM itself and thus obtain an AME-type formulation, which is denoted here as 'super-AME' (SAME). Several of the possible variants of the superparameterization concept are illustrated in Fig. 1.

2.4 A 1D AME Formulation

Having presented the rationale for the AME paradigm and the main elements of its implementation for a particular application, adaptations for other purposes are considered. As in the atmospheric flow application, a key consideration is the spatial dimensionality that is most cost-effective for a given application.

There are several applications for which a 1D formulation is advantageous in principle. One is wall-bounded flow, in which evolution of the wall-normal profile of flow properties embodies the dominant physics. Analogously, thin free shear flows (e.g. jets, wakes and mixing layers) are boundary-layer type flows whose representation based on property profiles along a lateral coordinate is common [19]. Another such application is horizontally homogeneous vertically stratified buoyant flow. 'Single-column models' (SCMs) are vertically oriented 1D formulations that are commonly applied to ABL flows of this type [17]. Finally, 'stationary laminar flamelet models' (SLFM) used in turbulent combustion simulations involve 1D (flame-normal) flow representations [20].

The 1D formulation described here has been applied to all these flow types. Representative results are discussed in Sect. 7. For now, the utility of a 1D formulation is assumed, deferring the question of how turbulence can be modelled in 1D until Sect. 5.

Given such a 1D model, there are several ways that it can be used in the construction of a 3D simulation. One way is to associate a 1D line segment, on which the model is implemented, with each control volume of an LES, analogous to one of the ways of incorporating CSRMs into a GCM (Sect. 2.3). This 1D analogue of SP might be termed 'semi-superparameterization' (SSP) because the sub-grid model dimensionality is half that of SP. Extending the analogy, assume a Cartesian mesh of cubic LES control volumes. Consider the rectangular volume formed by a linear stack of LES control volumes in any one of the three coordinate directions (analogous to a row or column of GCM control volumes). Now take each side-edge of each of these rectangular volumes to be a 1D domain for implementation of the 1D model of turbulence. Each of these domains then spans the flow in a given direction, and is presumed to resolve all relevant length scales. Thus it has the needed attributes for an all-scale AME-type formulation, subject to the specification of suitable rules for coupling the various domains. This is the 1D analogue of SAME (Sect. 2.3). For consistency with terminology used previously [21], this 1D methodology is denoted ODTLES, while SSP and related 1D formulations are denoted LESODT. Despite the terminology, ODTLES is an AME formulation that does not involve the advancement of LES-type equations, just as SAME (Sect. 2.3) dispenses with the GCM machinery.

2.5 Hybrid Formulations

Section 2.2 alludes to two of the many possible hybrid formulations that combine attributes of the approaches mentioned thus far. Description of all the promising possibilities is beyond the scope of this chapter, but a particular hybrid formulation that is based on the modelling approach discussed in Sect. 4.2 is mentioned here.

The distinction between parameterization, in which information is transferred between different scale ranges, and AME, in which all information transfer involves spatially resolved quantities, has been emphasized thus far. For some applications, it is advantageous to evolve some variables using parameterization and evolve others with full resolution. A variant of the domain geometry described in Sect. 2.4 with reference to SSP (nominally a parameterization) has in fact been implemented as a hybrid of this type [22, 23]. Namely, for combustion simulation, thermochemical information (species mass fractions and enthalpy) resides solely on the 1D line segments, while momentum and pressure reside on the coarse 3D mesh. The upscale information transfer consists of density changes that drive the mesh-resolved advancement of the continuity equation. The downscale transfer consists of velocities normal to control–volume faces. These velocities prescribe volume transfers,

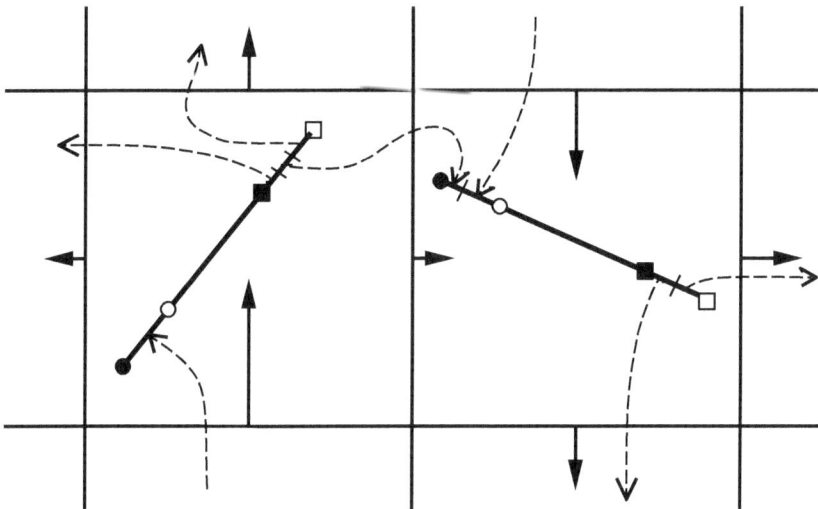

Fig. 2. Example of a hybrid formulation in which thermochemical information resides solely within 1D sub-grid domains (*tilted line* segments) but flow evolves on the coarse 3D mesh (Cartesian array of control volumes). *Solid lines* with arrows are flow velocities, evolved on the coarse mesh, that determine volume transfers between 1D sub-grid domains. The splicing mechanism that implements these transfers is illustrated. Each 1D domain has an input end (*circle*) and an output end (*square*). Open and filled symbols demarcate the 1D domains before and after splicing. Portions transferred during splicing are separated by tick marks. *Dashed curves* with *arrows* indicate transfers between 1D domains in different control volumes. For details, see [24]. An alternative to specified input and output locations is to use periodic 1D domains. Then there are no preferred locations at which to remove and insert domain segments

in a Lagrangian sense, between 1D segments associated with control volumes that share a common face. This Lagrangian transfer operation, termed 'splicing' (see Fig. 2), preserves chemical states, thereby mitigating the inherent artefact of reacting-flow parameterization (Sect. 2.2). It also preserves small-scale spatial structure, subject to an important caveat. In a receiver segment, newly spliced fluid from a donor segment contacts receiver fluid at one location, possibly creating an unphysical local configuration (e.g. cold fuel in contact with cold air under physical conditions that would require flame at all fuel–air interfaces). This is the first of several illustrations that dimensional reduction involves compromises and trade-offs, as in any modelling approach. Nevertheless, splicing is an advective transfer rather than a surrogate mixing operation, so it preserves local chemical states. The only species mixing in this formulation is by a physically accurate molecular mixing process, in contrast to models that are strictly parameterizations.

Among the various proposed model formulations encompassed by the rubric 'superparameterization' are some that would be termed hybrids in the present classification [18]. It can be anticipated that the distinct but overlapping interests of the geophysical and engineering fluid dynamics communities (as well as the astrophysics community, whose interests are discussed in the chapter "Turbulent Convection and Numerical Simulations in Solar and Stellar Astrophysics" and the chapter "Turbulent Combustion in Thermonuclear Supernovae" in this volume) will stimulate a productive cross-fertilization of modelling concepts as progress continues in these arenas.

3 Proposed Modelling Strategy

3.1 Overview

The goal of this chapter is to outline a turbulence simulation strategy in which ODT is a central element, and in so doing, to motivate as well as explain ODT. The strategy as outlined has not yet been implemented computationally, although an effort to do so is underway and development of several key components of the strategy has been completed.

Section 1 introduces the AME paradigm and explains its advantages for turbulent flow simulation. The strategy outlined here is designed with this in mind, subject to the inevitable compromises involved in modelling.

In Sect. 2.4, the 1D domain is defined geometrically as a line segment, specifically, a line segment corresponding to an edge of a linear stack of cubic control volumes (CVs). This is useful conceptually, but for numerical implementation, it is preferable to interpret 1D model evolution as occupying a volume of space, enabling a finite-volume numerical representation. For this purpose, the rectangular volume occupied by each stack of control volumes within the 3D domain is taken to be a 1D model domain.

In particular, assume that the 3D flow domain is itself rectangular, with coordinate bounds $0 \leq x \leq X$, $0 \leq y \leq Y$ and $0 \leq z \leq Z$. This geometry corresponds to a proof-of-principle application proposed in Sect. 7.1, but for now it is illustrative. Assume nominal CVs (whose role, in the absence of mesh-scale advancement, is as yet unexplained) that are cubic with edge length M. X, Y and Z are all assumed to be integer multiples of M, so the flow domain can be tiled with a cubic array of these CVs. For convenience, express length in units of M, so X, Y and Z are integers. Then the CVs in the array are indexed (i, j, k), where $1 \leq i \leq X$, $1 \leq j \leq Y$ and $1 \leq k \leq Z$, and the respective CVs are denoted C_{ijk}.

CVs stacks, each of which is a 1D model domain as defined above, are formally defined as $\bigcup_n C_{ijk}$, where n denotes either i, j, or k. For example, $S_x(j, k) = \bigcup_i C_{ijk}$ is the index-(j, k) stack oriented in the x direction, which is then the coordinate direction of the 1D model implemented on $S_x(j, k)$. $S_y(i, k)$ and $S_z(i, j)$ are defined analogously, yielding three arrays of stacks oriented in the respective coordinate directions. Each array fills the flow domain. Likewise, each CV C_{ijk} is contained in three stacks.

In the proposed formulation, three distinct flow solutions are time-advanced concurrently, each in one of the stack arrays. Each is a self-contained solution in that it does not exchange fluid or fluid properties with the other solutions, but the solutions are coupled in that each determines fluxes that are used to close the other two solutions. Each solution is designated by the corresponding subscript of S, i.e. x, y, or z.

Now consider the substructure of each stack, or 1D domain. (These terms are used interchangeably.) An x-oriented stack, or x-domain, is considered for illustration. (In general, statements about x-domains are likewise applicable to y-domains and z-domains.) By definition, each x-domain has a substructure consisting of a linear array of X cubic CVs, each of edge unity in the chosen scaled units. The first and last CVs each have one face interior to the x-domain, four contained in its respective side-faces, and one coinciding with an end-face of the x-domain. The other CVs in the x-domain each have two faces interior to the x-domain and four contained in its respective side-faces. The union of noninterior CV faces coincides with the surface of the x-domain.

The CVs are central to the coupling of the three concurrent flow solutions. Additional x-domain substructure needed to advance the individual solutions, illustrated in Fig. 3, is now introduced.

The x-domain is already partitioned into X CVs by the CV interior faces. A refinement of this partitioning is introduced. Parallel to those interior faces, additional faces are introduced so that the x-domain is partitioned into mX cells of identical shape, denoted 'wafers', where m is an integer. The x-domain is now a linear array of mX wafers of edge $1/m$ (in scaled units), such that the union of each successive set of m wafers coincides with a CV. Each wafer is a rectangle of dimensions $(1/m) \times 1 \times 1$, where in general $m \gg 1$, hence the terminology.

This x-domain refinement defines a mesh, resolving the length scale $\Delta x = 1/m$, on which the 1D model is implemented. This is the length scale at which the flow is resolved within the 1D treatment. 3D flow is captured explicitly at length scales above unity through the coupling of flow solutions. Below length scale unity, 3D flow is captured only to the extent that it is represented implicitly within the 1D model. (See Sect. 3.2.) If the CVs formed the mesh for an analogous explicitly 3D flow simulation, then the range of represented scales would be 1 through X. In the present formulation, an additional factor m of scale resolution is introduced through modelling. Because this additional resolution is introduced in 1D rather than 3D, the number of computational cells in the simulation is smaller, by a factor of order m^2, than the number required for equivalent 3D resolution. The attendant computational cost reduction is the benefit of the present formulation. The trade-off for this cost reduction is the use of a model, rather than the exact governing equations, to evolve the flow at scales smaller than unity.

Commensurate with the disparate scales at which the flow is resolved in 1D and in 3D, the time step for advancement of the 1D model on an x-domain is considerably shorter than the time step for coupling of the three flow

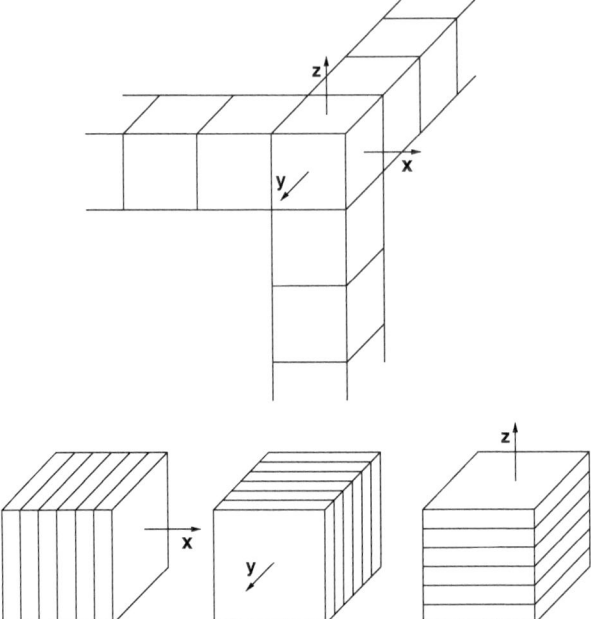

Fig. 3. *Top*: Three orthogonal 1D domains that contain a given control volume, here taken to reside at a corner of the 3D flow domain. *Bottom*: Within the control volume (expanded view), the substructure of the portion of each 1D domain that intersects the control volume is shown. In this example, the 1D spatial refinement factor relative to the 3D control–volume structure (*top*) is $m = 6$

solutions. Therefore, 1D model advancement is subcycled within an overall time-advancement cycle whose time increment corresponds to the solution-coupling time step. (If fractional-step advancement is used, there may be several coupling operations per time-advancement cycle. Numerical implementation is not considered here at this level of detail.) The advancement cycle is explained further in Sect. 3.3.

Several aspects of numerical implementation are noted. First, during 1D subcycling, each x-domain evolves autonomously. This provides an efficient domain decomposition for parallel implementation that should yield near-perfect scalability owing to the predominant cost of 1D subcycling relative to other operations during the advancement cycle. Second, spatial uniformity of the 1D refinement of the x-domain has been introduced for clarity of exposition. Though this meshing is used in 1D model applications reported to date, an adaptive-mesh formulation presently under development offers the possibility of substantial cost savings that will extend the range of applicability of this formulation. (See Sect. 6.1.) Third, the advancement cycle as outlined excludes any advancement subprocesses on the 3D mesh (union of CVs). If an incompressible formulation of the momentum equation were adopted, then enforcement of continuity would require an elliptic solution of the pressure Poisson equation, contravening the AME paradigm. A formulation of this type has in fact been implemented [21], as discussed in Sect. 6.3. For the purpose of formulating a model within the AME paradigm and noting its attributes, a compressible analogue of that incompressible formulation is proposed.

3.2 1D Advancement

The 1D subcycling on each x-domain is both the novel feature and the main physical content of the proposed formulation, so it is explained in detail in Sect. 5. Here, the 1D modelling concept is introduced briefly.

Within the proposed compressible-flow treatment, a natural context for 1D modelling is 1D gas dynamics [25]. In general, 1D gas dynamics is a steady-state formulation useful for analysis of shocks and other high-Ma phenomena. Here, compressibility is introduced in order to exploit its technical advantages for turbulent flow simulation, as in [26], rather than for investigation of high-Ma phenomena per se.

Starting from conventional 1D gas dynamics, possible extensions to represent turbulent flow effects are considered. Steady-state representation of compressible as well as incompressible turbulent flow is provided by 'Reynolds-averaged Navier–Stokes' (RANS) formulations. For compressible flow, the simplest of these formulations introduce an eddy viscosity and an eddy diffusivity (for temperature). Similarly, one can introduce eddy transport coefficients within (otherwise inviscid) 1D gas dynamics.

The advantages and limitations of such a formulation as a self-contained model of compressible turbulence are not of interest here. The purpose of a 1D compressible turbulence formulation in the present context is not to obtain a self-contained model, but rather, to obtain a submodel suitable for the proposed 3D AME framework.

Moreover, 1D gas dynamics with eddy transport does not in itself address the present need. Operationally, eddy transport is a diffusive process that smooths fluctuations rather than generating or sustaining them, contrary to the present goal of explicitly simulating small scale turbulent fluctuations.

In this regard, recall the discussion in Sect. 1.2 of the relationship between PDF evolution equations and SDEs. Turbulent transport, which is diffusive in an average sense (and is represented diffusively in PDF as well as in RANS turbulence models), can be represented in a fine-grained sense by SDEs. However, as noted in Sect. 1.2, if the SDEs are formulated in conformance to model-based PDF evolution equations rather than the exact governing equations, the resulting fine-grained representation may not be physical, and in fact, generally is not. (It is noted in passing that this caveat applies also to LGH. Although LGH solves equations that are exact at the continuum level, local particle fluctuations correspond in this instance to a postulated subcontinuum dynamics that is not intended to be an accurate representation of molecular fluctuation effects.)

In fact, there is a generic difficulty with the introduction of fine-grained structure using SDEs. SDEs are driven by noise fields that are difficult to constrain so that they obey global conservation laws, which require the constancy of spatial integrals over specified functions of the noise, the dependent variables, or both. For applications involving separation of length scales, such as the thermodynamic or hydrodynamic limit of statistical mechanics, this does not necessarily cause a problem. For example, if molecule numbers in a set of control volumes are allowed to fluctuate individually (e.g. Poisson shot noise reflecting local density fluctuations), the constancy of total molecule number in the whole system, which holds for a closed system with no chemical reactions, is not enforced. However, in the thermodynamic limit, the stochastic model describes a grand canonical ensemble that either converges to the behaviour of the physically correct canonical ensemble or can be used to infer properties of the true physical system.

There is no separation of length scales in turbulent flow and hence no freedom to deviate from global constraints, but there are ways to incorporate conservation constraints in particular cases. For example, an SDE for the stream function can be used to introduce velocity fluctuations while preserving continuity in 2D. (In 2D, velocity is the curl of the stream function, assuring that the flow remains solenoidal [27].) However, there is no obvious way to use SDEs to obtain a reasonable fine-grained 1D representation of turbulence that obeys applicable conservation laws. In this regard, the stochastically forced Burgers equation [28], though in many ways an illuminating 1D analogue of turbulence, is manifestly incapable of evolving fluid density in conformance to a specified equation of state.

Thus, the utility of SDEs for modelling the small scales of turbulence is not precluded, but an SDE formulation suitable for present purposes has not been identified. On physical grounds, there is an inherently more robust approach.

The compressible flows of interest here involve both solenoidal (divergence-free) and dilatational (curl-free) motions. The dilatational motions represented during 1D subcycling are governed by conventional 1D gas dynamics (Sect. 5.4). Dilatational motions not included within this representation are captured during solution coupling (Sect. 3.3). To be captured in a 1D formulation, solenoidal motions require special treatment, as follows.

Consider the advancement, for a time Δt, of the advective operator in the equation of motion for any property field $\theta(x, t)$, assuming numerical operator splitting so that other evolution processes (e.g. molecular transport) are omitted. This advancement is equivalent to a mapping $x \rightarrow x'(x)$ of each location x to a new location x'. The corresponding transformation of θ is $\theta(x) \rightarrow \theta(x'(x))$. This specifies the transformed θ field as a function of the coordinate x' at the new time by setting θ at new location x' equal to the θ value at the old time at the location x that is mapped to x' by the advancement operation.

This rather elaborate representation of advection, whose conventional representation is the $v \cdot \nabla$ operator, is introduced because the two are not equivalent, but rather, the former is a *generalization* of the latter. To see that the former includes the latter, integrate the Lagrangian advective equation $dx/dt = v(x, t)$ from t to $t + \Delta t$ to obtain the mapping $x \rightarrow x'(x)$ that is equivalent to $v \cdot \nabla$ advancement for a given $v(x, t)$ space–time history.

The advantage of the map representation of advection is that it can be used to formulate models that decompose the advection process into a sequence of discrete operations that advance property fields over any specified time interval, e.g. finite rather than infinitesimal. This decomposition can replicate physical advection exactly if the map is based on integration of the exact Lagrangian advective equation for given $v(x, t)$ over a finite time interval Δt. However, if $v(x, t)$ is not known a priori because fully resolved advancement of the exact 3D governing equations is unaffordable, then a map representation based on a postulated stochastic process can be used to model this advancement.

Stochastic iterated maps are in fact familiar tools of statistical mechanics modelling, including turbulence models [29, 30]. The noteworthy feature here is the application of the maps to the independent variable x rather than to the dependent variable θ. This approach allows incorporation of features of advection that are needed for physically sound flow simulation.

In VD (Sect. 1.2), the continuum process of vortical advection is spatially discretized but advanced in continuous time. The map representation of advection likewise enables discretization of a continuum process, in this case in the time domain. The specific map *ansatz* that is introduced is analogous to the individual vortex blob in VD in that it is applied to a finite spatial region and is intended to represent an elementary fluid motion ('eddy') in turbulence. However, a vortex blob can persist indefinitely (although some VD implementations allow blob merger) and execute any number of circulations, but each map is a one-shot event representing a particular displacement field ($x' - x$ as a function of x), e.g. one circulatory motion.

Map-based advection modelling, applied in 3D, yields novel mathematical insights as well as an efficient simulation method for a class of turbulent multiphase processes [31]. For present purposes, the key point is that map-based advection can be applied in 1D.

In 1D, the only solenoidal flow that can be generated by the $v \cdot \nabla$ operator is rigid translation. In map language, the solenoidal property can be stated as follows: $\int_{\sigma'} dx' = \int_{\sigma} dx$ for any subset σ of x, where σ' is the image of the subset σ obtained by the transformation $x \rightarrow x'$. (Henceforth, boldface is omitted in statements specialized to 1D, although in this and some other cases the validity of the statement is not restricted to 1D.) This is a statement of measure preservation by the map. It is more general than the usual solenoidal condition $\nabla \cdot v = 0$ because it encompasses a more general class of advection processes. The existence, within the map representation of advection, of nontrivial 1D motions that are measure preserving, and obey another essential property, is the key motivation for introducing map-based advection here (although it is likewise useful in 3D, as noted).

The other essential property is a particular form of continuity. It is different from adherence to the continuity equation, which reduces, for incompressible flow, to the solenoidal condition. Here, continuity refers to the relation

$$|x(x'_1) - x(x'_2)| \leq B\,|x'_1 - x'_2|,\tag{1}$$

where subscripts denote particular values of x', and B is a finite numerical constant; for the map *ansatz* adopted here (Sect. 4.2), $B = 3$. Equation (1) ensures that the map does not introduce spatial discontinuities into a continuous function, i.e. $h(x') \equiv g(x(x'))$ is continuous in x' if $g(x)$ is continuous in x.

It is important to enforce this form of continuity, not only because it is obeyed by the exact equations of motion (except for inviscid compressible flow, which is not considered here), but also because violations of this condition can introduce significant artefacts. Velocity discontinuities correspond to infinite local strain and thus, unphysically large local turbulence production. A possible anomaly resulting from species concentration discontinuities is noted in Sect. 2.5.

These artefacts can be remedied to some extent, but there is a more fundamental reason for enforcing (1). The coefficient B in (1) bounds the multiplicative decrease in separation that a map can induce between a pair of fluid elements (here meaning fluid states at particular points in space). Central to turbulent cascade phenomenology is the notion of locality of the turbulent cascade in scale space, i.e. individual fluid motions in turbulence (eddies) induce at most order-unity reduction of fluid-element separation [11]. Intermittency suggests deviations from this picture that can be interpreted within the present framework as locally large values of B. The mapping *ansatz* has been formulated in a way that accommodates this [32], but implementations to date conform to (1) with $B = 3$, and in one instance $B = 5$ [33].

Formally, a map represents a change of configuration corresponding to some time increment Δt. It would therefore appear that a time update should be associated with map implementation. However the formulation does not accommodate this for several reasons. First, maps are applied to finite spatial regions, representing the spatial extents of individual turbulent eddies within the 1D representation. In turbulence, many eddy motions are occurring at a given instant, implying multiple overlapping time increments, if a corresponding literal time advancement is triggered by each map. This leads to conceptual as well as computational difficulties. Second, the intent is to model all physical processes subsumed in the governing equations, not solely advection. There is no plausible way to time advance, e.g. diffusive transport, as a subcycling process within a map representation of an eddy because a map is inherently instantaneous. Hence, the finite time duration of an eddy motion cannot be represented operationally within the model.

The physics associated with eddy time scales is nevertheless contained in the model, albeit in an indirect way that does not fully capture turbulence phenomenology. Operationally, 1D advancement consists of conventional advancement of subprocesses other than solenoidal advection, punctuated by instantaneous maps representing the latter (with no associated time incrementation). This is equivalent to a sequence of initial-value problems, where the system state after a map is the initial state, which is advanced until the occurrence of the next map, which modifies the spatial structure of the dependent variables in some subregion, thereby establishing initial conditions for further time advancement. The statistics of the time intervals between maps in various size ranges are the model representation of the temporal character of eddy motions.

Thus, within the 1D advancement there is an operator splitting, reflecting the qualitatively different model representations of solenoidal advection, consisting of maps, and other subprocesses. Between each map and the next, the other subprocesses are subcycled; this might involve additional operator splitting based on numerical considerations. These other subprocesses correspond to 1D gas dynamics in the conventional sense, including dilatational flow aligned with the 1D domain (Sect. 5.4).

3.3 Advancement Cycle

A minimal description of the advancement cycle, omitting consideration of chemistry, output gathering and related issues, is presented. As noted in Sect. 3.1, three distinct coupled flow solutions are advanced concurrently. For a given dependent variable θ, e.g. density or a velocity component, its state at a given time t is specified, for a given solution (e.g. the solution labelled x), as $\{\theta_{jk}(x,t)\}$, where $0 \leq x \leq X$, $1 \leq j \leq Y$ and $1 \leq k \leq Z$. Here, x is any real number in the specified range of the continuum 1D domain. For the discrete finite-volume formulation based on wafers of width $1/m$ (in the units of Sect. 3.1), in which x corresponds to wafer centres, x takes the values $(2n-1)/(2m)$, where n is an integer in the range $1 \leq n \leq mX$. The integer indices j and k label the θ profiles in the corresponding domains $S_x(j,k)$.

For the various dependent variables θ, this prescription fully specifies the states of the three flow solutions at time t. Note that no variables associated with the coarse CVs C_{ijk} are needed to specify the solution states. This is the hallmark of an AME-type formulation.

Initial and boundary conditions are specified with reference to individual property profiles $\theta_{jk}(x,t)$. For illustration, Rayleigh convection, a suitable target case for initial model application (Sect. 7.1), is considered. This flow is generated by holding each boundary of the rectangular flow domain at fixed temperature so as to induce gravitational instability, e.g. taking the bottom boundary ($z = 0$, where z is the vertical coordinate) to be at a given temperature T_0, while the other boundaries are held at some common temperature $T_1 < T_0$. No-slip conditions are applied at all these boundaries.

The simulation is run until initial transients are relaxed, as measured, e.g. by velocity or temperature fluctuations at the centre of the enclosure.

Flow statistics are then gathered during the subsequent statistically steady advancement. Therefore initial conditions are irrelevant. A simple choice of initial conditions is uniform temperature T_1 and motionless fluid throughout the domain.

   Application of the boundary conditions is closely tied to the advancement cycle, which is now considered. Conceptually, though not necessarily in an efficient numerical implementation, the advancement cycle consists of two steps:

1. Subcycling, independently within all the domains $S_x(j, k)$, $S_y(i, k)$ and $S_z(i, j)$ of all three flow solutions, to advance the processes described in Sect. 3.2 from time t to time $t' = t + \Delta t$, where Δt is the advancement time step.
2. Property transfers across the boundaries of all the domains $S_x(j, k)$, $S_y(i, k)$, and $S_z(i, j)$ to enforce the equality of property fluxes across each CV face in the three flow solutions during the time interval Δt.

This advancement procedure is illustrated schematically in Fig. 4.

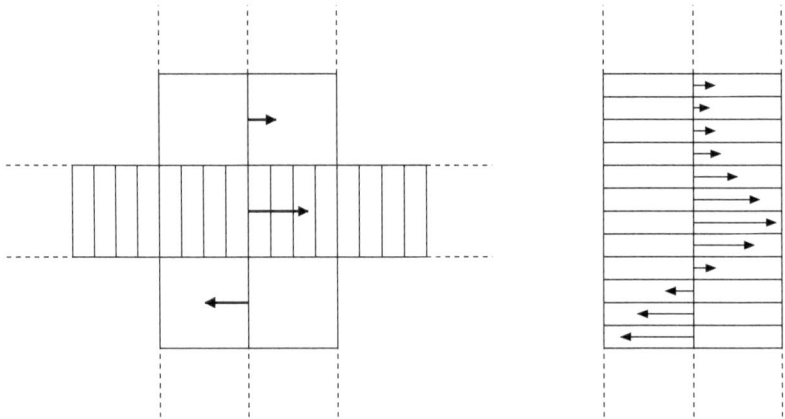

Fig. 4. Schematic illustration of the advancement procedure, here restricted to domain directions x (horizontal) and y (vertical). *Left*: A particular x-domain $S_x(j, k)$ is shown partitioned into wafers (here, $m = 4$). During step-1 subcycling within this domain, a property θ is fluxed across an interface on the common boundary of the two y-domains that are shown, denoted $S_y(i, k)$ and $S_y(i + 1, k)$ in the text. CV boundaries of these two domains are shown, but not their wafer substructures. The three arrows originate on the common boundary of the two y-domains. The *middle arrow* represents the θ flux due to step-1 processes on the x-domain that is shown. The arrows above and below it are fluxes generated by neighbouring x-domains (not shown). *Right*: The wafer substructures of the two y-domains are shown. The arrows are flux-conserving interpolants of the fluxes shown on the *left*. They prescribe θ transfers between wafer pairs across the wafer faces on which the arrows originate, thereby implementing lateral coupling of the two y-domains

The significance of step 2 is illustrated by considering the common face of CVs C_{ijk} and $C_{(i+1)jk}$. It is interior to $S_x(j,k)$ and on a lateral boundary of each of the domains $S_y(i,k)$, $S_y(i+1,k)$, $S_z(i,j)$ and $S_z(i+1,j)$. All three flow solutions require physically accurate property fluxes across this face during Δt. Step 1 induces property fluxes across surfaces interior to each 1D domain, but none across the lateral bounding surfaces of these domains, so these fluxes must be prescribed and implemented in some other way.

As the example illustrates, each CV face on a lateral boundary of a 1D domain is in the interior of a 1D domain of a different flow solution. Therefore properties are fluxed across that face during the step-1 advancement of the flow solution in which it is an interior face. These fluxes can be monitored during step-1 advancement of that flow solution.

The modelling assumption that closes the 3D formulation is that each property flux across a given CV face is the same in all three solutions. This implies, by Gauss' theorem, that all three solutions are the same at the mesh-filtered level unless property sources and sinks associated with 1D subcycling (step 1) are different within a CV for different solutions. This is possible in the present formulation, e.g. due to differing details of small scale mixing that affect chemical reaction rates locally. Mesh-filtered conserved properties evolve identically in the three solutions. (Here, filtering is a data-reduction technique rather than a part of the advancement algorithm.)

Thus, each property flux across a given CV face that is determined by one of the solutions during step 1, becomes a prescribed flux that, during step 2, governs transfer of the property across that face in the other two solutions. The step-2 transfers are between pairs of wafers in adjacent 1D domains, e.g. the 1D domain pair $S_y(i,k)$ and $S_y(i+1,k)$ in the example, and likewise, the 1D domain pair $S_z(i,j)$ and $S_z(i+1,j)$.

To specify the transfers in detail, consider the step-2 transfers of property θ across the common face of CVs C_{ijk} and $C_{(i+1)jk}$ in solution y. Let F be the θ flux across this face that is prescribed by solution x during step 1. Then for each integer n in the range $[1, m]$, the wafer values $\theta_{ik}(x,t)$ and $\theta_{(i+1)k}(x,t)$, where $x = i + (2n-1)/(2m)$, are incremented by $\pm f(x)\Delta t$, where $-$ and $+$ apply to the respective θ values and $f(x)$ is an interpolated flux constrained to obey $\sum_{n=1}^{m} f(x) = mF$. (Interpolants constrained in this manner have previously been used analogously [21, 34].)

The implementation of boundary conditions is essentially the same as the treatment of property fluxes at CV faces in the interior of the flow domain. Consider a CV face that is part of the flow boundary. The boundary condition at that face is applied during step-1 subcycling of the 1D domain that is bounded by that face and oriented normal to it. The flux of a given property θ through that face during step 1 is monitored, or it may have a known value specified by the boundary condition. During step 2, this flux value F is an imposed flux across that face for the other two 1D domains that are bounded by the face. The interpolant f is constructed and θ values are modified accordingly, analogous to flux implementation across faces interior to the flow domain.

3.4 Relationship to Conventional Methods

The advancement cycle outlined in Sect. 3.3 is applicable irrespective of the details of the 1D advancement (Sect. 3.2). The fluxes during step 1 reflect contributions by molecular and advective transport, where the solenoidal part of the advective contribution is due to fluid displacements by maps. Alternatively, the solenoidal part could be based on a postulated eddy diffusivity. This representation of solenoidal flow on the 1D domain would smooth rather than wrinkle property profiles, so 1D mesh refinement ($m \gg 1$) would become spurious. Nevertheless, this alternative indicates the formal analogy, as well as the key physical distinction, between the present framework and conventional LES of compressible flow. The distinction is the resolution and explicit evolution, rather than smoothing, of small scale processes in the present formulation.

Not only is the present formulation formally analogous to LES; it can be rendered equivalent to an LES through constraints on the implementation of maps. As explained in Sect. 5, the map sampling process generates a distribution of map sizes that generally conform to the eddy size distribution inferred from conventional turbulence phenomenology. For a given magnitude of property gradient, the map-induced flux depends primarily on the size-vs.-frequency distribution of maps.

To render the model formally equivalent to LES, one can deviate from this physically based prescription as follows. Characterize the overall magnitude of map-induced transport by an eddy diffusivity κ_e, which scales as ϕL^2, where ϕ and L are a representative frequency and size of the large eddies (i.e. the largest eddies implemented in 1D; see Sect. 6.2), which dominate transport. Assume that ϕ is increased and L is reduced so as to maintain constant κ_e, yielding many small eddies inducing the same transport as a smaller number of larger eddies. In the limit of diverging eddy frequency and vanishing L, the law of large numbers implies that fluctuations of property fluxes time averaged over the advancement step Δt vanish, so the stochastic model becomes deterministic. In this limit, the map sequence no longer induces physically relevant fine structure and its role is reduced to transport characterized by the diffusivity κ_e. Formally then, the model reduces to LES with an eddy-diffusivity closure, where the specific form of the closure depends, as in conventional LES, on how the dependence of κ_e on the flow state is specified. In this regard, the ODT eddy-selection process (Sect. 5.1) is closely analogous to conventional LES closure; see [35] for details.

Apart from its reduction to the physical modelling content of conventional LES, the present formulation requires an alternating-direction solution algorithm that differs from conventional LES numerics. Revision of the idealized advancement scheme of Sect. 3.3 can be anticipated as the algorithm is developed and tested.

This reduction to LES highlights the physical contribution of the map process when it has a realistic size-vs.-frequency distribution. As the scale L is dialed up from zero, fluctuations and associated fine-scale structure are introduced, but these properties do not necessarily enhance the realism of the model if the map distribution is not physically accurate. To benefit from this departure from an LES formulation, the induced fluctuations must be sufficiently accurate to provide a gain in model fidelity that justifies the computational cost of the method. This is best judged from the performance of the model. The particular formulation outlined here has not yet been implemented, but it is closely analogous to existing formulations, which are now considered in further detail in order to highlight the modelling concept and to assess how it might perform within the formulation proposed here.

4 Map-Based Advection Models

4.1 1D Models of Turbulent Premixed Combustion

Efforts by the author and co-workers to develop map-based methods for turbulent flow simulation in one or more spatial dimensions are summarized. The intent is to indicate the variety of possible formulations and the physics that is captured and omitted in particular instances.

The starting point was an effort to develop a minimal model of turbulent premixed combustion. The initial outcome was a formulation in which the instantaneous state of a turbulent flame is idealized as a bit vector (row of integers 0 or 1) in which each pair of adjacent bits interacts in two ways.

First, each 0 is converted into a 1 at a mean rate B times the number (0, 1, or 2) of adjacent bits in state 1. This process represents laminar burning with laminar flame speed BL, where L is the nominal spatial separation of adjacent bits. Note that there is some subtlety even at this level of description. The middle bit in a 101 configuration is deemed to burn twice as fast as in a 100 or 001 configuration because flames consume it from both sides, which is a reasonable but not uniquely plausible idealization of flame propagation. Also, this is a random process but could be plausibly formulated as a deterministic process.

Second, each pair of adjacent bits is exchanged (e.g. 01 to 10, 10 to 01, 00 and 11 unaffected) at a mean rate R, thus idealizing turbulent advection with eddy diffusivity RL^2. (Note that bits execute simple random walks with event rate $2R$.) Like laminar burning, this process is random in time, namely a Poisson process with mean event rate R for each bit pair. Model dynamics are governed by one nondimensional parameter, $\gamma = R/B$, which can be viewed as an idealization of the quantity u'/S that governs 3D turbulent combustion, where u' is the root-mean-square turbulent velocity fluctuation and S is the laminar flame speed. $1/B$ times the mean rate of 0-to-1 conversions is then the model analogue of u_T/S, where u_T is the turbulent burning velocity.

For a step-function initial bit profile, this process relaxes to statistically steady propagation that captures some qualitative features of turbulent premixed combustion [36]. It has been shown that the model analogue of u_T is governed by the Kolmogorov–Petrovsky–Piscounov velocity-selection principle in the large-γ limit [37]. To improve the physical realism of this formulation, it was extended by allowing exchanges of the positions of nonadjacent bit pairs, idealizing the effects of turbulent eddies of various sizes [38].

4.2 Linear-Eddy Model

Though bit-pair exchange over a range of bit separations reflects the range of eddy motions in turbulence, it does not reflect the coherence of eddy motions, meaning that a large eddy displaces a larger volume of fluid in a given direction than does a small eddy. Accordingly, an exchange process denoted block inversion was introduced, involving the reversal of the order of bits j through $j + l - 1$ to represent a size-l eddy [39]. This change was necessitated by the application of the 1D approach to diffusive scalar mixing rather than flame propagation; bit-pair exchange gives far too rapid length-scale reduction in this context. This artefact occurs also for flame propagation, but is less severe in that context because u_T is more sensitive to the distance and frequency of the largest bit displacements than to the amount of fluid transported.

Block inversion introduces scalar discontinuities at eddy endpoints. From a spectral viewpoint, this corresponds to transfer of scalar fluctuations from finite wave-number k to $k = \infty$, violating the spectral locality of length-scale reduction that is a hallmark of the inertial-range turbulent cascade [11].

To remedy this artefact, the scalar-mixing formulation, denoted the 'linear-eddy model' (LEM), was improved by introducing a new exchange process, termed the triplet map [40]. This is not a pair exchange, but rather, a permutation of cell indices j through $j + l - 1$. Taking the map range l to be a multiple of 3, the triplet map, illustrated in Fig. 5, permutes the cell indices

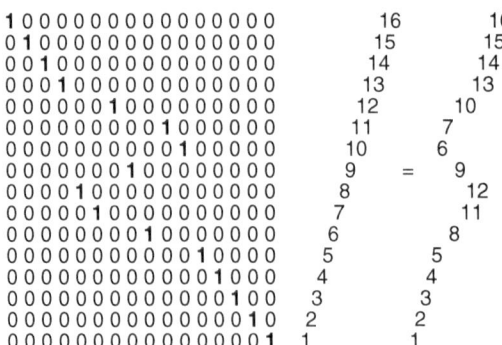

Fig. 5. Application of a triplet map, with $l = 9$, to a 16-element column vector with vertically increasing cell indices. For clarity, unity matrix elements are boldface and cells are shifted horizontally in proportion to their index values. The shifts are intended to suggest the 1D profile of the mapped variable

into the new order j, $j+3$, $j+6$, ... , $j+l-3$, $j+l-2$, $j+l-5$, $j+l-8$, ... , $j+4$, $j+1$, $j+2$, $j+5$, $j+8$, ... , $j+l-4$, $j+l-1$. This operation reduces the separation of any pair of cells by no more than a factor of three, thus satisfying the scale locality of length-scale reduction. It is the simplest of a family of permutations that preserve scale locality, and is optimal in that no other member of the family enforces a smaller bound B (Sect. 3.2) on the maximum scale-reduction factor. Because it is a permutation of equal-sized cells, the triplet map is measure preserving in the sense defined in Sect. 3.2.

LEM is parameterized by a Péclet number Pe, which is the eddy diffusivity associated with transport by the triplet-map sequence divided by the molecular diffusivity. On this basis, LEM has been used to study the dependencies of turbulent mixing and reaction processes on Pe and on the initial and boundary conditions imposed on one or more scalar profiles that evolve on the 1D domain [33, 41–46].

4.3 One-Dimensional Turbulence

LEM simulates mixing induced by parametrically specified turbulent advection. To obtain a model that, instead, predicts turbulent flow evolution, profiles of one or more velocity components were introduced on the 1D domain, and the random selection of individual eddies (here parameterized by j, l, and time of eddy occurrence) was generalized [32]. In LEM, the eddy rate is a prescribed function of l, reflecting the known inertial-range frequency-vs.-wavenumber scaling [11], and also depends on j if the flow is spatially inhomogeneous. In the predictive flow model, denoted 'one-dimensional turbulence' (ODT), the sampling rate for each eddy (parameterized by j and l) is a function of the instantaneous flow state, based on turbulence production and dissipation mechanisms that are conventionally used to estimate eddy time scales [47]. A key distinction here is that conventional estimation based on mixing-length phenomenology is typically applied to quantities subject to some form of averaging or filtering, but in ODT, mixing-length phenomenology is applied to instantaneous property profiles that are not subject to averaging or filtering.

In ODT, the key molecular process that evolves concurrently with eddy events (i.e. the analogue of laminar flame propagation in premixed combustion and molecular diffusivity in LEM) is molecular viscosity, as prescribed by the viscous-transport term of the momentum equation. The corresponding nondimensional parameter that governs constant-property flow evolution in ODT is a Reynolds number. In ODT, as in 3D flow simulation, the nominal Reynolds number is defined in terms of domain geometry and flow initial and boundary conditions, but the turbulent Reynolds number, defined in terms of u', the mean energy dissipation rate, and the kinematic viscosity, is an outcome of simulated flow evolution rather than an input.

Velocity profiles in incompressible ODT do not advect fluid (see Sect. 5.4 for discussion of compressible ODT), but they influence triplet-map advection through their role in determining eddy-sampling rates. In this sense they are auxiliary variables, but in addition, they are the flow observables. The tight two-way coupling between velocity-profile evolution and eddies (triplet maps advect velocity profiles) maintains overall consistency of velocity statistics and map-induced transport.

Buoyancy effects have been incorporated into ODT, and buoyant strati-fied flows have been studied extensively [32, 47–51]. In fact, buoyancy alone (velocity profiles omitted) is a sufficient input to eddy rate determination to provide a reasonable representation of some flows of interest (including the flow considered in Sect. 7.1), motivating a simplification of ODT that is termed 'density-profile evolution' (DPE) [32, 48]. ODT has also been used to study free shear flow [52–54], confined flow (Sect. 7.2) and combustion [55–57].

4.4 Higher-Dimensional Map-Based Methods

The triplet map generalizes straightforwardly to higher spatial dimensions. This generalization is found to be useful both theoretically and computa-tionally [31]. The relaxation of advective time-stepping constraints, and the option of a mesh-free Lagrangian algorithm (based on the spatial continuum definition of the triplet map, see Sect. 5), offer substantial computational ad-vantages even in 3D.

In higher dimensions, it is possible to define a deterministic map-based ad-vection protocol that is a useful representation of turbulence in some contexts. One such formulation, 'deterministic turbulent mixing' (DTM), has been used to study flame-front geometry in turbulent premixed combustion [58].

5 ODT Formulation of Substructure Advancement

5.1 Boussinesq Formulation

To date, compressible gas dynamics has not been incorporated into ODT. The existing formulation that has the features closest to those needed in a compressible formulation is one that is based on the general variable-density conservation equations (i.e. not specialized to small density fluctuations) [54]. This formulation involves mathematical intricacy that obscures the underly-ing modelling concepts, and a compressible formulation will be even more obscure in this regard. Therefore the formulation outlined here is based on the Boussinesq approximation [59], in which density variations are deemed negligible except in the gravitational forcing term. Gravity is included here both to illustrate the treatment of a dynamically active scalar property (here, density) and because the initial target application of the proposed ODTLES formulation is a buoyancy-driven flow. This formulation is roughly analogous to the ODT formulation in [47].

A mathematical statement of this illustrative formulation is presented. In Sect. 4.2, a spatially discrete definition of the triplet map was given. Henceforth, space and time variables are continuous unless stated otherwise, and the triplet map is defined on the spatial continuum.

The ODT formulation utilized here simulates the time evolution of velocity components u, v and w and density ρ defined on a 1D domain representing the vertical (z) coordinate. This evolution involves two processes: (1) a sequence of eddy events, which are instantaneous transformations that represent turbulent stirring and (2) intervening time advancement of conventional form. Each eddy event may be interpreted as the model analogue of an individual turbulent eddy. The location, length scale, and frequency of eddy events are governed by a stochastic process.

During the time interval between each eddy event and its successor, the time evolution of property profiles is governed by the equations

$$\left(\partial_t - \nu\partial_z^2\right) u(z,t) = 0 \tag{2}$$

$$\left(\partial_t - \nu\partial_z^2\right) v(z,t) = 0 \tag{3}$$

$$\left(\partial_t - \nu\partial_z^2\right) w(z,t) = 0 \tag{4}$$

$$\left(\partial_t - \gamma\partial_z^2\right) \rho(z,t) = 0. \tag{5}$$

Here ν is viscosity and γ is diffusivity of the scalar, temperature, that controls the density. For simulation of Rayleigh convection, discussed in Sect. 7.1, these equations are solved on a vertical domain $[0, H]$, where H is the height of the convection cell. Boundary conditions applied to the velocity at $z = 0$ and H are $u = v = w = 0$. Density boundary conditions are $\rho(0,t) = \rho_1$ and $\rho(H,t) = \rho_2$, where $\rho_2 > \rho_1$ to enforce unstable stratification, which drives the flow.

Each eddy event consists of two mathematical operations. One is a triplet map representing the fluid displacements associated with a notional turbulent eddy. The other is a modification of the velocity profiles in order to implement pressure-induced energy redistribution among velocity components and net kinetic-energy gain or loss due to equal-and-opposite changes of the gravitational potential energy. These operations are represented symbolically as

$$\begin{aligned} \rho(z) &\rightarrow & \rho(M(z)) \\ u(z) &\rightarrow & u(M(z)) + c_u K(z) \\ v(z) &\rightarrow & v(M(z)) + c_v K(z) \\ w(z) &\rightarrow & w(M(z)) + c_w K(z). \end{aligned} \tag{6}$$

According to this prescription, fluid at location $M(z)$ is moved to location z by the mapping operation, thus defining the map in terms of its inverse $M(z)$. This mapping is applied to all fluid properties. The additive term $c_s K(z)$, where $s = u$, v, or w, affects only the velocity components. It implements the aforementioned kinetic-energy changes. Potential-energy change is inherent in the mapping-induced vertical redistribution of the ρ profile; see (10).

In the spatial continuum, the triplet map is defined as

$$M(z) \equiv z_0 + \begin{cases} 3(z - z_0) & \text{if } z_0 \le z \le z_0 + \frac{1}{3}l, \\ 2l - 3(z - z_0) & \text{if } z_0 + \frac{1}{3}l \le z \le z_0 + \frac{2}{3}l, \\ 3(z - z_0) - 2l & \text{if } z_0 + \frac{2}{3}l \le z \le z_0 + l, \\ z - z_0 & \text{otherwise.} \end{cases} \tag{7}$$

This mapping takes a line segment $[z_0, z_0+l]$, shrinks it to a third of its original length, and then places three copies on the original domain. The middle copy is reversed, which maintains the continuity of advected fields and introduces the rotational folding effect of turbulent eddy motion. Property fields outside the size-l segment are unaffected.

The spatially discrete numerical implementation of the triplet map, illustrated in Fig. 5, transparently obeys conservation properties because it is implemented as a permutation of equal-sized cells, as noted in Sect. 4.2. In the continuum limit that is approached by increasing the spatial refinement, the map definition (7) is recovered. Figure 5 reflects key features of the continuum definition, notably the increase of property gradients in the compressed copies and gradient reversal in the middle copy.

In (7), the parameters z_0 and l are the continuum analogues of the integer quantities j and l in the discrete definition of the triplet map. Here, z_0 specifies the location, and l the size, of the eddy event.

In (6), K is a kernel function that is defined as $K(z) = z - M(z)$, i.e. its value is equal to the distance the local fluid element is displaced. It is nonzero only within the eddy interval, and it integrates to zero so that the process does not change the total (z-integrated) momentum of individual velocity components. It provides a mechanism for energy redistribution among velocity components, enabling the model to simulate the tendency of turbulent eddies to drive the flow towards isotropy, constrained by the requirement of total (kinetic plus potential) energy conservation during the eddy event (which is nondissipative).

To quantify these features of eddy energetics, and thereby specify the coefficients c_s in (6), it is convenient to introduce the quantities

$$s_K \equiv \frac{1}{l^2} \int s(M(z))K(z)\,\mathrm{d}z, \tag{8}$$

where $s = u, v, w,$ or ρ. Substitution of the definition of $K(z)$ into (8) yields

$$s_K = \frac{1}{l^2} \int [zs(M(z)) - M(z)s(M(z))]\,\mathrm{d}z = \frac{1}{l^2} \int [s(M(z)) - s(z)]z\,\mathrm{d}z. \tag{9}$$

Because $M(z)$ is a measure-preserving map of the z domain onto itself, the domain integral of any function of $M(z)$ is equal to the domain integral of the same function with argument z. This allows the substitutions of z for $M(z)$

that yield the final result in (9). For $s = \rho$, this expression is proportional to the potential-energy change induced by the triplet map. The energy change Δ caused by an eddy event can then be expressed as

$$\Delta = \rho_0 l^2 \left(c_u u_K + c_v v_K + c_w w_K \right) + \frac{2}{27} \rho_0 l^3 \left(c_u^2 + c_v^2 + c_w^2 \right) + g l^2 \rho_K, \quad (10)$$

where a reference density ρ_0 (defined here as mass per unit height, based on a nominal column cross-section) is introduced (i.e. the standard Boussinesq prescription), as well as the gravitational acceleration g.

The representation of both the potential and kinetic energy contributions in (10) using (8) is a consequence of the definition chosen for K. Based on this definition, another equivalent form of (8),

$$s_K \equiv \frac{4}{9l^2} \int_{z_0}^{z_0+l} s(z)[l - 2(z - z_0)] \, \mathrm{d}z, \quad (11)$$

which is useful for numerical implementation, is readily obtained.

Overall energy conservation requires $\Delta = 0$. Two additional conditions are required to specify the coefficients c_s. These are based on a representation of the tendency for eddies to induce isotropy. For this purpose, it is noted that there is a maximum amount $Q_s = (27/8)\rho_0 l s_K^2$ of kinetic energy that can be extracted from a given velocity component s during an eddy event [53]. (The amount of energy actually extracted or deposited depends on c_s.) Q_s is thus the 'available energy' in component s prior to event implementation. The tendency towards isotropy is introduced by requiring the available energies of the three velocity components to be equal upon completion of the eddy event. This provides the additional needed conditions and yields the following expression determining c_s:

$$c_s = \frac{27}{4l} \left[-s_K \pm \sqrt{\frac{1}{3} \left(u_K^2 + v_K^2 + w_K^2 - \frac{8gl}{27} \frac{\rho_K}{\rho_0} \right)} \right]. \quad (12)$$

The physical criterion that resolves the sign ambiguity is explained in [53]. Note that the last term in (12) is the square root of a quantity proportional to the net available energy $Q_u + Q_v + Q_w - P$, where the quantities Q_s are the component available energies prior to event implementation and P is the gravitational potential energy change caused by triplet-mapping of the ρ profile, requiring equal-and-opposite change of available energy during eddy implementation, as enforced by the condition $\Delta = 0$. If P is positive (stable stratification) and larger than the available energy, then the eddy is energetically prohibited. In this case, the argument of the square root in (12) is negative and the eddy event is not implemented (see below).

Although the formulation of an individual eddy event incorporates several important features of turbulent eddies, the key to the overall performance of the model is the procedure for determining the sequence of eddy events during a simulated flow realization. The expected number of eddies occurring during a time interval dt, whose parameter values are within dz of z_0 and within dl of l, is denoted the 'eddy rate distribution' $\lambda(z_0, l; t) \, dz_0 \, dl \, dt$, where λ has units of (length$^2 \times$time)$^{-1}$. Eddies are randomly sampled from this distribution. Mathematically, this generates a marked Poisson process [60] whose mean rate as a function of the 'mark' (parameter) values z_0 and l varies with time. The physical content of the eddy selection process is embodied in the expression for λ that is adopted,

$$\lambda = \frac{C\nu}{l^4} \sqrt{\left(\frac{u_K l}{\nu}\right)^2 + \left(\frac{v_K l}{\nu}\right)^2 + \left(\frac{w_K l}{\nu}\right)^2 - \frac{8gl^3}{27\nu^2}\frac{\rho_K}{\rho_0}} - Z. \qquad (13)$$

This expression involves two free parameters, C and Z, whose roles are explained in Sect. 6.2. λ is set equal to zero if the argument of the square root is negative, indicating an energetically prohibited event; see the discussion of (12).

For $Z = 0$, the argument of the square root is a scaled form of the net available energy. Thus, for given z_0 and l, (13) with $Z = 0$ is simply the dimensionally consistent relation between the net available energy and the length and time scales of eddy motion, where the associated time scale is the inverse of the (appropriately normalized) eddy rate λ. Thus, (13) may be viewed as a representation of mixing-length phenomenology within the ODT framework. This phenomenology is the basis of many turbulence modelling approaches. In particular, it is central to LES closures based on eddy viscosity, hence the analogy between conventional LES and the proposed ODTLES methodology (Sect. 3.4). However, the present approach, which does not involve averaging, differs from the typical use of mixing-length concepts to close averaged equations in several respects:

1. Rather than assigning a unique l value at each spatial location, ODT allows eddies of all sizes throughout the spatial domain, with their relative frequencies of occurrence at different locations specified by (13).
2. Quantities on the right-hand side of (13) depend on the instantaneous flow state rather than an average state, so eddy occurrences are responsive to unsteadiness resulting from transient forcing or statistical fluctuations inherent in the eddy-sampling process.
3. Eddy occurrences thus depend on the effects of prior eddies and affect future eddy occurrences. These dependencies induce spatio-temporal correlations among eddy events, leading to a physically based representation of turbulence intermittency.

These attributes of ODT are the basis of its detailed representation of turbulent cascade dynamics coupled to boundary conditions, shear and buoyant forcing, etc. In particular, the stochastic variability of simulated ODT realizations arises from a physically based representation of turbulent eddy statistics, and thus enables a conceptually sound and mathematically consistent assessment of the effects of stochastic variability on the variability of, and correlations among, output statistics.

If two of the three velocity components are removed from the model, (13) reduces to the eddy rate distribution used in [49]. If the buoyancy term is omitted, (13) resembles the expression for λ that appears in [53], except that here, λ is based on the total available energy (including contributions from all three velocity components) rather than the available energy associated with vertical motion. Use of the total available energy is advantageous here because it gives the correct critical Richardson number, $Ri_c = 1/4$ [61], for the onset of instability (in the present context, eddy events). Another distinction from [53] is that the procedure that was used previously to suppress occasional unphysically large eddy events is omitted here. For ODTLES implementation, a bound on eddy sizes follows from consideration, in Sect. 6.2, of the complementary roles of steps 1 and 2 of the advancement cycle.

5.2 Numerical Implementation of Eddy Sampling

The unsteadiness of the rate distribution λ suggests the need to reconstruct this distribution continually as the flow state evolves. This prohibitively costly procedure is avoided by an application of the rejection method [62], involving eddy sampling based on an arbitrary sampling distribution that is designed to over-sample all eddies. True rates are computed only for sampled eddies, and are used to determine eddy acceptance probabilities. The resulting procedure adequately approximates the desired sampling from λ [32], and is exact in the limit of infinite over-sampling. The choice of the arbitrary sampling distribution affects the efficiency of the sampling procedure, but not the statistics of the eddies that are selected for implementation.

This implies modification of the split-operator cycling during 1D advancement that is outlined at the end of Sect. 3.2. Denoting the arbitrary joint PDF used to sample z_0 and l values as $h(z_0, l)$, and choosing a sufficiently small eddy-sampling time-step Δt_s, the advancement cycle during step 1 of the overall advancement (Sect. 3.3) is

1. Advance the concurrent processes such as viscous transport (Sect. 3.3) for a time interval Δt_s.
2. Sample z_0 and l values from $h(z_0, l)$.
3. For these values, compute $\lambda(z_0, l)$ based on the current flow state.
4. Compute the ratio P of the rate $\lambda(z_0, l)$ of occurrence of an eddy with these z_0 and l values as given by the model to the rate $h(z_0, l)/\Delta t_s$ resulting from the sampling procedure.
5. Implement the selected eddy with probability P based on a Bernoulli trial, i.e. implement the eddy if $P = \lambda(z_0, l)\Delta t_s/h(z_0, l)$ is larger than a random variable sampled from the uniform distribution over $[0, 1]$.

Δt_s must be assigned a value small enough so that P never exceeds unity. For numerical accuracy, $P \ll 1$ should be obeyed with at most rare exceptions.

For an evolving flow, it is efficient to adjust Δt_s during advancement in order to direct the P values towards a target range, typically of order 0.01.

There is a theoretical advantage to sampling Δt_s values so that eddy-sampling occurrences obey Poisson statistics. The rejection procedure then corresponds to the thinning algorithm for generating nonstationary Poisson processes [63]. This approach has been used in recent ODT simulations.

5.3 Planar Free-Shear Flows

The ODT representation of a time-developing Kelvin–Helmholtz instability, illustrated in Fig. 6, indicates some of the flow features captured by the model. This illustration is based on the ODT formulation of [53], which includes the large-eddy suppression procedure that is needed for stand-alone ODT simulation of unbounded flows (Sect. 5.1).

The rendering shows that the width of the active mixing zone grows primarily by the relatively infrequent occurrence of a large event extending beyond the current range of the mixing zone, with some additional contribution by the more numerous small events. This process is consistent with the dominant role of large engulfing motions and the secondary role of small-scale nibbling in turbulent entraining flows under neutral buoyancy conditions. (The effect of density stratification on the ODT representation of turbulent entrainment has been investigated [32, 54].)

Bunching of events, especially after the occurrence of a large event, reflects the interactions between the eddy events and the evolving velocity profile that induce the model analogue of the turbulent cascade. Each eddy event

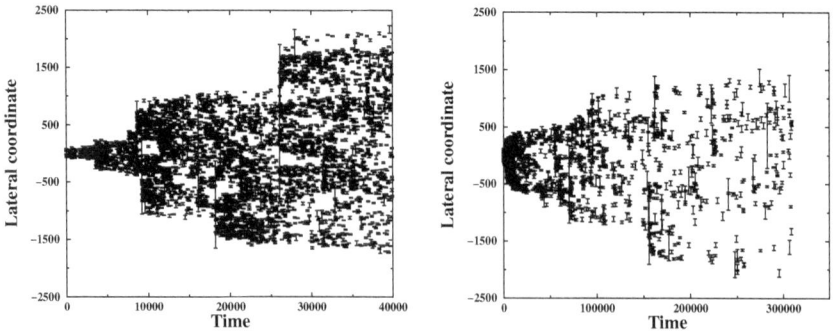

Fig. 6. Graphical representation of the sequence of eddy events during a simulated ODT realization of a time-developing Kelvin–Helmholtz instability (*left panel*) and a time-developing planar wake (*right panel*) [53]. The Kelvin–Helmholtz and wake simulations are initialized using step-function and top-hat initial velocity profiles, respectively. The space and time units in this illustration are arbitrary. In the plots, each eddy is represented by an error bar whose vertical span corresponds to the eddy range $[z_0, z_0 + l]$, and whose horizontal location corresponds to the time of eddy occurrence

compresses and folds the velocity profile within the range of the event. This increases the local shear that contributes to mechanical turbulence production in the relation (13) governing the frequency of subsequent events within that range. (In stratified flows, the buoyant production is also affected.) A feedback process is thus induced that promotes the occurrence of successively smaller events. Eventually, velocity fluctuation length scales are reduced sufficiently so that damping of the fluctuations by concurrent viscous transport dominates the production of fluctuations by eddy events. Viscous damping thus terminates the local burst of eddy activity.

A planar wake simulation is also shown in Fig. 6. In the Kelvin–Helmholtz simulation, vigorous turbulence, indicated by the number and size range of eddies as the flow evolves, is sustained by the shear imposed on the flow by the free-stream conditions (far-field velocity difference). The wake, however, evolves in a uniform background. As the initial velocity perturbation is dispersed by eddies and dissipated by concurrent viscous evolution, the turbulence intensity decreases, affecting the eddy frequency and size range and slowing the growth of the turbulent zone. These qualitative impressions are supported by the quantitative consistency of ODT simulation statistics with the known similarity scalings for these flows [53].

5.4 Proposed Compressible Formulation

The main modifications of the Boussinesq ODT formulation that are required in order to incorporate compressible gas dynamics are:

1. During 1D advancement, the w velocity now advects all properties, so introduce a $w\partial_z$ term on the left-hand side of (2–4) and in the additional evolution equations mentioned below (item 4).
2. Adopt the general variable-density formulation of [54], which generalizes the momentum equations (2–4) and the energy-redistribution step during eddy implementation.
3. In the w equation, introduce a pressure-gradient $(\mathrm{d}p/\mathrm{d}z)$ term.
4. Introduce 1D continuity and energy equations, and an equation of state (e.g. ideal gas) that determines the pressure locally (in each wafer) from density, temperature and composition (for multi-species mixtures).
5. Generalize the potential-energy contribution in (13) to reflect the equivalence of the gravitational body force and $\mathrm{d}w/\mathrm{d}t$ acceleration.
6. Generalize the viscous stress terms in (2–4) to compressible form, introducing (manageable) complications that are not elaborated here.

This scheme introduces acoustic time scales, which are very short relative to other time scales at low Ma. As in other low-Ma compressible simulations, a pseudo-compressible scheme based on an artificially low sound speed [1] can improve the efficiency of this formulation at low Ma.

A poorly understood feature of compressible turbulence is the coupling of acoustic and vortical motions. Because w is an advecting velocity rather than an auxiliary variable in the proposed compressible formulation, the triplet map introduces such a coupling. For example, if the property profile in Fig. 5 is taken to be the w profile, then in this instance the map, representing vortical motion, converts a pure expansion into an alternating expansion–compression–expansion, i.e. an acoustic source. Given the limited state of understanding of vortical-acoustic coupling in compressible turbulence, it is difficult to ascertain whether this is a good representation of this coupling.

This question is best addressed by implementing the proposed formulation and evaluating its predictive capabilities. Unlike incompressible ODT (Sect. 5.1), this compressible formulation is not intended for use as a standalone model. Its incorporation into 3D ODTLES is now considered.

6 ODTLES

6.1 Features

For implementation within ODTLES, an important feature of the formulation of Sect. 5.4 is the distinguished role of w in the z solution (and likewise of u and v in the x and y solutions respectively). Here it is convenient to introduce the alternate notation $v_{j,k}$ denoting velocity component j in solution k. The component $v_{i,i}$ now advects properties in solution i, but for $j \neq i$, components $v_{j,i}$ are auxiliary variables in solution i, as in Sect. 5.1. In addition to their usual role in determining the eddy sampling rate, these components, like all other flow properties, are fluxed through CV faces, thus prescribing inter-domain transfers of these velocity components in the $j \neq i$ flow solutions (step 2 of the advancement cycle; see Sect. 3.3). In this manner they influence the evolution of the advecting components $v_{j,j}$ in the $j \neq i$ solutions.

This highlights the multi-faceted relationships among the three velocity components, the three flow solutions, and the two steps of the advancement cycle. A related consideration is the manner in which the simulated evolution communicates pressure effects in 3D. If the pressure is locally high in one of the solutions, it is likely to be high in the same vicinity in the other solutions because they are all subject to the same fluxes at CV face locations (enforced by advancement step 2). Then step-1 (1D compressible) advancement in each solution will generate flow directed away from this vicinity in one coordinate direction. Step 2 will then communicate these outward-directed flows among the solutions so as to yield an approximate representation of radial outflow from the high-pressure region in all the solutions.

Step 2 transfers properties over a distance unity, rather than the 1D resolution scale $1/m$, using first differences. This is a diffusive representation of fluxes that are primarily advective, and therefore induces numerical dissipation of kinetic energy. Conservation of total energy, which is obeyed exactly, implies conversion of the lost kinetic energy into heat. For conventional eddy-diffusivity closure, incorporation of a sub-grid energy evolution equation can recast this dissipation as conversion of mesh-resolved kinetic energy into subgrid kinetic energy, with subsequent conversion to heat by viscous dissipation. The analogous mechanism in ODT is the use of the kernel operation to deposit the numerically dissipated energy into the velocity fields. Using the method of [54], this can be done in conformance with momentum conservation for variable-density flows. To distinguish numerically dissipated energy from true viscous dissipation, separate accounting of advective and viscous fluxes across CV faces is needed. This can be done by straightforward generalization of the procedure used in [53] to variable-density flow.

Because step 2 involves diffusive representation of advective transport, it is subject to some of the same limitations as parameterizations in which the representation of advective transport below 3D mesh-resolved scales is solely diffusive. Nevertheless, owing to the 1D subcycling during step 1, salient characteristics of the small scale flow structure are preserved, as has been demonstrated using the formulation described in Sect. 6.3 [21]. Splicing is a different method for implementing step-2 property transfers that is not diffusive in character, but is subject to other limitations, as noted in Sect. 2.5.

Because step 2 of the advancement cycle applies fluxes to a given flow solution that are interpolants of fluxes in a different flow solution, it is mathematically possible to violate realizability. Namely, it is possible to flux more of a non-negative quantity such as mass out of a wafer than it contains. This establishes a CFL-type constraint on the time step Δt. The allowed magnitude of Δt is of the same order for ODT closure as for eddy-diffusivity closure in which there is no spatial refinement below the CV scale, though for ODT the constraint is slightly more restrictive due to stochastic variability. Use of a small value of Δt incurs no significant cost penalty because the 1D subcycling using smaller time steps is the most costly part of the computation.

As noted in Sect. 3.3, Gauss' theorem constrains the evolution of the three distinct solutions. For each flow solution, it implies that the change of the CV integral of a conserved property during one advancement cycle is equal to the sum of the transfers of those properties through CV faces. The flux across each face is operationally defined as the corresponding face transfer divided by the time step Δt. (Recall that the face area is unity in scaled units.) This identity motivates a definition of mesh-scale output statistics that conserves the property exactly. Consider total CV mass, denoted $\bar{\rho}$ because the scaled volume is unity. Gauss's theorem implies that the quantities $\bar{\rho}$ and the face fluxes of mass, i.e. momenta $\langle \rho v \rangle_{\text{face}}$ (where v is the face-normal velocity), form a conservative set of output variables. An additional assumption or definition is needed to define the mesh-scale face velocity V. A natural choice is $\langle \rho v \rangle_{\text{face}} = \langle \rho \rangle V$, where $\langle \rho \rangle$ is an interpolant of the $\bar{\rho}$ array evaluated at the face centre at the midpoint $\Delta t/2$ of the advancement cycle (corresponding to the midpoint of the time integration that determines mass transfers across faces).

An output protocol of this sort is needed because the 3D conservation laws are applicable only to CVs and only with reference to state changes from the beginning to the end of the advancement cycle. Analogous considerations arise in conventional advancement schemes.

Discrete and continuum definitions of the triplet map are provided in Sects. 4.2 and 5.1 respectively, where the former is applicable in the Eulerian uniform-mesh (Sect. 3.1) implementation described thus far. In ongoing work, an alternative Lagrangian-mesh ODT implementation has been developed in which wafer faces are advected by the w velocity (if the flow is compressible) and by triplet maps. Here, the continuum map definition is used, resulting in tripling of the number of wafer faces within the mapped interval. A mesh-management scheme is used to suppress the excessive proliferation of wafers. This formulation will be particularly advantageous for wall-bounded flows in which high spatial resolution is needed only in near-wall regions, as in the applications discussed in Sect. 7.

6.2 Parameter Assignment

The model parameters C and Z are introduced in the formulation of eddy sampling in Sect. 5.1. C scales the eddy event rate, and hence the simulated turbulence intensity, for a given flow configuration. The role of Z is to impose a threshold eddy Reynolds number that must be exceeded to allow eddy occurrence [47]. In near-wall flow, the transition from the viscous layer to the buffer layer is sensitive to this threshold and hence to Z [64]. For $Z > 0$, eddies are suppressed entirely when local values of the eddy Reynolds number are sufficiently small. The circumstances under which this occurs in ODTLES are considered. This question is closely tied to the upper bound on the range of allowed eddy sizes l.

As noted in Sect. 6.1, the 3D character of the flow is captured above the CV scale by step 2 of the advancement cycle. Therefore it would be redundant to allow l values greatly exceeding unity. Likewise, the bound on l should not be much less than unity, because this would omit representation of eddies larger than the bound but smaller than unity. The signature of either of these artefacts would be apparent in the 1D energy spectrum, which can be extracted from ODT simulations [21, 32, 35]. Examination of energy spectra from simulations of representative flows therefore allows empirical determination of a bound on l.

As noted, the bound will be of order unity. Therefore the largest Reynolds number of a 1D eddy event that will occur is of the order of the Reynolds number of the largest eddy that is not resolved at the CV scale. If the mesh is increasingly refined (decreasing CV size in physical units) for a given flow configuration, then the Reynolds number of the largest unresolved eddy decreases until it is below the threshold value corresponding to the assigned value of Z. At this mesh refinement, eddies are entirely suppressed during 1D subcycling, so no fine structure is generated and additional 1D refinement below the CV scale (i.e. $m \gg 1$) becomes superfluous. At this point, the role of physical modelling is eliminated and the ODTLES simulation reduces to DNS. This and the considerations of Sect. 3.4 highlight the nature of the assumptions and approximations on which ODTLES is based.

6.3 Comparison Case: Incompressible Formulation

An incompressible analogue of the formulation of Sect. 6.1 has been developed and applied to homogeneous decaying turbulence [21]. The main differences between the two formulations are summarized.

Because the incompressible formulation precludes dilatational flow, continuity must be enforced on a time-accurate basis. Therefore a two-step advancement cycle, with similarities to that described in Sect. 3.3, is implemented. However, in step one ODT lines are fully coupled through the introduction of additional transport terms that account for LES-scale multi-dimensional momentum transport. Also, each 1D domain evolves only the two velocity components $v_{j,k}$ for $j \neq k$. The $j = k$ component that advects fluid in the compressible formulation is omitted. Instead, fluid is advected along the 1D domain by a separately defined 'advecting velocity' that is determined by continuity (here, the solenoidal condition), based on 'fluxing velocities' that govern inter-domain transfers (the analogue of step 2 in Sect. 3.3). The fluxing velocities are moving averages, in time, of the velocity components evolved on the 1D domain. (This is an incomplete description because it omits consideration of the staggered mesh on which the simulation is implemented, and its algorithmic implications.)

In step two of the overall advancement cycle, a pressure projection is performed to enforce continuity of the mesh-scale filtered velocity field. The resulting velocity corrections are passed down to the 1D level using an adjustment scheme involving a momentum-conserving interpolant of the mesh-scale corrections. The interpolant is slightly dissipative, but this can be corrected where it degrades the flow solution using the kernel operation, as in Sect. 6.1.

It seems likely that this and the compressible formulation will exhibit comparable performance for flow regimes to which both can be applied, but this remains to be demonstrated. Computational costs are also likely to be comparable. Other perspectives on ODT-based 3D simulation of incompressible turbulent flow are provided in [34] and [65].

7 Illustrative ODT Applications

7.1 Rayleigh Convection

Rayleigh convection is a suitable initial application for compressible ODTLES. Here, previously reported results for this flow [51] obtained using the Boussinesq ODT formulation of Sect. 5.1 are summarized in order to illustrate the performance of ODT and the additional predictive capability that might be provided by compressible ODTLES.

The ODT formulation used to simulate Rayleigh convection was simpler than that of Sect. 5.1 in that one instead of three velocity components was evolved. An even simpler ODT-type formulation was previously used to simulate this flow [32]. Termed 'density profile evolution' (DPE), it evolves only density or a density surrogate (temperature), but no velocity components. Equation (13) indicates that in a gravitationally unstable state, gravitational potential energy, in the absence of fluid motion, is sufficient to generate eddy motion, consistent with the physical occurrence of spontaneous onset of motion under such conditions (e.g. the Rayleigh–Taylor instability, which has also been simulated using DPE [32]). Other buoyant-stratified-flow applications of both DPE [48] and one-component ODT [13, 49, 50] have been reported.

The ODT representation of Rayleigh convection corresponds to the ideal configuration of horizontally homogeneous flow between horizontal plates of infinite extent, but computed results are compared to measurements in convection cells that are necessarily laterally bounded. The dimensional parameters governing this flow are plate separation, buoyant forcing, viscosity ν and thermal conductivity κ. They are grouped into two nondimensional parameters, the Rayleigh number Ra, which quantifies the strength of the gravitational instability, and the Prandtl number $Pr = \nu/\kappa$, a fluid property that controls the relative thicknesses of the near-wall viscous and thermal layers. At high Ra, the thin near-wall layers strongly influence flow dynamics, as demonstrated by the significant observed Pr dependence of flow structure [66]. ODT is an efficient method for resolving these thin layers.

In Fig. 7, an instantaneous density profile from an ODT simulation of Rayleigh convection highlights the analogies between the model and physical processes in the flow. Large localized deviations from the time-averaged profile are the signatures of map-induced displacements of near-wall fluid into the

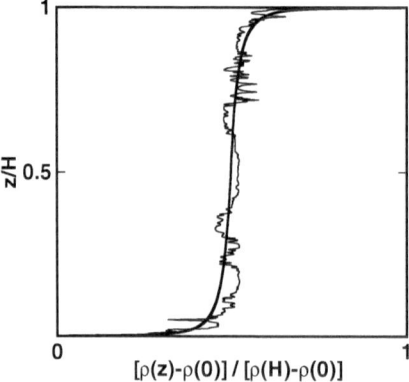

Fig. 7. Instantaneous (*thin line*) and time-averaged (*thick line*) vertical profiles of normalized density from an ODT simulation of Rayleigh convection for Ra = 1.4×10^9 and Pr = 0.7

bulk flow region. Though these displacements do not capture the persistence in time of buoyant plumes, they emulate the mechanism of entrainment of near-wall fluid into the bulk flow. Smaller eddy events subdivide and compress entrained parcels (upper region of the profile). In conjunction with molecular transport, this leads to smoothing of the fluctuations (e.g. smooth regions in the central and lower regions of the profile that deviate from the time average). Over time, these processes communicate wall forcings to the centre plane, as indicated by the nonzero density gradient at the centre of the profile. Using ODT, these flow mechanisms have been quantitatively characterized [51], which in turn has motivated another innovative modelling approach [67].

The behaviour of greatest interest is the dependence of the turbulent enhancement of mean heat flux, denoted Nusselt number (Nu), on Ra and Pr. A choice of the model parameters C and Z is identified that yields good agreement with measured Nu values over a wide range of Ra and Pr values. Without further parameter adjustment, ODT yields accurate predictions of centre-plane fluctuations, including PDFs of velocity and temperature [51].

Comparison of near-wall simulated PDFs with measurements [68] indicates large discrepancies that may reflect the inability of ODT to capture the 'wind,' a symmetry-breaking large scale circulation [69]. Given the good performance of the model in other respects, an ODTLES formulation that incorporates 3D boundary conditions and emulates 3D large scale motions might reproduce the wind and its influence on fluctuation statistics. This will be a useful initial test of the compressible formulation, both because conventional methods cannot affordably capture the relevant small scale near-wall phenomena and because copious experimental data, exhibiting nontrivial parameter dependencies, are available.

The confinement of influential small scale phenomena to the near-wall region implies a strong preference for the Lagrangian numerical scheme, which by construction provides high resolution only where needed. It can therefore be anticipated that this application will be no more costly computationally than a previously demonstrated near-wall ODT closure for confined flows (Sect. 7.2), which was used to simulate high-Re channel flow on a single processor [64].

7.2 Channel Flow

Channel flow corresponds to the same geometry as idealized Rayleigh convection, i.e. flow between parallel plates with no-slip boundary conditions, but the flow is forced by a pressure gradient parallel to the plates rather than gravitation normal to the plates. Like Rayleigh convection, channel flow relaxes to a statistically steady state. It is a canonical test case for conventional LES [70, 71].

To address the near-wall closure difficulties described in Sect. 2.1, an ODT-based near-wall sub-grid closure for LES was implemented and applied to channel flow [64]. The closure is similar in structure to, and in fact was the precursor of, the formulation described in Sect. 6.3. In this regard, it might appear that the full-flow closure is superfluous away from the near-wall region. (As in Rayleigh convection, small scale motion and transport are disproportionately influential in the near-wall region.) For channel flow specifically, this may be correct, but the compressible formulation with Lagrangian 1D implementation may be comparable in cost, as is expected for Rayleigh convection. Moreover, the near-wall closure involves potentially problematic parameterization in the region of transition between the ODT near-wall treatment and conventional closure in the bulk flow.

As in the application to Rayleigh convection, ODT parameters were adjusted to match a mean flow property, in this case, the mean velocity profile. Here, Z controls the height of the transition from the viscous to the buffer region. The LES with ODT sub-grid closure reproduced the friction law and wall-normal profiles of velocity fluctuations with good accuracy. Stand-alone ODT yielded less accurate near-wall fluctuation statistics, indicating the need for a 3D bulk-flow representation in order to represent accurately the bulk forcing that drives near-wall fluctuations. It is anticipated (Sect. 7.1) that ODTLES may likewise capture the wind effect in Rayleigh convection, yielding comparable performance improvements relative to stand-alone ODT.

8 Discussion

High-fidelity simulation of turbulent flows and their interaction with other processes ultimately requires local (in space and time) resolution of all relevant processes. Because this is unaffordable in 3D DNS, a modelling strategy involving resolution of small scales in 1D is proposed. The drawbacks of two-way information transfer between resolved and coarse-grained treatments suggests that an all-scale 1D formulation should be adopted, with large scale 3D motion captured through suitable couplings within and among arrays of 1D domains rather than through a separate coarse-grained treatment. A proposed formulation within this 'autonomous microstructure evolution' paradigm has been outlined. The underlying 1D methodology, 'one-dimensional turbulence', has been described, with emphasis on the gain in fidelity when a resolved 1D representation of relevant flow phenomenology is introduced.

Representation of turbulent fluid motion in 1D is enabled by generalization of the usual mathematical representation of advection through the introduction of a map-based representation. It is noted that this concept is not specific to 1D, and has potentially useful 3D applications.

Although the specific 3D turbulence simulation method outlined here has not yet been implemented or demonstrated, steps in its development that are indicative of its ultimate form and performance have been described. An analogue of this engineering-focused approach that is under development by the atmospheric science community has been noted. The potential for fruitful cross-fertilization of ideas across disciplines is plainly evident.

Acknowledgements

The author thanks R. Ecke for sharing unpublished results, and R. McDermott, R. Schmidt, and S. Wunsch for efforts and insights that were largely responsible for the progress reported here. This research was supported by the U.S. Department of Energy, Office of Basic Energy Sciences, Division of Chemical Sciences, Geosciences, and Biosciences. Sandia National Laboratories is a multi-program laboratory operated by Sandia Corporation, a Lockheed Martin Company, for the United States Department of Energy under contract DE-AC04-94-AL85000.

References

1. Toro, E.F.: Riemann Solvers and Numerical Methods for Fluid Dynamics, 2nd edn. Springer, Berlin Heidelberg New York (1999)
2. Alder, B.J., Ladd, A.J.C.: Simulation by molecular dynamics. In: Trigg, G.L. (ed.) Encyclopedia of Applied Physics, vol 18, p. 281 VCH, New York (1997)
3. Liu, G.R., Liu, M.B.: Smoothed Particle Hydrodynamics: A Meshfree Particle Method. World Scientific, Singapore (2003)
4. Flekkøy, E.G., Coveney, P.V., De Fabritiis, G.: Phys. Rev. E **62**, 2140 (2000)
5. Succi, S.: The Lattice Boltzmann Equation for Fluid Dynamics and Beyond. Oxford Univ. Press, Oxford (2001)
6. Pope, S.B.: Prog. Energy Combust. Sci. **11**, 119 (1985)
7. Leonard, A.: J. Comput. Phys. **37**, 289 (1980)
8. Leonard, A.: Annu. Rev. Fluid Mech. **17**, 523 (1985)
9. Winckelmans, G.S., Leonard, A.: J. Comput. Phys. **109**, 247 (1993)
10. Sagaut, P.: Large Eddy Simulation for Incompressible Flows, 3rd edn. Springer, Berlin Heidelberg New York (2006)
11. Frisch, U.: Turbulence: The Legacy of A.N. Kolmogorov. Cambridge Univ. Press, Cambridge (1995)
12. Yakhot, V., Sreenivasan, K.R.: J. Stat. Phys. **121**, 823 (2006)
13. Dreeben, T.D., Kerstein, A.R.: Int. J. Heat Mass Transf. **43**, 3823 (2000)
14. Randall, D.A., Khairoutdinov, M., Arakawa, A., Grabowski, W.: Bull. Amer. Meteor. Soc. **84**, 1547 (2003)
15. Krueger, S.K.: J. Atmos. Sci. **45**, 2221 (1988)
16. Xu, K.-M., Krueger, S.K.: Mon. Wea. Rev. **119**, 342 (1991)
17. Randall, D.A., Xu, K.-M., Somerville, R.C.J., Iacobellis, S.F.: J. Climate **9**, 1683 (1996)
18. Jung, J.-H., Arakawa, A.: Mon. Wea. Rev. **133**, 649 (2005)
19. Townsend, A.A.: The Structure of Turbulent Shear Flow, 2nd edn. Cambridge Univ. Press, Cambridge (1976)
20. Peters, N.: Turbulent Combustion. Cambridge Univ. Press, Cambridge (2000)
21. Schmidt, R.C., Kerstein, A.R., McDermott, R.: Comput. Meth. Appl. Mech. Eng., in press (2008)
22. Chakravarthy, V.K., Menon, S.: Flow Turb. Combust. **65**, 133 (2000)
23. Chakravarthy, V.K., Menon, S.: Combust. Sci. Tech. **162**, 175 (2001)

24. Sankaran, V.: Subgrid combustion modeling for compressible two-phase reacting flows. Ph.D. Thesis, Georgia Institute of Technology, Atlanta (2003)
25. Vincenti, W.G., Kruger, C.H.: Introduction to Physical Gas Dynamics. Krieger, Melbourne (1975)
26. Xu, X., Lee, J.S., Pletcher, R.H.: J. Comput. Phys. **203**, 22 (2005)
27. Batchelor, G.K.: An Introduction to Fluid Dynamics. Cambridge University Press, New York (1977)
28. Chekhlov, A., Yakhot, V.: Phys. Rev. E **51**, R2739 (1995)
29. Juneja, A., Lathrop, D.P., Sreenivasan, K.R., Stolovitzky, G.: Phys. Rev. E **49**, 5179 (1994)
30. Vicsek, T., Barabási, A.-L.: J. Phys. A: Math. Gen. **24**, L845 (1991)
31. Kerstein, A.R., Krueger, S.K.: Phys. Rev. E **73**, 025302(R) (2006)
32. Kerstein, A.R.: J. Fluid Mech. **392**, 277 (1999)
33. Kerstein, A.R.: J. Fluid Mech. **231**, 361 (1991)
34. McDermott, R.J.: Toward one-dimensional turbulence subgrid closure for large-eddy simulation. Ph.D. Thesis, University of Utah, Salt Lake City (2005)
35. McDermott, R.J., Kerstein, A.R., Schmidt, R.C., Smith, P.J.: J. Turbul. **6**, 1 (2005)
36. Kerstein, A.R.: J. Stat. Phys. **45**, 921 (1986)
37. Bramson, M., Calderoni, P., De Masi, A., Ferrari, P.A., Lebowitz, J.L., Schonmann, R.H.: J. Stat. Phys. **45**, 905 (1986)
38. Kerstein, A.R.: Proc. Combust. Inst. **21**, 1281 (1988)
39. Kerstein, A.R.: Combust. Sci. Tech. **60**, 391 (1988)
40. Kerstein, A.R.: Combust. Sci. Tech. **81**, 75 (1992)
41. Kerstein, A.R.: Phys. Fluids A **3**, 1110 (1991)
42. Kerstein, A.R.: J. Fluid Mech. **240**, 289 (1992)
43. McMurtry, P.A., Gansauge, T.C., Kerstein, A.R., Krueger, S.K.: Phys. Fluids A **5**, 1023 (1993)
44. Cremer, M.A., McMurtry, P.A., Kerstein, A.R.: Phys. Fluids **6**, 2143 (1994)
45. Kerstein, A.R., McMurtry, P.A.: Phys. Rev. E **50**, 2057 (1994)
46. Kerstein, A.R., Cremer, M.A., McMurtry, P.A.: Phys. Fluids **7**, 1999 (1995)
47. Kerstein, A.R., Wunsch, S.: Bound. Layer Meteorol. **118**, 325 (2006)
48. Kerstein, A.R.: Dyn. Atmos. Oceans **30**, 25 (1999)
49. Wunsch, S., Kerstein, A.R.: Phys. Fluids **13**, 702 (2001)
50. Wunsch, S.: Phys. Fluids **15**, 1442 (2003)
51. Wunsch, S., Kerstein, A.R.: J. Fluid Mech. **528**, 173 (2005)
52. Kerstein, A.R., Dreeben, T.D.: Phys. Fluids **12**, 418 (2000)
53. Kerstein, A.R., Ashurst, Wm.T., Wunsch, S., Nilsen, V.: J. Fluid Mech. **447**, 85 (2001)
54. Ashurst, Wm.T., Kerstein, A.R.: Phys. Fluids **17**, 025107 (2005)
55. Echekki, T., Kerstein, A.R., Chen, J.-Y., Dreeben, T.D.: Combust. Flame **125**, 1083 (2001)
56. Hewson, J.C., Kerstein, A.R.: Combust. Theor. Model. **5**, 669 (2001)
57. Hewson, J.C., Kerstein, A.R.: Combust. Sci. Tech. **174**, 35 (2002)
58. Niemeyer, J.C., Kerstein, A.R.: Combust. Sci. Tech. **128**, 343 (1997)
59. Emanuel, K.A.: Atmospheric Convection. Oxford Univ. Press, New York (1994)
60. Snyder, D.L., Miller, M.I.: Random Point Processes in Time and Space, 2nd edn. Springer-Verlag, New York (1991)

61. Turner, J.S.: Buoyancy Effects in Fluids, 2nd edn. Cambridge Univ. Press, Cambridge (1979)
62. L'Ecuyer, P.: Random number generation. In: Gentle, J.E., Haerdle, W., Mori, Y., (eds.) Handbook of Computational Statistics. Springer–Verlag, Berlin (2004) Chap. 2
63. Law, A.M., Kelton, W.D.: Simulation Modeling and Analysis, 3rd edn. McGraw-Hill, New York (2000)
64. Schmidt, R.C., Kerstein, A.R., Wunsch, S., Nilsen, V.: J. Comput. Phys. **186**, 317 (2003)
65. Kerstein, A.R.: Computer Phys. Commun. **148**, 1 (2002)
66. Siggia, E.D.: Annu. Rev. Fluid Mech. **26**, 137 (1994)
67. Grossmann, S., Lohse, D.: Phys. Fluids **16**, 4462 (2004)
68. Ecke, R.E.: unpublished
69. Sreenivasan, K.R., Bershadskii, A., Niemela, J.J.: Phys. Rev. E **65**, 056306 (2002)
70. Cabot, W., Moin, P.: Flow Turb. Combust. **63**, 269 (1999)
71. Piomelli, U., Balaras, E.: Annu. Rev. Fluid Mech. **34**, 349 (2002)

Index